Y0-BDY-798

HARVARD HISTORICAL
MONOGRAPHS

XLVII

Published under the Direction
of the Department of History
from the income of
The Robert Louis Stroock Fund

CHŌSHŪ

in the

MEIJI RESTORATION

ALBERT M. CRAIG

Harvard University Press
Cambridge, Massachusetts
London, England

Third Printing 1978

Library of Congress Catalog Card Number 61–8839

ISBN 0–674–12850–8

Printed in the United States of America

TO MY MOTHER

Adda Clendenin Craig

Note to the Third Printing

As this book was being translated into Japanese by Mrs. Makiko Hasegawa and Professor Seisaburō Satō, a number of minor errors came to light. They have been corrected in this printing. I deeply appreciate the aid, patience, and meticulous attention to detail which the translators gave to the work. I also would like to express my appreciation to Mr. W. Mark Fruin, who, while a Stanford graduate student, pointed out an error in a footnote citation in Chapter Eight. The footnote dealt with the social composition of the *Chinsei kaigiin,* an important neutralist group during the Chōshū civil war. I corrected the citation while revisiting the Chōshū archives of the Yamaguchi-ken Bunshokan. Using domain records for name changes of samurai, which occurred frequently, I was able to identify more members of the group than I had originally. The larger sample corroborated my original finding: the samurai members of the neutralist clique were of higher status, as measured by stipendiary income, than the leaders of either of the contending cliques.

Acknowledgments

My debts to individuals for help received during the writing of this book are manifold, more than can simply be listed here. I should, however, particularly like to thank Professor Edwin O. Reischauer, who guided this work in its initial dissertation phase, for his kindness, his careful page by page corrections of the manuscript, and for his many valuable interpretive suggestions. I am also grateful to Dr. Robert N. Bellah for our many discussions about the nature of modern Japanese society as well as for specific criticisms of my work; to Professors John K. Fairbank, Roger F. Hackett, Marius B. Jansen, and Benjamin I. Schwartz, who have read the entire manuscript and have made suggestions incorporated in the text; and to Mr. Tanaka Akira and Mr. Seki Junya, both for their expert guidance in matters pertaining to Chōshū and for their personal kindness during my stay in Japan in 1955–1956. Lastly I should like to express thanks to my wife Teruko. Besides proofreading the entire manuscript and typing it in its earlier stages, she has made a number of substantive suggestions and was at all times a source of encouragement.

CONTENTS

TABLES

CHART

PART ONE

Background:

The Anatomy of a Han

Introduction

In July 1853 four American ships of war entered the harbor of Uraga in Japan. They were commanded by Commodore Perry who, bearing a letter from President Fillmore, demanded that Japan sign a treaty of friendship with the United States. Fifteen years later, years of debate, political turmoil, assassinations, and wars, there occurred the Meiji Restoration and Japan embarked on its course of modernization and empire. This book is a study of Japan, and more particularly, of one area, Chōshū, during this critical fifteen-year period.

Until 1853 Japan had existed for 250 years under a system of government that is often termed centralized feudalism. It is referred to as feudal since Japan was divided into over 260 han[1] (baronies). Chōshū was one of these han. Each han was ruled by its daimyo (feudal lord). The daimyo was aided and cherished by his samurai retainers, a hereditary military class, who, living on the income from their fiefs or on a hereditary stipend, resided in the castle town of their lord. Below the samurai, ruled by the han government, were the nonfeudal classes: the peasants, artisans, and merchants.

Each han government was staffed by a bureaucracy of rank and each functioned administratively as an autonomous unit. The finances of the han, its judiciary, its military organization and training, its fiscal policies, its educational system, and virtually all other

[1] I am using "han" in this work to refer to the feudal state *within its Tokugawa framework,* thus distinguishing it from its freer condition before the establishment of the Tokugawa peace. This follows its use by Japanese scholars. Strictly speaking, of course, "han" only came into common use as the designation for the daimyos' domains, people, and government, during the course of the Edo period—being borrowed from an earlier Chinese use by Japanese Confucian scholars. It was first employed as the official name for administrative units only after the Restoration, in 1869, referring to the areas under the control of the governors (daimyo).

aspects of its internal affairs were controlled and administered by the han government alone. It is necessary to stress that the han were, within the Tokugawa framework, autonomous, since the deceptive ease with which the system was overthrown after 1868 has led many to conclude that their apparent autonomy was merely a façade masking a highly centralized reality. This was not the case. De Tocqueville has argued that the subjection of all departments in prerevolutionary France to a single authority, that of the intendants and subdelegates, created a uniformity of practice, a populace conditioned to respond to bureaucratic control, and a *de facto* centralization of power which laid the groundwork for the centralized institutions of the revolutionary era that followed. Such an argument cannot be applied to pre-Restoration Japan. Certain areas of han life were regulated by the controls of the central authorities in Edo, but these were few and for the most part external.

The han were also different from one another both in their gross proportions and qualitatively. Some han were small with a total agricultural product of only ten thousand or so *koku*[2] of rice. In contrast to this, at least one, and probably several, of the larger han produced the equivalent of over one million *koku* of rice. Most han were territorially united, but a few of the smaller han of central Japan were divided into scattered fiefs—which limited, for example, the ability of those han governments to impose economic controls. The autonomy of the han led to significant variations in matters such as fiscal policies, the pattern of administration, and education. Other differences were historical in character. Similarities between the han are often explained in terms of their common feudal past, and it is true that none escaped the molding influence of the period of Warring States; yet it was this continuation of the past into the Tokugawa period (1600–1868) that also accounts for crucial differences in traditions, laws, class structure, and in the relative positions of the various han within the Tokugawa polity. By the eighteenth century the differences among the han in Japan were at least as pronounced as those existing between Prussia and some of the smaller

[2] One *koku* is equivalent to 4.96 bushels.

counties and duchies in pre-Napoleonic Germany. It is necessary to stress these differences since to a large extent the politics of the Bakumatsu period (1853 to 1868) hinged upon them.

The system of feudal states is spoken of as centralized since all the han were under the hegemony of the Tokugawa house. The head of the Tokugawa house, the shogun, exercised his hegemony through his "house government" which was known as the Bakufu. Literally, Bakufu means "tent government" or military headquarters; it suggests a contrast with the civil government under the Emperor in Kyoto. In fact, of course, the Imperial establishment in Kyoto was a captive court kept in existence merely to legitimate shogunal rule. The character of the Bakufu was dual. First, it was merely a larger han government which ruled the domains under its immediate control in the same fashion as did the governments of the various han. Second, the Bakufu was the organ through which the Tokugawa maintained their controls over the many other han. Established at Edo, it was the national center to which daimyo journeyed for their forced biennial residence, from which orders were issued for the periodic financial levies exacted from the han, and at which hostages were kept.

Thus there existed, on the one hand, the han maintaining a considerable amount of private power, and on the other, the Bakufu wielding the right to regulate certain of the han's external affairs. This equilibrium, first established in the early decades after 1600, was kept up for 250 years. Any untoward move by a han to extend its power or to free itself from the Bakufu control system would invite instant retaliation, and any move by the Bakufu to extend its controls or to interfere with the internal affairs of the han would run the risk of arousing fear and distrust among the han, and so disturb what was to the Bakufu a satisfactory *status quo*. The internal balance of the elements of this equilibrium was protected from outside interference by the Tokugawa policy of cocoon-like seclusion. Japanese were not permitted to leave Japan nor were foreigners permitted, with rare exception, to enter. Changes did occur in the Tokugawa scene: the expansion of the commercial sector of the economy, the development of a brilliant urban culture, a burst of new learning which included ideas introduced from the West, and so on; but, politically at least, the equilibrium was maintained.

It was upon this circumscribed feudal world of sword-bearing samurai, upstart merchants, and tax-ridden peasants that Perry impinged in 1853, forcing Japan to abandon its time-hallowed policy of seclusion. What was striking in the reaction that followed was that only 2 of the 260-odd han of Tokugawa Japan emerged to play vigorous roles in the overthrow of the Bakufu. The rest, with occasional exceptions, were quiescent; they were moved in the current of events, but did not move of themselves. The 2 that acted were Chōshū and Satsuma, han whose samurai were subsequently to carry out, as the dominant center of the new Meiji government, the revolution of the early Meiji period. The central question in this book is why one of them, Chōshū, could act when other han could not.

This is, at the present stage of our knowledge, perhaps the most important type of question that can be asked concerning the formation of modern Japan. It is important because of past neglect: several very general studies have been made of the Meiji Restoration, excellent monographs are available concerning the Tokugawa economy, the Tokugawa value system, the agrarian change which preceded the Restoration, and certain aspects of the economic and political changes of the early Meiji period. There are also a number of studies of the Bismarcks who collectively led Japan during its early phase of modernization. But as yet, in the Western literature on Japan, virtually nothing has been written on the two han that played a role in the "unification" of Japan analogous to that played by Prussia in Germany.[3] This book is an attempt to fill that hiatus in the case of Chōshū.

More concretely, the han was the central level of samurai identification during the Tokugawa period. Because of the hierarchical nature of their loyalty the samurai saw themselves, unlike English gentlemen or Chinese literati during the same period, not primarily as

[3] The single exception that enters my mind is Robert Sakai, "Feudal society and modern leadership in Satsuma-han," *Journal of Asian Studies,* 16:365–376 (May, 1957).

members of a social class, but as members of a particular han. Within the han samurai were organized into horizontal groupings by rank, and they were aware that they were objectively members of a single class: this can be seen in Fukuzawa Yukichi's experiment in which he mimicked the attitudes of other classes, or even more clearly in statements by samurai of virtues which they possessed and commoners did not. But, this did not lead them to regard samurai from other han as their fellows. On the contrary, these were viewed as members of potentially hostile groups, as samurai whose pride was in conflict with their own, almost as aliens. The vertical social cleavage of which this was an effect was strikingly evident in the pattern of Bakumatsu politics.[4]

A second reason for studying the han is, curiously, to illuminate questions of individual motivation. The historian is concerned professionally with abstractions, with the particular aspects of past reality that someone, for whatever purpose, has chosen to record. A considerable portion of his labor is spent in attempting to reconstruct the meanings which these abstractions held for those who lived in the past. That one set of abstractions rather than another was recorded is in itself an indication of interests, yet reconstruction is difficult since even fairly obvious actions by individuals took place, in Lewin's sense of the term, in a structured field. Since most of this field was taken for granted, it is difficult to survey using biographical materials alone. In the "field" of samurai of Tokugawa Japan the chief realities were certain intermediate groups—vassal groups, schools, military units, and bureaucratic cliques—centering on the han. Only as members of such groups or collectivities could individual samurai act.[5] To inquire into the structure which groups possessed, to see the way in which frustrations, ambitions, duties,

[4] The apparent exception to this rule is the case of the loyalist ideologues whose common doctrine enabled them to work together from time to time. Yet even here cooperation invariably took the form of associations of han cliques and evidences of hostilities and in-group feelings are abundant.

[5] The Bakumatsu activities of *rōnin* or masterless samurai seem to furnish an exception here. Yet it must be noted that during the Bakumatsu period most "masterless samurai" were not masterless, but were samurai who had been given permission to leave their han. Moreover, most effective action by such *rōnin* took place within the framework of action of some other han, within a *rōnin* militia unit in Chōshū, for example.

and ideals were joined in such groups, and to view the evolving relations among them, is to comprehend for the first time the field in which individual action took place. To ignore such infra-han structure is to view the activist samurai and commoners of the Bakumatsu period as individual, self-contained bees buzzing in a hollow hive. On the other extreme, to combine into a single class amalgam the diverse characters which such groups possessed and to use this as a philosophers' stone to explain all social change, not only ignores the vertical cleavage of that society, but obscures, even within a single han, which groups were important and which not.

To understand why Chōshū (and Satsuma) could act in the Restoration period, it is obviously necessary to explain why other han could not. A definitive answer to this question must be comparative. Unfortunately, at the present time there are few monographic studies, even in Japanese, which can be used for this purpose. As a result, most of this study is fairly concrete. In the first half, which I have called "The Anatomy of a Han," I have attempted to block out certain large areas of Chōshū organization and function up until 1858, the year when the decision was made to involve Chōshū in national politics. Under this rubric is included an account of han politics to 1858, as well as items such as the structure of the Chōshū samurai class, the size of the han, its finances, its government, and its position in Tokugawa Japan. In each case I have stressed those aspects of the Chōshū background that appear to have contributed to its role in Restoration history. I have attempted to show in some detail how these background factors worked out in Bakumatsu politics between 1858 and 1868 in the second half of the book.

CHAPTER I

Chōshū and the Tokugawa Polity

Gross Productivity and Samurai Manpower

Chōshū was big. Its domains were extensive; its samurai class was one of the largest in Japan. Any discussion of the factors contributing to Chōshū's ability to act in Bakumatsu politics might well begin with a consideration of size. It may be argued that the product of a han and the size of its samurai class are only gross factors and as such are not important. What was done with the income of a han was obviously more important than its gross product. Similarly, equipment, training, morale, leadership, and so on were more important considerations than gross numerical strength in determining the military power of a han. Nevertheless, gross factors were significant; even though the largest han were not always the most important politically, the most important were always among the largest. Moreover, a consideration of such gross factors is a convenient point at which to begin since comparative materials are relatively abundant.

During the Tokugawa period, the size of a daimyo's domain was ranked according to its productive capacity. This was known as the *kokudaka* of the han, and was expressed in terms of so many *koku* of rice. Actually, the figure included some crops other than rice, some commercial taxes, some taxes on other products (though not those under the han monopoly system), some "house taxes" on city dwellings, taxes on salt, and in the case of Chōshū, even a tax on bathhouses in Yamaguchi.[1] Table 1 compares the productive capacity of the sixteen largest han in Bakumatsu Japan; all are han whose officially listed productive capacities (see left-hand column) were more than 300,000 *koku*. In terms of these official listings, Chōshū was the ninth largest han in Japan.[2]

The official listings (*omotedaka*), however, were fictions which had been established during the early years of the Tokugawa period, and were only rarely revised. In almost every case, the actual product (*naidaka* or *jitsudaka*) was different. In a very few cases, the actual product of a han was less than its officially recognized capacity. Mito, for example, was officially listed as possessing domains of 350,000 *koku* but its actual product was less than this.[3] In most cases, the real product of a han exceeded its official listing. Chōshū, for example, had an officially listed product of 369,000 *koku,* but its actual product was, according to Table 1, 713,600 *koku.* (This does not include the domains of the four Chōshū branch han which would raise this figure to about 895,000 *koku.* One authority estimates the total product of Chōshū at over 1,000,000 *koku.*[4]

The source of this discrepancy between the official listing and the actual productive capacity can be seen clearly in the case of Chōshū.

[1] Misaka Keiji, *Hagi han no zaisei to buiku* (Tokyo, 1944), pp. 35–42.

[2] The 300,000-*koku* figure which I have chosen for the purpose of comparison is obviously an arbitrary limit. It would have been more realistic to base the comparison on the *jitsudaka* rather than on the *omotedaka,* but few of the latter are readily available. However, Tosa, whose *omotedaka* was 242,000 *koku* but whose *jitsudaka* was 442,700 *koku* might have been included on the list.

[3] *Ishin shi* (Tokyo, 1939), I, 261–262.

[4] Misaka, *Hagi han no zaisei to buiku,* p. 169. The official listing (*omotedaka*) of the branch han (*shihan*) vis-à-vis the main han was fixed in 1625 at a total of about 183,000 *koku,* just as that of the main han vis-à-vis the Bakufu was fixed in 1610. Misaka suggests that subsequent gains in the productivity of the branch han were probably proportionally as great as the recorded gains made by the main han. This would almost put Chōshū over the 1,000,000 *koku* mark. Subsequent research tends to confirm this. Iwakuni han, one of the branch han, whose *kokudaka* was cut from 120,000 *koku* to 30,000 *koku* after Sekigahara, had increased its production to over 73,000 *koku* by the 1790's. See Koyama Yasuhiro, "Iwakuni han no hansei kaikaku," *Shigaku kenkyū,* 76:35 (May, 1960).

TABLE 1. The great han of Tokugawa Japan

Name of han: Castle town	Name of han: Area name	Name of daimyo	Officially listed productive capacity (*omotedaka*) (in *koku*)	Actual recorded product (*jitsudaka*) (in *koku*)	Status
Kanazawa	(Kaga)	Maeda	1,022,700	1,353,300	*tozama*
Kagoshima	(Satsuma)	Shimazu	770,000	869,500	*tozama*
Sendai	(Mutsu)	Date	625,600	958,400	*tozama*
Nagoya	(Owari)	Tokugawa	619,500	. . .	*shimpan*
Wakayama	(Kii)	Tokugawa	555,000	539,400	*shimpan*
Kumamoto	(Higo)	Hosokawa	540,000	721,000	*tozama*
Fukuoka	(Chikuzen)	Kuroda	520,000	. . .	*tozama*
Hiroshima	(Aki)	Asano	426,000	488,000	*tozama*
Hagi	(Chōshū)	Mōri	369,000	713,600	*tozama*
Saga	(Hizen)	Nabeshima	357,000	. . .	*tozama*
Mito	(Hitachi)	Tokugawa	350,000	. . .	*shimpan*
Hikone	(Ōmi)	Ii	350,000	. . .	*fudai*
Tottori	(Inaba)	Ikeda	325,000	428,100	*tozama*
Tsu	(Ise)	Tōdō	323,000	. . .	*tozama*
Fukui	(Echizen)	Matsudaira	320,000	. . .	*shimpan*
Okayama	(Bizen)	Ikeda	315,000	469,100	*tozama*

Source: This table is a composite of materials taken from several sources. The statistics for those han for which the *jitsudaka* is given were taken from Toba Masao, "Edo jidai no rinsei," *Iwanami kōza Nihon rekishi* (Tokyo, 1934), IX, 24-25. I have used all of the *jitsudaka* given by Toba without substituting, in one or two cases, apparently more accurate figures, in the hope of preserving internal consistency among the figures which he has given. Footnote 5 below contains one such figure. The various Chōshū *jitsudaka* in Chapter II are also somewhat different from those given in this table. In general *jitsudaka* are extremely difficult to obtain. Ideally, of course, each *jitsudaka* should be identified as the *daka* of such and such a year. In the absence of such identification the discrepancies between the figures given by different writers may not be taken simply as evidence of inaccurate scholarship. Other figures in this table were taken from one or the other of the following: Itō Tasaburō, *Bakuhan taisei,* (Tokyo, 1956); Nihonshi jiten, (Osaka, 1954).

After the battle of Sekigahara, Chōshū's domains, or rather, the domains of the house of Mōri, were cut to the two provinces of Suō and Nagato, which had an assessed production (according to a cadastral survey made in 1600) of only 298,480 *koku*. This was less than one-fourth of their previous domains. Moreover, since the Mōri had already collected and used the taxes for the year 1600 from the domains which subsequently had been confiscated, they were also faced with the task of repaying the taxes to the new lords of these domains. Their only source of income from which this could be repaid was their greatly reduced domain (the Chōshū of the Tokugawa period). They were also encumbered with many other financial burdens, such as the establishment of a new castle town and the resettling of their retainers in Chōshū. Therefore, in 1610 another land survey was carried out in which the assessed productivity of Chōshū rose to 525,435 *koku*. A fraction of the gain of 226,955 *koku* may represent an increase in productivity; most of it, however, was undoubtedly due to the looseness of the earlier survey when the dependence of the daimyo on his vassals prevented too close a scrutiny of the productive capacity of their fiefs. The Chōshū daimyo, Mōri Terumoto, apprised of the results of the new survey, was going to report the increase in production to the Bakufu. His Elders, the chief ministers, however, suggested that since the Bakufu levies would increase in proportion to the gain in production they should follow the example of the neighboring Fukushima Masanori of Hiroshima and report an assessed product of 369,411 *koku*.[5] This was done, and as a consequence, in spite of continuous increases in Chōshū's actual production, the 369,411-*koku* figure remained the official Chōshū listing until after the Meiji Restoration. It would appear to have been in the interests of the Bakufu to have forced the han to bring their official listings in line with their actual production, since Bakufu levies for men and money were proportional to the official listings. The Bakufu was aware that the revenues of the han were increasing, yet it lacked the means by which to ascertain the actual product of a han or to compel the han to change its official listing. In this, as in most other areas concerning internal administration, the autonomy of the han was inviolable. The levies, however, constituted only one

[5] Misaka, *Hagi han no zaisei to buiku*, pp. 38–39. Another example of the same phenomenon was the case of Kumamoto. When the Hosokawa received it as a fief in 1634, its actual production had already increased from 750,000 *koku* to 960,000 *koku*, yet its *omotedaka* was intentionally kept at 540,000 *koku*. See Itō, *Bakuhan taisei*, p. 43.

small part of the total Tokugawa control system. The inequities resulting from the discrepancy between official listings and actual production were not as serious as the antagonisms which would have been created by direct Bakufu interference in the internal finances of the han.

Figures of the actual products are not available for all the han listed in Table 1. Judging, however, from those at hand, Chōshū was probably about the fourth or fifth largest han in Tokugawa Japan. Satsuma, the only other han able to take independent political action under the conditions that emerged in Japan after 1853, was probably second or third. Considering their size, it is not at all surprising that they were able to act when so many others could not.

Moreover, both Chōshū and Satsuma were favored with an unusually large number of samurai in proportion to the size of their domains. In general, in Tokugawa Japan, big han tended to maintain large numbers of samurai, medium han had medium-sized samurai classes, and so on. This "natural tendency" was reinforced by Bakufu orders promulgated during the 1648–1652 period, which stipulated the number of warriors that were to be maintained by its retainers. This official scale was as follows:[6]

Size of domains (*koku*)	Number of samurai
200	5
1,000	21
5,000	103
10,000	235
100,000	2,155

This scale was intended primarily for the *fudai* and *hatamoto* retainers of the shogun, and as such it provides a convenient yardstick in terms of which comparisons can be made between these and the "outside" han. When this is done one of the most significant differences to emerge is that those han which were great independent baronies before Sekigahara tended to have many more vassals in relation to their officially assessed productive capacities than those han formed after 1600, such as the majority of the Tokugawa col-

[6] Itō, *Bakuhan taisei*, p. 17.

lateral or branch houses. This was especially true of Satsuma and Chōshū.

At the time of Hideyoshi's conquest of Kyūshū in 1587, the domains of the Lord Shimazu, the Satsuma daimyo, which had originally covered most of the island, were cut to two and one-half provinces. Most of the Shimazu vassals crowded into the newly reduced domain, giving it a much greater density of samurai population than would have obtained under normal circumstances. As it was not financially feasible to support all of them as castle town samurai, some, those already living in villages, were kept there, continuing into the Tokugawa period an earlier form of samurai dominated villages. Other samurai moving to the new domains as well as some already living in the castle town were also forced, it appears, to reside in the countryside. The majority of these constituted the famous Satsuma *gōshi* class, most of whom were farmers with swords and the status of samurai.[7] As a consequence, Satsuma, which with 777,000 *koku* of domains should have had only about 14,000 samurai according to the ratios given in the tabulation above, in reality had almost twice that number. They could be supported only because its revenues were higher than its official listing, and even then, only by using the *gōshi* system; yet Satsuma was clearly stronger militarily because of its samurai manpower potential.[8]

The case of Chōshū was quite similar. For siding against the Tokugawa at Sekigahara, Mōri Terumoto saw his domains reduced from nine to two provinces. Many of the Mōri vassals who could no longer be supported were dropped, but about 5657 were retained, and, since many of these were of fairly high rank, they brought with them an unusually large number of rear vassals: over 5000 in all.

[7] In 1708 there were in Satsuma 4,110 castle town samurai, 20,297 samurai living outside the castle town (i.e., in the country), and 3,146 whose rank was that of *ashigaru* or below, who also lived outside the castle town. It should be noted that while many of those living outside the castle town were, financially at least, little more than farmers, some received stipends. Eight received more than 10,000 *koku*, 26 received more than 1,000 *koku*, 336 received more than 100 *koku*, and 1,588 received *kirimai* (stipends payable in grain or coin). It should also be noted that in Satsuma, the term *gōshi* referred to *shi* (upper samurai) and not *sotsu* (lower samurai). See *Kagoshima-ken shi*, II, 18–26.

[8] Itō, *Bakuhan taisei*, pp. 54–55.

Therefore, Chōshū had about 11,000 samurai rather than the 6000 that the Bakufu would require of a house with domains of a similar assessed productive capacity. These could be maintained because its domains were far richer than the official assessment.[9]

In contrast to Chōshū or Satsuma, Owari and Aizu, which were both Tokugawa branch houses, had considerably smaller samurai classes in proportion to their officially listed productive capacity. Owari, with 619,500 *koku* should have had over 12,000 samurai; it had only 5910.[10] Aizu, with 230,000 *koku* should have had over 4000 samurai but had only 3089. As a result, in proportion to their size, they were much weaker in samurai manpower than either Chōshū or Satsuma.[11]

Yet, the most striking example of a disproportion between land (or productive capacity) and samurai manpower was that of the Bakufu domains. On the eve of the Restoration the Bakufu held domains with an assessed production of about 7,060,000 *koku*—or about 23 per cent of the official estimate for all of Japan, which was 30,550,000. Of this 7,060,000 *koku,* 57 per cent, or 4,000,000 *koku,* was land whose revenues went directly to the Bakufu (this was called *kurairichi*). The remaining 3,060,000 *koku* or 43 per cent was distributed in *hatamoto* fiefs. The largest of these was a fief of over 9000 *koku* which differed very little from the fief of a small *fudai* daimyo, and among the lesser *hatamoto* there were 766 *hatamoto* with fiefs whose productive capacity was less than 100 *koku.* These smaller *hatamoto* differed very little from the upper ranks of the *gokenin,* the third echelon of hereditary Bakufu retainers, except that in principle the former possessed fiefs (which they managed personally only if over 3000 *koku*) whereas the latter were given *kirimai* or a direct payment in rice. The *gokenin* received their revenues from the income of the 4,000,000 *koku* of land which went directly to the Bakufu.

With over 7,000,000 *koku* of domains, the Bakufu should by its

[9] Umetani Noboru, "Meiji ishin shi ni okeru kiheitai no mondai," *Jimbun gakuhō,* 3:44 (Mar., 1953).

[10] *Ishin shi,* I, 275–276.

[11] Itō, *Bakuhan taisei,* p. 56.

own scale have had over 140,000 samurai. Actually, it neither had nor pretended to have anything near this number. Rather, it spoke of its military collectively as the "80,000 mounted horsemen." [12] Even this figure or image is deceptive and substantially overstates its true strength. In the 1713–1735 period, for example, there were only 5204 *hatamoto* and 17,309 *gokenin,* a total of 22,513 men. In addition to these, there were also soldiers (*sotsu: ashigaru* and below) among the direct vassals of the shogun; the exact number of such low-ranking vassals is not known but one authority has estimated that there were several thousand of them. This would bring the number of samurai, although not mounted samurai, under the direct control of the shogun to about 30,000. To this, of course, must be added the vassals of the *hatamoto* who were rear vassals vis-à-vis the shogun. Had all of the *hatamoto* maintained the requisite number of vassals in proportion to the size of their fiefs, the total number of rear vassals, soldiers, *gokenin,* and *hatamoto* would have reached the 80,000 figure. Actually, however, the *hatamoto* often maintained fewer than the required number of vassals, or they maintained them in such a condition that they could never be effectively mobilized. Because of this and because the rear vassals were not for the most part "mounted" warriors, one can only conclude that when the "80,000 mounted horsemen" are spoken of, it must be taken to include the troops of the *fudai* and *shimpan* daimyo of which more will be said presently.[13]

From this one can conclude that the Bakufu was faced with the formidable task of supporting its far-flung and diverse domains, and staffing its apparatus of government and scattered officialdom, with a basic force of only 20,000 or 30,000 vassals. This could be done. It is impossible, however, to imagine that such a puny force could suffice to control and keep in order the powerful daimyo on the perimeter of the Bakufu domains. To preserve its hegemony over these outside lords, the Bakufu was dependent on its institutional system of controls, on custom, and ultimately, on the cooperation and loyalty of the military forces of the *fudai* and *shimpan* daimyo.

[12] Itō, *Bakuhan taisei,* p. 17.
[13] *Ibid.,* pp. 9–18.

From this point of view, it would not be inconceivable to write the history of the Bakumatsu period in terms of the process of gradual estrangement between the Bakufu and its traditional supporters. Like a bank, the Bakufu hegemony lived on credit. During the Bakumatsu period, as its controls broke down and it became isolated from its allies, its credit began to disappear, and finally in 1867 there occurred a simultaneous withdrawal of accounts. When this occurred, its weakness of numbers ensured its military defeat.

Structural Tensions in the Tokugawa Polity

And yet, granting the weakness of the Bakufu, the question still remains: why did Chōshū and Satsuma rather than others among the sixteen largest han play the leading roles in toppling its age-heavy structure? A partial answer to this question is that five of the sixteen were *shimpan* or *fudai* han: han who by their very nature were closely associated with the Bakufu, and, if in some cases they were unwilling to support it at its moment of crisis, they proved equally unwilling to contribute actively to its overthrow.

Daimyo were divided into three classes depending on the relation of their house to the house of Tokugawa. We will first examine the case of the *fudai* daimyo, those daimyo whose ancestors had already become the vassals of the Tokugawa before the battle of Sekigahara in 1600. Throughout the Edo period, these daimyo played a dual role; they ruled in their own han, as did other daimyo, but at the same time they staffed the top positions in the Bakufu hierarchy as retainers in the house government of the shogun. The latter role served as a channel through which the more ambitious and able among them could advance in position and rank. It gave them a voice in national affairs, and from time to time it enabled them to become the lords of larger domains in reward for their services. Such position within the Bakufu government seems to have created within the most able *fudai* daimyo, who were potentially the most likely to become disaffected, a positive sense of bureaucratic identification with the Bakufu. This, rationalized within the lord-vassal loyalty ethic, overshadowed their identity as daimyo of autonomous areas.

This tendency was further strengthened by their size. Of one hundred and forty-five *fudai* daimyo, only one merited inclusion in our list of large daimyo, and ninety-four were lords of domains whose listed productive capacities were less than 50,000 *koku*, that is to say domains which were politically and militarily insignificant. Yet size alone did not determine their dependence on the Bakufu; even the house of Ii of Hikone han, which possessed men and resources enough for independent action, remained a solid bulwark of the Bakufu to the end. In fact, it seems to have remained a stauncher supporter of the system than the last shogun himself.

In contrast to these *fudai* daimyo, the *tozama* daimyo whose ancestors had first sworn fealty to the Tokugawa after Sekigahara ruled only their own han. They could not hold positions in the Bakufu nor could they voice opinions in matters concerning national affairs. Moreover, while the Bakufu tended to be lenient in its treatment of *fudai* han, its treatment of the *tozama* han was harsh. For the slightest infraction of the "Laws Governing the Military Houses" (*Buke shohatto*), their domains would be reduced or confiscated. The severity with which they were treated can be seen in the diminution of their numbers: out of one hundred and eighty *tozama* han existing in 1598, only ninety-eight survived in Bakumatsu times.[14] Of course, of the eighty-two *tozama* lords who did not survive the Edo period, the vast majority lost their domains either immediately after Sekigahara or during the first century of Tokugawa rule. By the early part of the eighteenth century the situation had become more or less stabilized and the Bakufu grew less and less inclined to challenge the *status quo*. Nevertheless, the earlier, distant attitude towards the *tozama* daimyo persisted; the political potentials of the *tozama* remained circumscribed, and consequently, during the Bakumatsu period it was these han that tended to become the focuses of anti-Bakufu activity.

The third type of han, the *shimpan* or branch han of the Tokugawa house, were in a more complex and sometimes even ambivalent relation to the Bakufu. Like the *tozama* daimyo, the *shimpan* daimyo ruled only their own han, and even the ablest could not hope to

[14] *Ibid.*, p. 26.

obtain official position within the Bakufu. On the other hand, insomuch as they could give advice to the shogun, they resembled the *fudai* lords. But, they differed from both *tozama* and *fudai* daimyo in that a daimyo of certain *shimpan* houses might himself become the shogun should the shogun die without an heir. Consequently, when their advice was heeded and their favored candidate appointed shogun, they were among the most ardent supporters of the Bakufu, but when their advice was ignored and their candidate rejected, they could become quite critical. There was always, however, a limit to their criticism. As Tokugawa branch houses, no matter how bitter their attacks, as in the case of Nariaki of Mito, they were always leveled against an administration and never against the institution of the Bakufu itself. Even the loyalist movements in the branch han were affected by this restriction. For example, in Mito, the ideological birthplace of the *sonnō jōi* movement, the samurai extremists always felt constrained to maintain the fiction of acting in the *sonnō jōi* spirit of Ieyasu.[15] Consequently, the *shimpan* role in the Bakumatsu political struggles was important only during the early years of the period, when Bakufu policies and not the institution of the Bakufu were under attack. Moreover, within the *shimpan,* tradition and the ties of custom were always such as to hinder rather than help their extremist *sonnō* factions, which, in every case, were crushed or effectively suppressed during the early phase of Bakumatsu politics. Therefore, as in the case of the *fudai* han, in spite of their size the *shimpan* were unlikely to lead a movement against the Bakufu. This enables us to eliminate as potential rebels the *fudai* han, Hikone, and the four *shimpan,* Mito, Owari, Kii, and Fukui, from the ranks of the sixteen large han with sufficiently strong material bases to carry on autonomous political activity during the Bakumatsu period.

Traditional Hostilities and Obligations in Tokugawa Society

What then of the remaining eleven *tozama* han? Why were only two active; for what reasons were the other nine not? A complete and satisfactory answer to these questions must await the appearance

[15] Tōyama Shigeki, *Meiji ishin* (Tokyo, 1951), pp. 78–79.

of further monographic studies on these han during the Bakumatsu period, but some light may be cast on the subject by a consideration of the traditional hostilities or traditional feelings of obligation that characterized the attitudes of different *tozama* han toward the Tokugawa.

Among the *tozama* han one can distinguish two types: those who fought against the Tokugawa at Sekigahara, and those who sided with the Tokugawa or remained neutral. Western historians of Japan have sometimes incorrectly assumed that the *tozama* han were those who fought against the Tokugawa and that the *fudai* han were those who were their allies. Actually, the distinction between *tozama* and *fudai* hinged on the question of vassalage and not on the question of who fought on which side.[16] Most of those who were vassals of the Tokugawa before 1600 (later known as *fudai* daimyo, *hatamoto, gokenin,* and so on) fought on the Tokugawa side, but so did many other lords who later were classified as *tozama* daimyo. Of the eleven *tozama* han to be found among the sixteen largest han, only three, Chōshū, Satsuma, and Saga, were ranged against the Tokugawa at Sekigahara; the other eight either sided with the Tokugawa or remained neutral.

The distinction between these two types of *tozama* han can be clearly seen in the difference in the treatment meted out to them by Ieyasu after Sekigahara. Those who fought against him had their domains reduced or confiscated, or at best, were left untouched. Those who aided him were generously rewarded. Among the eleven largest *tozama* houses in the Bakumatsu period, those of Maeda, Date, Hosokawa, Asano, Tōdō, and Kuroda were given enlarged domains after the battle of Sekigahara in reward for their services to Ieyasu. Not having been vassals, they were not allowed to participate in the house government of the Tokugawa, the Bakufu; nevertheless, they were regarded by the Tokugawa as trustworthy allies. And trust in their loyalty was translated into practical terms: the

[16] It would be even more accurate to define a *fudai* daimyo as one recognized by the Bakufu as a *fudai* daimyo. This definition would encompass even the house of Satake which was given honorary *fudai* status by the Tokugawa in spite of not having been a vassal before Sekigahara. At times history is so illogical that only tautologies are completely true. See Itō, *Bakuhan taisei,* pp. 23–24.

"friendly *tozama*" were strategically placed by the Tokugawa to hedge in their former enemies in the outlying areas of Japan in much the same manner that the *shimpan* and *fudai* han were used to guard the approaches to the Tokugawa heartland in Kantō. For example, Lord Asano was placed at Hiroshima and Lord Ikeda at Okayama to surround hostile Chōshū; Lord Hosokawa was placed at Kumamoto and Lord Kuroda at Fukuoka to form a northern tier against the anti-Tokugawa potential of Satsuma and Saga.[17] In fact, Ieyasu's first action following Sekigahara was to summon Kuroda Nagamasa and to present him with his new fief. Before his assembled generals, Ieyasu praised him as follows: "Today's victory is entirely due to your loyalty and care, and as a reward for your great merit, as long as my house shall last the interests of the house of Kuroda shall never be allowed to suffer."[18]

In sharp contrast to this, the house of Mōri (of Chōshū) which until the time of Sekigahara had been the second greatest feudal power in the country, had its domains of 1,205,000 *koku* cut to the two provinces of Nagato and Suō with an assessed production of only 298,480 *koku*. Satsuma, which had suffered a drastic reduction of its domains some years earlier, and Saga were not diminished at the time of Sekigahara. All three han, however, seem to have maintained a strong anti-Tokugawa bias throughout the Tokugawa period. In this sense, the history of Chōshū in the Meiji Restoration can be said to begin from the time of its defeat at Sekigahara.[19]

To the Mōri and their vassals, the Bakufu, towards whom most of the great *tozama* han felt an obligation, represented the enemy who had laid low the once great house of Mōri. Even the meager stipends of the Mōri vassals were, to some extent, seen as a consequence of their reduced domains. Most of this anti-Tokugawa bias was formless, a part of the Chōshū heritage, but in a few cases it seems to have crystalized into specific practices. One ceremony embodying this animus was held annually on the first day of the new year. Early in the morning when the first cock crowed, the Elders

[17] *Ibid.*, pp. 27–28.

[18] A. L. Sadler, *The Maker of Modern Japan: The Life of Tokugawa Ieyasu* (London, 1937), p. 210.

[19] Misaka, *Hagi han no zaisei to buiku*, p. 26.

and Direct Inspectors would go to the daimyo and ask, "Has the time come to begin the subjugation of the Bakufu?" The daimyo would then reply, "It is still too early; the time is not yet come."[20] While obviously secret, this ceremony was considered one of the most important rituals of the han. Another comparable custom in a more domestic setting has also been recorded. Mothers in Chōshū would have their boys sleep with their feet to the east, a form of insult to the Bakufu, and tell them "never to forget the defeat at Sekigahara even in their dreams."[21] In the case of Satsuma, every year on the fourteenth day of the ninth month the castle town samurai would don their armor and go to Myōenji, a temple near Kagoshima, to meditate on the battle of Sekigahara. And on the following day, they would return to the castle town to listen to "The Military Record of the Battle of Sekigahara" (*Sekigahara gunki*).[22]

After more than two hundred years of peace under the Tokugawa hegemony, how important were these traditional feelings? Even in a society such as the Tokugawa, where revenge was ethically sanctioned and the repayment of obligation (*hōon*) a matter of honor, it is extremely difficult to judge the extent to which such formalized historical elements weighed on the feelings of the vassals in the governments of the han. Here one can only suggest that these traditional hostilities formed one of several potentially subversive elements in the ideology of the samurai of Chōshū, Satsuma, and Saga; whereas in the case of the pro-Tokugawa *tozama* han, their awareness of past (and therefore present) obligations formed an element that had to be rationalized and resolved before they could join in any anti-Bakufu movement.

The case of Tosa illustrates admirably that these obligations

[20] Hirano Shūrai, *Chōshū no tenka* (1912), p. 26. Almost everyone who has dealt with Chōshū, historians, public officials, and so on, seems to know of this custom, but no one seems able to document it. I am indebted to Ishikawa Takumi, librarian of the Chōshū archives at the Yamaguchi-ken Library for the above citation. At the same time, I must state that *Chōshū no tenka* is not an academic work, and that it is probably little more reliable than oral sources. But in view of the blatant anti-Tokugawa nature of this custom, a lack of documentation does not necessarily disprove it.

[21] *Ibid.*, pp. 26–27.

[22] *Ishin shi*, III, 59.

incurred after Sekigahara continued even in Bakumatsu times. Before Sekigahara, the house of Yamanouchi had held a small fief with domains of only 60,000 *koku*. Then, in spite of having played a very minor part in the Tokugawa victory, it was given Tosa as a fief, with domains officially assessed at 242,000 *koku*. Its gratitude for this, at least as a formal element, continued even through the Bakumatsu period. In a letter from Yamanouchi Yōdō to Matsudaira Keiei written in 19/5/62[23] concerning national affairs, Yōdō reveals his attitude: "Though I am of little ability, from the time of my ancestors, [the house of Yamanouchi] has in many ways received favors from the Tokugawa. Because of this, although I have retired [he was forced to retire by the Bakufu], I voluntarily express to you my innermost thoughts. I do this out of gratitude [towards the Bakufu]."[24] This special relation between Tosa and the Bakufu was recognized by others as well.

In a letter dated 7/59 Date Munenari, in speaking of Tosa to Matsudaira Keiei of Fukui, wrote that their special gratitude to the Tokugawa for their 200,000 *koku* set them apart from the other *tozama* han.[25] In contrast to these feelings on the part of the Yamanouchi house, their hereditary vassals, the Tosa *gōshi,* who were originally the vassals of the Chōsokabe house deposed by Ieyasu, were hostile toward the Bakufu and formed the nucleus of the Tosa extremist anti-Bakufu movement.[26] Here as elsewhere this element of traditional hostility was probably not as immediate as their material frustrations as *gōshi* or the difficulties that cropped up from time to time during the Tokugawa period between them and the hereditary upper vassals of the Yamanouchi. It is, nevertheless, a

[23] All dates given in this book, unless indicated otherwise, will follow the Japanese calendar. Thus, 19/5/62 refers to the nineteenth day of the fifth month in 1862 (the second year of the Bunkyū period); 7/59 refers to the seventh month in 1859 (the sixth year of Ansei). Now and then I will also use the form 18/8, referring to the eighteenth day of the eighth month, without mentioning the year when it is obvious. Of course, 1862 is not perfectly congruent with the second year of Bunkyū; it does, however, indicate chronological sequence with greater clarity. Intercalary months will be indicated by "i" before the number of the month.

[24] *Ishin shi, III,* pp. 208–209.

[25] *Ibid.,* pp. 208–209.

[26] *Ibid.,* pp. 209–212.

not unimportant factor, to which too little significance has been attached in recent years.[27]

Finally, in discussing traditional relations and their influence on the actions of the han in Bakumatsu Japan, it is worth noting the unique historical relation of the house of Mōri to the Imperial Court. The *Bōchō kaiten shi* refers to this as one "great cause" explaining why "the Mōri of Chōshū alone, surpassing the other lords, energetic and daring, could finally achieve the great deed of the Restoration":[28]

The Mōri were originally of the Kyoto official class. For generations they frequented the palace; they were close to the jeweled throne receiving from it many favors. From the time of [Ōe no] Hiromoto [the 38th generation in the line from which the Mōri traced their descent] the Mōri became a military house. Yet because of these circumstances its feelings of longing for the imperial city were almost those of a distant traveller longing for home, feelings incomparably greater than those of the ordinary warrior of a military house.[29]

This was followed by several centuries of ups and downs until the house of Mōri was finally established following the demise of Ōuchi. Yet even in this period of glory the Mōri did not forget their debt to the Court. When the fortunes of the Court were at their lowest ebb during the period of Warring States, Mōri Motonari donated money to the Emperor Ōgimachi to pay for his enthronement ceremonies. And, after this for three successive generations it was the custom to contribute each year to the Court the income from certain silver mines. As a consequence of these early ties, so the *Bōchō kaiten shi* relates, during the Tokugawa period when all direct relations between the military houses and Court nobles were interdicted, the

[27] A recent exception to this rule is the article by Marius Jansen on Takechi Zuizan in which he speaks, on the one hand, of the Chōsokabe origins of the Tosa *gōshi* and, on the other, stresses the importance of the traditional differences between Tosa and Chōshū: "The ruling houses of the two areas stood in a very different relationship to the house of Tokugawa, as the Yamauchi of Tosa bore it gratitude for good treatment while the Mōri of Chōshū had a long-standing grudge to settle." See Marius B. Jansen, "Takechi Zuizan and the Tosa Loyalist Party," *The Journal of Asian Studies,* 18:199–200 (Feb., 1959).

[28] Suematsu Kenchō, *Bōchō kaiten shi* (Tokyo, 1929), I, 29.

[29] *Ibid.,* pp. 29–30. This and all other translations unless otherwise noted are mine.

daimyo of Chōshū alone was permitted to visit the noble houses on his way to and from Edo.[30]

Elements of the above history are obviously open to question, but at least during the Tokugawa period Chōshū does seem to have had a particularly close relation to the Court. How strong an influence this had on its Bakumatsu action is problematic, but Chōshū samurai were at least not unaware of it. Pleading for stronger han participation in national politics, Kusaka Genzui wrote on 8/4/62: "From the first the house of Mōri has been different from other han and for the last several hundred years it has been the stronghold of the Imperial (*sonnō*) cause."[31] At least, as in the case of the traditional hostility to the Bakufu mentioned above, this relation to the Court provided an excuse for those advocating stronger participation in national affairs and proved an obstacle to those supporting a neutral, passive role.

[30] *Ibid.*, p. 34.

[31] *Ishin shi*, III, 4. Not only were the Mōri originally descended from a Kyoto noble house, they also, according to Chōshū tradition, were an offshoot of the Imperial line: allegedly the twenty-seventh lord in the line from which the Mōri descended received as a wife a lady-in-waiting of the Nakatomi family who was already pregnant by the eldest son of the Emperor Heizei (806–809). As a consequence, the ancestors of the Mōri who by a still older tradition were descended from the gods (*shimbetsu*), were from the twenty-eighth generation also of Imperial stock. See Misaka, *Hagi han no zaisei to buiku*, pp. 17–18.

CHAPTER II

Chōshū's Finances and the Tokugawa Economy

The Tokugawa Economy

Still another factor contributing to Chōshū's ability to act during the Bakumatsu period, when most han were of necessity quiescent, was its financial solvency. Too few studies, even of the larger han, are available for us to say with confidence that Chōshū (and perhaps Satsuma) was unique. Yet it can be said that Chōshū's solvency contrasts strongly with the reputed insolvency of most other han, and that it was this solvency that not only enabled Chōshū to buy rifles, cannons, and ships, but also gave to Chōshū the leeway needed to plan and carry out various reform programs before and during its involvement in national political affairs. The system by which Chōshū was able to obtain this solvency was a peculiar one in which, on the one hand, savings were accumulated and cash reserves built up, while, on the other hand, the han borrowed money and went deeply in debt. In certain respects, this appears as an early form of deficit financing in which the right hand did not know, or was pledged not to interfere in, what the left hand was doing. Before treating this relatively complex system in greater detail, however, it is necessary first to sketch briefly the relevant aspects of the Tokugawa economy in order to see the context within which the Chōshū policy had evolved.

The Tokugawa economy has been fairly well studied. It is a field in which only fragmentary evidence exists in many areas, yet from these fragments many of the leading economic institutions, especially those of the major cities, have been carefully reconstructed. Many of these studies emerged from the great historical debate of the 1930's concerning the nature of early Japanese capitalism, and many more have been made during the modified form of this debate which has continued since the end of the Second World War. In contrast to studies of economic structure, very little work has been done on the dynamics of the Tokugawa economy. Even fairly elementary questions such as the relative prices of staples, manufactured products, and services over the course of the Tokugawa period, questions for which historical materials do certainly exist, have not been satisfactorily answered. Keeping in mind, then, the tenuous monographic base on which all accounts of Tokugawa economic process rest, let us turn to several general explanations of the rise of han debt.

The earliest theory, and one that is still widely held, explained han debt in terms of the incompatibility of fiscal practices geared to a "natural" economy and the encroachment of a money economy. According to this theory Japan had an almost natural or moneyless economy at the beginning of the Tokugawa period. Then, supposedly, about 1700 a money economy was ushered in, which gradually spread until by the end of the Edo period it had permeated almost every area of life. Accordingly, the expenses of the han increased, but, since their incomes were fixed, they became increasingly dependent on loans and sank deeper and deeper into debt. Interest rates were usurious; therefore, at a certain point the circle closed and escape became economically impossible. This in sum is the "monetary revolution" or "inflationary theory" of han debt.

Actually, an equally strong case can be made for the opposite contention: that han indebtedness gave rise to the money economy. This is the *"sankin kōtai* theory," which in recent years has become justly popular.[1] According to this theory, the rise of han debts began

[1] This *sankin kōtai* theory is held, in various forms, by proponents of diverse schools of thought. On the left Naramoto Tatsuya writes: "The special character of the society of the Tokugawa feudal system was that the financial systems of the various han were from their very inception forced into the sphere of activity of merchant capital by the unique control forms of the Tokugawa Bakufu, thus accelerat-

early in the Tokugawa period and was due primarily to the demands of the Bakufu control system, that is, to the periodic Bakufu levies for money or labor and the *sankin kōtai* system which required the families of daimyo to live in Edo and the daimyo themselves to reside in Edo every other year. There seems to be little doubt that it was these additional expenses entailed by the Bakufu control system that gave rise to the enormous debts characteristic of the period.

This also seems to have been the most important factor contributing to the rise of the Tokugawa money economy. The daimyo were forced to obtain specie with which to pay for these "control expenses"; their only source of specie was the sale of the "surplus production" of the han, and consequently, from early in the seventeenth century, the daimyo attempted to develop cash crops within their domains. The sale of these cash crops in Osaka and other market centers, and the subsequent use of the specie to meet han expenses in these same centers, led to a rising volume of commercial transactions, a nationwide distribution net, and a vastly increased use of money.

In a large measure, this *sankin kōtai* theory helps to explain the vigor of the Tokugawa economy, a vigor which in many ways was not to be expected in the transition from the open economy of pre-Tokugawa times to the secluded economy of the Edo period. Even before the unification of Japan by Nobunaga and Hideyoshi a considerable degree of commercial practice had risen in Japan. The economy was not as moneyless as some have assumed. The *za* or early guilds and the *toimaru* (that in some cases developed into the *toiya* or wholesale houses of the Edo period) had both existed since Ashikaga times. Even if each *za* and each *toimaru* is viewed as an

---

ing the development of the fundamental contradictions which cut across the roots of the feudal society." He then goes on to specify these control forms as the hierarchical social pyramid, the *sankin kōtai* system, and the Bakufu levies. See Naramoto Tatsuya, *Kinsei hōken shakai shiron*, pp. 26–27. More towards the center, Sakata Yoshio writes: "The main contradiction was the financial poverty of the han, which was due not to the development of a merchant economy but to the Tokugawa control system, a contradiction present from 1700, by which time the han were completely at the end of their tethers." See Sakata Yoshio, "Meiji ishin to Tempō kaikaku," *Jimbun gakuhō*, 2:2 (Mar., 1952). Undoubtedly the best treatment in English of this system is that by George Toshio Tsukahira, "The Sankin Kōtai System of Tokugawa Japan, 1600–1868" (diss. Harvard University, 1951).

island of monetary usage in a sea of barter payments in kind, by the time of Nobunaga these islands had spread to include almost all commercial transactions, and an exceedingly complex network of trade had grown up between them. The castle town of each area served as the center of both consumption and handicraft production, and a considerable trade flourished between the different areas, the craft and crop specialties of each being sold in areas in which they were lacking. By the middle of the sixteenth century several semi-autonomous towns had risen that were quite similar to the free cities of late medieval Europe; these functioned as distribution centers for the surrounding feudal areas and were also actively engaged in foreign trade.

Then, with the political centralization of the country, and more particularly, with the adoption of the policy of seclusion, the Japanese economy was forced to turn in upon itself. The semiautonomous towns, for the most part, were taken over by the Bakufu in the form of *daikansho,* areas under the direct control of the Bakufu, and deprived of their foreign trade, they soon dwindled and declined.

But, at the point at which it appeared that the commercial economy of Japan might, stagnating, drop to even lower depths, it rose and began to expand. The castle towns, which earlier had been the primary centers of production, became, with the institution of the *sankin kōtai* system, secondary to Edo and Osaka. Like iron filings on paper above the ends of a magnet, the trade of the country formed into an almost magnetic field converging on these two poles of the economy. Edo, which was filled with the samurai of the various han and the families of the daimyo, became the chief consumption center of the country. On alternate years the daimyo with their huge retinues would proceed to Edo. Han residences of great size and magnificence were established. Each daimyo attempted to outdo his fellows in pomp and ostentation. Businesses catering to their needs and wants grew up and prospered, and other businesses were begun in turn to care for the city's swelling population. Edo was the most important center of consumption in Japan, and Osaka, the "kitchen" of Tokugawa Japan, became the distribution center servicing its needs. Prices rose, and the daimyo were forced to spend in Edo all that they could wrest from their domains. It should be noted that the highest tax rates of the period of Warring States, "one-half [of the han's agricultural production] for the lord and one-half for the people," were continued, and that in some cases they were even raised during the Tokugawa period.[2] Within this context the *sankin kōtai* system might be viewed as a pump-priming device—actually it acted more as a pump than a primer—to counteract the depressing effects of seclusion, although it was certainly not set up with any such end in mind.

An index of the concentration of consumption in Edo was its sudden growth. Before Sekigahara, Edo had been merely a small village used as a military headquarters by Tokugawa Ieyasu. Only one hundred years later, it boasted a population of 800,000, of which 470,000 were merchants. In contrast to this, London at the beginning of the eighteenth century had a population of 674,000 and Paris only 550,000. Berlin in 1740 had a population of only 90,000.

And, to further extend the metaphor of the *sankin kōtai* system as a pump, I would suggest that it was one the very mechanism of which limited the capacity of the economy for further expansion. Most of the elements determining the cost of *sankin kōtai* procedures were fixed by political authority and custom: ceremonies in Edo, the permanent staff in Edo, the size and number of gardens in the Edo mansions, the size of the daimyos' retinues, standards of dress and equipment, and so on. Once fixed in the late seventeenth century these institutionalized forms, on the one hand, kept the economy going at a certain level, and on the other, prevented it from going beyond that level. By the early eighteenth century the limit of growth within this system had been reached (population might be taken as an index to this). As "government cities" Edo and (derivatively) Osaka could expand no further.

Beyond this limit further growth could come about only by qualitative change within—through a better technology, or by economic rationalization of one sort or another—or as a response to new economic forces outside of the *sankin kōtai* mechanism. Change

[2] Sakata, "Meiji ishin to Tempō kaikaku," p. 3.

within, however, was particularly difficult. In the early eighteenth century the Bakufu accorded monopolistic privileges to merchant associations that had functioned freely without such controls since before 1600. This had the effect of encapsulating and fixing the commercial *status quo*. Another factor inhibiting further growth was the conservatism of the merchants themselves. The forms and practices that had led to such astonishing growth in the period of initial commercial expansion from 1600 to 1700 were hallowed by success and difficult to change. This tendency was abetted by Confucian traditionalism, by the paternalistic structure of the merchant houses, and by the moralistic "house codes." Some restructuring did take place after 1700, but there was little dramatic innovation.

Most of such restructuring seems to have come as a response to increased commercial relations with local trade centers. But one must not overemphasize the degree of such changes. Like Edo, the castle towns were basically "government cities." The forced residence of samurai made of the towns local consumption centers and much of their commerce was concerned with supplying the needs of their warrior population. The castle towns were also centers for trade with Osaka, one component in the *sankin kōtai* net. As such they had achieved most of their Tokugawa growth by the end of the seventeenth century. And, for the same reasons given earlier in the case of Edo and Osaka, castle town merchants tended to be conservative. Yet the castle town merchant class seems to have gradually increased throughout the eighteenth century, when there also appeared some noncastle town commercial cities. In Chōshū, Shimonoseki furnishes the best example of this, but evidences of increased commercial activity are available for other Inland Sea ports as well. One factor distinguishing han merchants from those of Osaka and Edo was their greater dependence on han power (even during their phase of initial growth). In the eighteenth century when han finances were increasingly troubled and han governments were searching for profitable expedients, this dependence may have made some han merchants with semiofficial functions less resistant to change than their Edo counterparts.

Another factor, and one especially important in the case of noncastle town commercial towns, was that of commercial growth in rural areas—a type of growth usually associated with, though not limited to, the late eighteenth and early nineteenth centuries. To mention only a few of the varieties of rural enterprise, there were: factories begun by well-to-do peasants (producing sake, soy sauce, and the like), rural handicraft operations, rural commercial ventures: by the mid-nineteenth century peasants were dependent for many necessities on the purchase of goods which they no longer produced (or had not previously regarded as necessities), rural production organized by city merchants, and rural merchant houses handling certain aspects of the han's monopolies. Both the degree of development and the type of rural activity varied widely from area to area. Chōshū, even in those areas along the Inland Sea, was probably much less developed than most of central Japan. It is also difficult to generalize regarding the meaning of such local developments. At times they were organized by city merchants, at times they were parallel to, but not competitive with, city enterprises, and at other times they competed directly with city guilds.

Most of the commercial forms that have been described above, whether castle town guilds, port warehouses, or the great wholesale distributors, were called *nakama. Nakama* is a very vague term meaning a group engaging in some form of business. In the course of the eighteenth century, particularly after 1721, the *nakama* were given official recognition—or at least those in the Bakufu domains were—and so became *kabunakama*. The *nakama* were recognized so that they might be controlled and taxed. The annual tax was known as the *myōgakin*. In exchange for this payment each *nakama* was given monopoly rights in its particular field. This tended to limit the number of businesses, but also made business more profitable, particularly since the annual tax appears to have been nominal. In modern Japan *kabu* means stock. During the Tokugawa period, however, the *kabunakama* were not joint stock companies in any sense. Rather, in its most developed form, which was found chiefly in certain Osaka and Edo *toiya,* the possession of *kabu* merely signified the right to deal in certain commodities within a monopolistic framework. Most *kabu* were obtained through inheritance, although some were purchased; buying such a *kabu* may be compared to the purchase of a seat in a commodity exchange, but even this may be an overstatement.

It is tempting to see the orderly evolution of the Japanese economy as roughly parallel with the European pattern of development. Examining solely the purely economic side of the picture, parallels do exist that appear to be more than superficial. As in the West early periodic markets were replaced in Japan by market guilds; these in turn gave way subsequently to a freer economy within which there appeared great commercial houses, the domestic or "putting out" system of production, and even factories engaged in handicraft production. Yet when we attempt to see this evolutionary process in relation to certain characteristic political forms, the parallels break down and we are forced to wonder whether cognate economic structures have, in fact, the same historical meaning.

For example, in Europe the Warring States phase of feudalism (say, France between the tenth and twelfth centuries) was one of self-sufficient feudal territories whose patchwork character and endemic warfare are often adduced to explain the absence of commercial development. In contrast to this the Japanese period of Warring States (which politically had many similar features: the culmination of the process by which estates became transformed into regional fiefs, the emergence of primogeniture, and warfare unhindered by the restraints of some central power) was coupled with the guild phase (the Ashikaga *za*) of its economy.

A disjunction can also be seen in the period that followed. In Europe commercial capitalism in its heyday was associated with the rising monarchies of the sixteenth and seventeenth centuries; it is usually viewed, in part at least, as the internal response to the stimulation of oceanic trade. In Japan, however, the rise of a vigorous maritime commerce occurred earlier during the period of regional guilds, while the development of a comparable commercial capitalism took place during the period of Tokugawa seclusion. It took place in a society whose basic political and social structure, to the extent that comparisons can be made at all, was not more advanced than that of France in the fourteenth century (although there existed a far greater degree of centralized control if not administrative centralization).

Not to grasp this, to assume that regular stages do exist—and Japanese historians as well as Western students of the Japanese scene are often guilty of this error—is to make incomprehensible many aspects of Tokugawa history. One such phenomenon that can only be understood by taking into account this "disjunction" is the political consciousness of the Edo merchant which, in spite of the relatively advanced level of the Tokugawa economy, remained low.

Yet can this sort of "disjunction" between the economy and polity have actually existed? Most Western historians have usually assumed that advanced economic forms could not develop within a restrictive "feudal" framework, or, that, once formed, such an economy could not but corrode and ultimately destroy such a society.[3] In many ways this view is obviously true. Many were the changes to which the Tokugawa economy contributed during the several hundred years of the period. Since most surveys of the economy have been previously concerned to document these, it would be superfluous to list them here.

Rather, I would emphasize that in spite of these changes the main outlines of the Japanese society of the late seventeenth century were still easily recognizable in the mid-nineteenth century, and, moreover, the old society was still surprisingly functional. The commercial economy was somehow less destructive than it should have been in terms of a European model of history.

In part this can be explained by the effective compartmentalization or containment of the Edo-centered economy. While the Tokugawa economy was national in scope it possessed a sort of imbalance: local areas were drained by heavy taxes which, used in the cities, did not benefit the taxpayers. There were areas of monetary exchange, both locally and among different regions; but the level was low. Until the end of the period more taxes were collected in kind than in cash.

[3] Of course we are speaking here at such a high level of generalization and about phenomena so complex that any statement is both true and false at the same time. It is easy, for example, to find in European history instances where the commercialization of agriculture resulted in the strengthening of the manorial ties rather than the opposite. See, for example, G. O. Sayles, *The Medieval Foundations of England* (London, 1956), p. 435. Still, on the whole, it would be extremely difficult to find in Europe an example of the juxtaposition of advanced commercial practice with the type of feudal society found in Tokugawa Japan.

Wide areas of economic self-sufficiency exist within any peasant society, yet it is always thought-provoking to consider the ways (money taxes, a greater degree of landlordism, etc.) in which the Chinese rural economy in the nineteenth century was more advanced than that of Japan.

The gulf between rural (or provincial) and city life was social as well as economic. Despite the often mentioned predilection of samurai for the forbidden delights of *chōnin* dissipations a fairly effective separation of classes was maintained. A few officials of the han, such as the father of Fukuzawa Yukichi, were stationed in Osaka to handle the sale of han products, and a larger contingent of han troops would accompany the daimyo on his biennial trek to the capital, but the majority stayed at home. Ideas and individuals traveled about but they had little impress on han life. Saikaku and the writings of han moralists seem to display the antipodality of the two types of society. This sort of comparison, however, is unfair for recent studies have shown that the merchants were, in their own way as "Confucian" as their samurai opposites: the dreary writings of *shingaku* illustrate this well enough. Still, the stricter controls of the castle towns and the poverty of most peasants made smaller what Benedict has termed the circle of human feeling and more onerous that of duty.[4]

The inability of the Tokugawa commercial economy to destroy its semifeudal society may suggest that even in Europe the corrosiveness of money has been overrated. War with other states and the rationalizing (if not economically beneficial) effects of war on political relations combined with money to break down European feudalism. The money economy alone, also present in Japan, may not have been enough. Thus seclusion and the Tokugawa peace may have enabled Japan to preserve the local autonomy necessary for the preservation of its "backward" social institutions. The wars of medieval and late medieval Europe, for example, forced the feudal barons and rising monarchs into greater dependence on the merchants than was the case in Japan. Daimyo did go into debt, and in extreme cases the control of the finances of certain han seems to have fallen into the hands of Osaka merchants, but the very political

[4] Ruth Benedict, *The Chrysanthemum and the Sword,* pp. 177–195.

existence of a daimyo was never balanced precariously on a merchant's decision to finance a war or not. Consequently the *chōnin* had much less leverage than their European counterparts and one never saw the emergence in Japan of middle-class bureaucracies or "nobles of the robe."

Differences in population density also help to explain the particular Japanese pattern of evolution. France in its period of Warring States had a population of about two million. Japan in the late Ashikaga period had a population at least seven or eight times as great on a more compact base of intensive agriculture. Perhaps the larger base and the greater wealth that could be extracted from it in taxes made it possible to support a larger warrior class and a higher level of demand even within a society in which vertical feudal loyalties were still dominant. This would explain, on the one hand, the great armies of Warring States Japan—which in their numerical strength resemble those of the era of bastard feudalism in Europe—and, on the other, the early development of commerce in the regionally segmented pre-1568 society. It would also help explain how during the Tokugawa period urban populations greater than those of Europe could exist on a commercial base that structurally was less differentiated and less rationalized than its European counterparts. More accurate parallels might be drawn with China. Professor Sakata Yoshio has spoken of the Tokugawa economy as one which, owing to the nature of its consumers, saw the development of industrial arts rather than industrial production.[5] Certainly, the higher degree of political centralization, as well as the existence of external markets, created in European countries a greater volume of trade in some industrial commodities than existed in Japan.

But it is not enough merely to say that, while essentially antagonistic to the old society and culture, the economy was unable to destroy it. In some ways one must go further and say that the older type of society and culture positively contributed to the establishment of the advanced economy. A few examples will illustrate that this means more than the truism that whatever is present in a society will in some fashion be joined. The *sankin kōtai* system, the economic consequences of which we have already considered, was in

[5] Sakata, "Meiji ishin to Tempō kaikaku," p. 4.

its dependence on a formalized version of the lord-vassal relation and a system of hostages (rather than some system of intendants) basically feudal in character. Only in connection with this system of controls can one explain the two hundred and fifty year long Tokugawa peace, so important to economic growth. Perhaps one can compare the economic unity of the Bakufu domain (*tenryō*) with the Five Great Tax Farms established in France by Colbert in 1664. Barriers within this area, unlike those of the greater han, were primarily political, not economic, in function.

A second positive contribution of the old society to the Tokugawa economy (one which, like the *sankin kōtai* system, acted as a limiting factor as well) was in the area of class structure. The inalienability of land, stemming from the feudal principle that all land belonged to the lord, and the hereditary character of the ruling class meant that, unlike China, and with occasional exceptions, city merchants were unable to become landowners, and landowners were unable to enter the ruling class. This channelized their profits into either high interest loans or reinvestment. Why this did not result in pyramiding and increasingly higher levels of economic activity is difficult to say. It would appear that the fixed level of demand and the relatively stable technology limited growth on the one hand, while the repudiation of debts and occasional levies (a sort of unofficial taxation, often at confiscatory levels) ate into capital on the other. Still, great merchant fortunes were amassed which, within a different social setting, might have been put to noncommercial ends. In rural areas the *de facto* transfer of land was more frequent but even here there were difficulties: the tendency for wealthy peasants to convert their surpluses to high interest or investment capital was not dissimilar to that of the *chōnin*.

Traditional Confucian ideas also fed into the economy: the great stress on frugality and the almost physiocratic emphasis on agriculture which predisposed the Tokugawa rulers to avoid merchant taxes furnish two examples of this. The basic Confucian denigration of profit, however, seems contradicted by much of the merchant tradition; this is reflected in *chōnin* literature, much of which is preoccupied with misers, or obversely, with profligates. This has led many writers, both during the Tokugawa period and at the present, to describe the *chōnin* ethos as basically different from that of the samurai. Actually, however, as rationalized within the city culture, profits were not selfish, they were not for the individual, but for the family (or house), and so, by transposition, became the fulfillment of filial piety. This emphasis on the house and the duties due to it is not unlike the Calvinist idea of stewardship. Certainly this concept helped to hold together the merchant fortunes of the period: the fate of a merchant house whose head had lost sight of these duties was a constant theme of the half didactic sketches of the Tokugawa writers.[6]

Comparisons are at best hazardous. Yet if the premises which they involve are not made explicitly they will be present as assumptions even in the most factual account: the inflation theory discussed earlier is an example of the dangers of transferring implicit assumptions based on European experience to the Japanese milieu, and the application of Marxist formulas to the problem runs the same risk. In any case, seen within a comparative context, the factors influencing the rise of the Tokugawa economy—and the course of han finances—are many; it cannot be explained in terms of any simple theory of inflation or in terms of any given element, such as the *sankin kōtai* system. The influence of some of these factors can be observed in the case of Chōshū.

Chōshū Finances

As we have suggested earlier, the financial imbalance began in the han not during the eighteenth century but early in the seventeenth. In 1643 Chōshū's Edo expenses came to 1260 *kan* of silver and those in Kyoto, Osaka, and the han itself to 1326 *kan,* a total of 2586 *kan*. The income of the han after the subtraction of fixed expenses was only 1253 *kan,* leaving a deficit of 1333 *kan*. This was the equivalent of 53,320 *koku* of rice at the prevailing 1643 market price. From these figures it is clear that the deficit was due to expenses incurred outside

[6] The most thorough and penetrating study of the larger question of religion and the Tokugawa economy is that of Robert N. Bellah, *Tokugawa Religion: The Values of Pre-industrial Japan,* pp. 107–177.

the han; these were mainly related to the *sankin kōtai* system.[7] The other han in Tokugawa Japan fared no better; many, in fact, were in even worse condition. By 1700, over half of the han could manage their finances only by supplementing their fixed incomes with money borrowed from the merchants and moneylenders of Osaka and Kyoto.[8]

Why was the income of the Chōshū government in 1643 only a meager 53,320 *koku* of rice after the deduction of fixed expenses? This seems a very small amount when the total assessed productive capacity for the han was 658,299 *koku* at this time. What was the nature of the fixed expenses which consumed the major part of the han's revenues? Unfortunately a complete set of budget figures is not available for the early years of the period; therefore, for purposes of illustration, the budget proposed at the time of the Tempō Reform in 1840—or that part of it dealing with revenues—will be given. This can be used to clarify the 1643 figures since the ratio between the estimated total production and the amount available as income was substantially the same in 1840 as in the early seventeenth century. The Chōshū budget in 1840 was:[9]

Amount in the domains of the Mōri branch han[10]	183,022	*koku*
Amount in fief held by Mōri retainers, etc.	205,557+	*koku*
Land taxed for stipends (*kirimai, fuchimai, ukimai*)	285,468+	*koku*
Buikukyoku (Savings Fund) revenues	41,414+	*koku*
Compensation for decline in production	7,471+	*koku*
Public works, peasant officials, etc.	26,062+	*koku*
The number of *koku* remaining to be taxed as a source of income to the han government	146,161+	*koku*
Total estimated Chōshū agricultural product	895,158+	*koku*

Regular han government tax income	28,518 *koku* and 870	*kan*
Supplementary han government tax income	13,826 *koku* and 803	*kan*
Total taxes expressed in *koku*	75,800	*koku*
Total taxes expressed in silver *kan*	3,790	*kan*

[7] Misaka Keiji, *Hagi han no zaisei to buiku*, pp. 54–57.

[8] Sakata, "Meiji ishin to Tempō kaikaku," p. 4.

[9] Suematsu Kenchō, *Bōchō kaiten shi*, I, pp. 283–288.

[10] The four Mōri branch han are discussed at greater length in Chapters IV and VIII.

It is evident from the preceding budget that the *kokudaka* of a han was not a statement of the revenues available to its government. In the case of Chōshū, 20 per cent of the han product went to the branch han which were autonomous financially. The main han, Chōshū, had absolutely no control over any of this amount. About 54 per cent of the han product, that is to say, the tax revenues on about 54 per cent, went to the vassals in the form of fiefs and stipends. The stipends were administered by the han as were certain of the fiefs. Other fiefs, particularly the larger ones, were administered by the samurai holding the fief—usually with the aid of rear vassals and peasant officials. The form in which the fiefs were held and their management, however, were subject to gradual change throughout the Tokugawa period. The taxes on the remainder of the han product in both 1643 and 1840 came to only 8 per cent of the total gross product. This was far short of the amount needed to meet the growing expenses of the han government.

Throughout the Tokugawa period Chōshū's Edo expenses seem to have been roughly equivalent to the expenses of the government in the han itself. As in 1643, the revenues of the han government did not suffice to meet these expenses. The han debt, therefore, grew relentlessly as shown here.[11]

Year	Debt expressed in *kan* of silver (One *kan* equals 8.27 lbs.)	Year	Debt expressed in *kan* of silver (One *kan* equals 8.27 lbs.)
1603	. . .	1704	11,613
1622	4,000	1705	15,822
1632	. . .	1707	10,000
1638	. . .	1708	13,000
1644	3,682	1712	50,000
1646	6,200	1724	12,547
1649	8,439	1730	15,076
1652	7,430	1731	15,000
1658	3,912	1758	41,300
1676	12,000	1838	92,026
1682	22,000	1840	85,252

[11] Tanaka Akira, "Chōshū han no Tempō kaikaku," *Historia*, 18:26 (June, 1957).

The enormity of the debt will be better appreciated when these figures are compared with those in p. 39, showing the total han income in terms of silver.

And yet, considering that in 1643 the han's income was less than one half of the han's expenses, and that a deficit of 1333 *kan* of silver had occurred, one might legitimately query why the rise of the han debt was not even more rapid. First, let us assume that a deficit occurred only in those years in which the daimyo was in attendance in Edo. There were fewer ceremonial expenses and fewer samurai at the Chōshū residence in Edo during the absence of the daimyo. But the family of the daimyo was always in Edo, and the permanent staff was probably more or less constant, so that the difference may not have been significant. In any case let us assume that it was. Second, let us assume that even in those years in which the daimyo resided in Edo,[12] the deficit came to only 1000 *kan* of silver—this is clearly an overly conservative figure since the deficit in 1643 was 1333 *kan*, and this must have at least doubled or tripled by the end of the Edo period owing to inflation. Even then, the han debt should have, over a period of two hundred and forty years, exceeded 100,000 *kan*. And the periodic Bakufu levies should have driven the debt even higher. Instead, the han debt in 1840 was less than this amount. It was, in fact, even less than the 85,252-*kan* figure given in the above tabulation, since it was counterbalanced by a "han savings account," of which more will be said later in the chapter.

The relatively slow rise of the debt becomes even more of a puzzle when one considers the annual increment represented by interest on the debt. The 1838 debt was 92,026 *kan* of silver. Payments had been discontinued on both the interest and the principal of 32,842 *kan* of this amount, leaving a *de facto* debt of 59,184 *kan* of silver. The annual payments of interest and principal on this amount came

[12] This is not to suggest that the stay of the Chōshū daimyo in Edo corresponded with the calendar year. It did not: both the Chōshū daimyo and those of the Chōshū branch han were, during the 1850's, scheduled to arrive in Edo during the fourth month of odd numbered years and to leave during the same month of even numbered years. All that my argument requires at this point is that the daimyo had spent half of his time in the han and the other half in Edo.

to 12,175 *kan* of silver, an amount in excess of the annual han income which, in turn, was not sufficient to pay for the annual expenses of the han.[13]

The explanation for the paradoxically slow rise in the debt, in spite of the above, is that the han benefited from increases in productivity, from the increased value which inflation gave to that part of its tax revenues collected in kind, and that it had other invisible sources of income, invisible since they were not included among the revenues in the draft budget given on p. 39.

The following tabulation[14] illustrates the gradual growth of the Chōshū *kokudaka* as disclosed by periodic land surveys.

Year of the Survey	Assessed Production Expressed in *koku*
1600	298,480
1610	525,435
1625	658,299
1686-87	818,487
1761-62	892,100
1840	895,158

A similar expansion in the actual productivity of other han can be seen by comparing their officially listed productivity (*omotedaka*) with their actual recorded productivity (*jitsudaka*) as given in Table 1 in the first chapter. Chōshū seems to have made far greater gains than any of the other han for which comparative figures are available. Chōshū's most conspicuous gains were made during the first twenty-five years following 1600; by the latter part of the seventeenth century the growth in its productivity had all but come to a halt. As was suggested in the first chapter, it seems plausible to assume that the abrupt rise in assessed productivity during the early years was not so much the result of a real increase in productivity as it was of gains uncovered by more thorough surveys. During the period of Warring States, many fields, such as those in the fiefs of lesser lords who had voluntarily become the vassals of Mōri, may have been exempted from land surveys in order to insure the loyalty of these lesser lords. After the establishment of the "Tokugawa peace," such political

[13] Misaka, *Hagi han no zaisei to buiku,* pp. 270–271.
[14] Tanaka, "Tempō kaikaku," p. 16.

considerations were no longer necessary, and, since even the most aggrieved vassal could not switch his allegiance, his fiefs were duly surveyed along with other lands in the han.

The influence of inflation on the total han economy cannot really be estimated until a breakdown of han expenses over the years can be obtained. The influence of inflation on han revenues, however, was beneficial, since even at the end of the Tokugawa period, more taxes were collected in kind than in cash. Thus, in 1840 the revenues of the han were equivalent to 75,800 *koku* of rice; this is only 50 per cent greater than the 50,120 *koku* revenue in 1643. The silver equivalent of the 1840 figure, however, was three times greater than that of the 1643 figure, and by Bakumatsu times the value—in terms of silver—of the han revenues was even greater. The inverse effect of inflation on the han debt is obvious.

Finally, invisible revenues augmented the real income available to the Chōshū government. One such income was that from the han-controlled monopolies which was not included in the budget given above; these will be discussed at length in the following chapter. Another even more important item was the amount "borrowed" from the fief revenues and stipends of the samurai. As we saw in the Chōshū budget, the total estimated production of the han in 1840, excluding that portion in the domains of the branch han, was 712,136 *koku*. Of this, the income from 491,025 *koku* went to the samurai of the han in the form of either fiefs or payment in rice. In the face of the steady deterioration of its finances, the temptation represented by this amount was too great to resist and early in the seventeenth century the han began to borrow a certain percentage of each samurai's income. The amount gradually increased until 1704, when for the first time the han was forced to borrow half of the samurai's income. This amounted to 70,000 *koku*, almost as much as the total fixed income derived from taxes on the remainder of its domains.[15]

[15] Considering the amount of the han product that went to the samurai in the form of fiefs and stipends, 70,000 *koku* may appear to be too small a figure. This will appear more reasonable, however, if two points are kept in mind: first, a samurai with a fief of 100 *koku* actually received only 40 *koku* (at the prevailing tax rate) and that one-half of this came to only 20 *koku*. Second, a full 50 per cent was borrowed only from those with stipends of 100 *koku* and above. See Misaka, *Hagi han no zaisei to buiku*, pp. 97–103.

The practice of taking half, or almost half, of the samurai's income persisted through the remainder of the Tokugawa period and probably constituted the most important single factor restraining the rise of the han debt.

And yet, lest such "invisible income" and increases in production paint too rosy a picture, it should also be mentioned that these were counterbalanced by "extraordinary expenses": the destruction by fire of one of the Chōshū Edo residences (*yashiki*), the costs of the marriage ceremonies for members of the daimyo's family, Bakufu levies, natural damages to crops by flood, drought or storm, and so on. Very few years passed in which one or another of these "extraordinary expenses" did not crop up, and so they were extraordinary in name alone. Time and again during the Edo period, a plan would be drawn up by which Chōshū's budget could be balanced and its debt repaid, only to have it founder and collapse with the appearance of one or several of these extraordinary expenses. Therefore, in spite of the mitigating factors discussed above, the debt continued to spiral upwards.

In the face of this rising debt, how did Chōshū emerge to play a dominant role in Bakumatsu politics? Power in these stormy years, while a function of many factors, was ultimately based on the military potential of a han which, in turn, meant cannon, modern rifles, and warships, which had to be purchased from abroad with specie. It was also expensive maintaining in the field even the relatively small armies that fought in the Bakumatsu campaigns. Staggering with debt, Chōshū was able to buy arms and maintain its armies. How? The answer to this question is of singular interest since it represents an institutional response of a sort that could have very well miscarried but nevertheless succeeded: a response of a totally different nature from that of Satsuma whose financial achievements of the Tempō period were largely based on geographical fortune.

The institution that gave Chōshū its financial leeway in the Bakumatsu period was called the *buikukyoku;* in effect it was a special office or bureau of the han government for savings and investment. Since the *buikukyoku* grew out of an earlier scheme called treasury-money (*hōzōkin*) we will first discuss the latter. By the early part of the eighteenth century the han had become adjusted to the idea of a managed debt (a practical if not a moral adjustment). The need to borrow occasionally was recognized. The problem faced by han officials was how to avoid borrowing in times of emergency or natural disasters when interest rates were hiked to usurious heights. Their answer was to set aside each year a certain amount of the han's income to use as a reserve fund for such times. The idea was sound but the reserve never accumulated, for each year, as soon as it was collected, it was promptly borrowed by the office in charge of han finance to augment the regular revenues of the han; this was easily managed since the treasury-money was under the control of the ministers of finance who argued with some logic that it was better to borrow from themselves than from outsiders, as the interest on the loan would then be conserved for the han. Consequently, when emergency expenses arose, no savings were available and the han continued to borrow at the same high rates of interest as before.[16]

From the failure of this treasury-money system was born in 1762 the *buikukyoku,* a special bureau of the han for the purpose of savings and investments, whose funds were not to be touched except in the direst emergencies. At the time of its formation, in order to ensure a regular income for effective functioning, the administrators decided to carry out a new land survey; the revenues from newly discovered taxable land were to go to the new bureau. The survey, while no different from previous ones, was called the "survey for the correction of tax inequalities" for fear that the Bakufu might burden Chōshū with further levies should it discover that an ordinary survey to find increased revenues was being made. The survey disclosed 63,373 *koku* of production previously untaxed. From this was deducted 21,764 *koku* representing formerly productive land no longer under cultivation; the net increase in production amounted to 41,609 *koku*. The revenues from this newly found production, about 20,000 *koku,* together with the revenues from a few other special taxes totaled about 30,000 *koku,* which then became the fixed

[16] *Ibid.*, pp. 70–77.

annual income of the *buikukyoku*. At the time of its formation, this was the equivalent of about 1000 *kan* of silver a year.[17]

We stated earlier that the function of the *buikukyoku* was primarily that of a bureau for savings and investment. Through savings, and by investment in areas that would both benefit the han and augment the capital of this office, it was hoped to obtain a reservoir that could be drawn on in time of emergency and from which relief could be given periodically to the penurious samurai of the han. Thus, from its inception the *buikukyoku* was cast against a background of rising debt and official depredations of the samurai's income. The name *buikukyoku* itself can be thought of as a public relations device to reassure the samurai that they were ever in the hearts of their leaders. *Buiku* means "to cherish" or "to rear with care" and it referred, of course, to the samurai, the class cherished by the daimyo. The distribution of these relief funds, a very small per cent of the total *buikukyoku* expenditures, was administered to the samurai in a truly Machiavellian fashion. The ministers in charge of the han's finances would first appropriate one half or nearly that much of the retainers' incomes to supplement the han finances, as had become customary by the middle of the eighteenth century. It would subsequently be announced that the daimyo, bearing in mind the welfare of his vassals, had decided to accord them a special grant; whereupon a sum very small in proportion to the amount taken previously from their incomes would be distributed to the samurai from *buiku* funds in the name of the daimyo.

Buiku investment took place in three fields: land, harbor development, and commerce. It was able to make a profit in each by combining capital resources with governmental authority. It would finance the reclamation of agricultural land that had gone to waste due to natural disasters. The reclaimed lands would then be sold cheaply and their taxes added to the annual income of the bureau. It built harbors along the Inland Sea, and it also invested in the commercial monopolies of the han, of which more will be said in the following chapter. Finally, from the beginning of the nineteenth

[17] The description of the *buiku* system given here is based almost entirely on the researches of Misaka Keiji. See *ibid.*, pp. 162–277.

century, the *buikukyoku* established at Shimonoseki a government *toiya*, or a combination warehouse-wholesale house, which afforded storage and lent money on security to the merchants of more distant han on their way to Osaka and also handled the sale of the Chōshū tax-rice. This *toiya*, known as the *koshinigata*, is of especial interest in that in Tempō times it alone was expanded when all other forms of government commercial enterprise were being abandoned.

How successful was the *buikukyoku* in preserving its autonomy in the face of the rising han debt? Profiting from their experience in the past with the treasury-money, the reformers who instituted the *buiku* office hedged it about with a number of checks to prevent its funds from being drawn into the regular accounts whenever a deficit occurred. It was established as an office completely separate from those offices which handled the han's regular finances. Placed under the direct supervision of the daimyo, his permission was required for any use of the funds. It was also announced in the name of the daimyo that other offices were to request *buiku* grants only when all other expedients had failed, and that for this to happen would be a sure sign of an official's incompetence. Furthermore, the precedent was established that each time the han was beset by a financial crisis, the daimyo would warn his ministers not to touch the *buiku* funds.

On the whole this system of checks worked fairly well. Between 1762 and 1784, 23,795 *kan* of *buiku* silver were used; of this only 12,125 *kan* were used to supplement the regular finances or as relief for the samurai. The remainder was used for emergencies, as originally intended, and for various extraordinary expenses, and several thousand *kan* were always kept on hand for further emergencies. Our knowledge of the *buiku* operations from 1784 on is largely a matter of conjecture since most of its records, unfortunately, have not been preserved. It is known that the system began to decline somewhat after 1784, until finally in 1800 all the savings were taken to fill a deficit in the regular finances of the han. This breakdown, apparently, provoked a reaction, in which new checks were established and a new and even more rigorous attitude was adopted towards the use of its funds. As a result, the *buiku* funds were left to accumulate almost untouched until Bakumatsu times. One historian attributes this post-1800 saving to the frequent injunctions by the daimyo, though he leaves unexplained why they should have become suddenly more effective after 1800, particularly since this was the period which saw the steepest rise in the han debt.[18] However implausible such a continuous accumulation of savings may seem, one is compelled to accept it, since, in spite of its debt, Chōshū was a comparatively wealthy han in the Bakumatsu period.

The rifles, cannon, and warships purchased from the West by Chōshū during the Bakumatsu period were paid for, almost entirely, with *buiku* money. The campaigns of the Chōshū troops during the last years of the Bakufu were also financed chiefly by *buiku* funds, although the han was later reimbursed for these by the new Meiji government. Furthermore, after the Restoration, Chōshū apparently still had 1,000,000 *ryō* of gold remaining in the *buiku* account; this was the equivalent of 71,600 *kan* of silver at the 1859 rate of exchange. This is to say that Chōshū's savings at the time were greater than its debt. Of this amount it was said that 700,000 *ryō* had been presented to the Court and 300,000 retained for the use of the han.[19] These figures, however, seem incredibly large, and since the 700,000-*ryō* gift to the Court was not officially recognized by the new regime until 1883 (Chōshū sources attribute the tardy acknowledgment to the confusion of the early Meiji period), an attitude of skepticism is perhaps justified. Considering the number of Chōshū personnel in the new government, it is somewhat odd that the recognition of such a gift should have come so late. Whatever the truth may be concerning the gift, there is no doubt that the financial role of the *buikukyoku* in Chōshū history was crucial. Without its savings Chōshū would have been unable to play a leading role in the Restoration. If the *buiku* savings were, in fact, as great as 71,600 *kan* of silver then one must surmise that the positive factors mentioned earlier more than counterbalanced the *sankin kōtai* expenses, the Bakufu levies, and the extraordinary expenses of the han. If this was so,

[18] *Ibid.*, pp. 247–249.
[19] *Ibid.*, pp. 325–328.

then even with a *jitsudaka* three times the size of its *omotedaka,* Chōshū was something of a phenomenon.

Can this phenomenon be explained in terms of the *buikukyoku* itself, or must more fundamental, underlying reasons be sought? If the answer is to be found in the *buikukyoku* itself, then Chōshū's financial strength may have stemmed from its skillful use of inflation and deficit financing. By offsetting borrowing with economically profitable investment, Chōshū could take full advantage of a century of gradual inflation. Its debts depreciated relatively in value, while its investments appreciated. This appears to have resulted in a much greater financial benefit than if the surplus had simply been used to reduce the han debt. As far as it goes this interpretation of the success of the *buikukyoku* appears to be sound. Inflation certainly decreased the real value of the han debt and the *buikukyoku* investments undoubtedly appreciated over time. Yet two doubts remain: first, were the gains from the effect of inflation on han debt and on the sales of han products not more than offset by the rise of han expenses, in the han and in Edo? At the present this question cannot be answered; further studies, however, of the prices of staples, handicraft products, and services as they changed throughout the Edo period may suggest that the han lost more than it gained by inflation. Second, while *buikukyoku* investments were profitable, they appear to have been relatively limited in scope. These alone cannot explain the Bakumatsu wealth of Chōshū. Several other factors may have been more important. One of these was the continuing stream of cash crops produced in the Chōshū domain. Another was the use of repudiation by the managers of Chōshū's debt. By 1838, one third of the entire han debt was, for all practical purposes, unacknowledged. Yet another factor that must be considered was the ability of the Chōshū government to impose on its samurai an effective program of frugality and to effectively curtail its own expenses. This manifestation of han strength, this driving concern with governmental goals, which seems to have been basic to the success of the *buiku* system, reflects the vitality which was also apparent in many other aspects of the han government.

Only further studies of Chōshū's total economy and of the finances of other han will enable us to evaluate the relative contributions of the above factors to the success of Chōshū's plan for *buiku* savings. There is no doubt, however, that it was these savings which made financially possible Chōshū's role in the Bakumatsu period.

CHAPTER III

Chōshū and the Tempō Reform

The Problem of the Reform

In recent years the thesis has been advanced among Japanese historians of the Marxist school that the reforms of the Tempō period (1830–1843) mark the starting point of Restoration history.[1] By this they mean that the particular class alliance which accomplished the Meiji Restoration first appeared on the scene in Tempō times. The Meiji Restoration, for most historians of this school, signifies the establishment of "absolutism" in Japan. "Absolutism," according to

[1] The great emphasis placed on the Tempō Reforms at the present time is a development that came only after the Second World War. The pioneer work on this subject was contained in two articles by Naramoto Tatsuya: "Kinsei hōken shakai ni okeru shōgyō shihon no mondai," *Nihon shi kenkyū,* V (Sept., 1947); "Bakumatsu ni okeru gōshi: chūnōsō no sekkyokuteki igi," *Rekishi hyōron,* X (Aug., 1947), both of which were later incorporated in his *Kinsei hōken shakai shiron* (Tokyo, 1952). Other more recent developments on this theme are: Tōyoma Shigeki, *Meiji ishin* (Tokyo, 1951); Inoue Kiyoshi, *Nihon gendai shi, I: Meiji ishin* (Tokyo, 1951); Horie Hideichi, *Meiji ishin no shakai kōzō* (Tokyo, 1954); Horie Hideichi, *Hansei kaikaku no kenkyū* (Tokyo, 1955). In addition to these works, which have provided the theoretical framework within which subsequent research has been done, monographic studies have also been made. Of special note are the studies on Chōshū by Seki Junya and Tanaka Akira, on Satsuma by Yamamoto Hirobumi, and on Saga by Gotō Yasushi. Most of the material in this chapter, if not otherwise designated, was taken from the work by Naramoto or from the monographs of Tanaka and Seki.

the Marxist interpretation, is not so much a theory of government as it is a particular type of class structure: a compromise stage between feudalism and capitalism in which an alliance is formed between certain strata of each class. As applied to the Meiji Restoration, the absolutist alliance is seen as having formed between lower samurai and an emerging class of local producers who represent the first stage of industrial capitalism. Opposed to this new absolutist alliance are the high-ranking (*mombatsu*) samurai, under whose aegis monopolistic merchant capitalists had been permitted to grow. Thus, on one hand, emergent local production (early industrial capitalism) confronts commercial capitalism; on the other, low-ranking samurai challenge high-ranking samurai. In the eyes of this school, the gradual overthrow of the old by the new is the essence of Restoration history: this is the rise of absolutism.

Since this new absolutist alliance of classes, reflecting in turn a fundamental evolutionary development in the productive process itself, first appeared in Tempō times, the Tempō Reforms are viewed by this school as fundamentally different from those of the Kyōhō and Kansei periods. The latter reforms were feudal reaction; the Tempō Reform, however, represented in its class basis the wave of the future, and therefore was progressive. Yet, in most han, even in Tempō times, the progressive forces were too weak, and were crushed. (In some fringe areas the component of early industrial capital was insufficient; in some central areas the progressive elements in the feudal class were too weak to override commercial capital.) Only in a few han, such as Mito, Chōshū, and Satsuma, were the progressive forces actually able to carry out their reform. The success, or partial success, of the reform in these areas contributed to the positive role played by these han in the Restoration movement. And, since the Tempō Reform was thus a type of reform, "Tempō Reform" is sometimes used as a generic name applicable to reforms which did not even take place during the Tempō period. Tosa and Echizen, for example, are said to have carried out their "Tempō Reform" after 1853.

Concretely, what form is this absolutist alliance said to have taken at the time of the Tempō Reforms; how were the rural producers and lower samurai supposed to have united? The answer is that a union of interests was formed in the items of the reform itself. The lower samurai were represented by a bureaucratic reform clique which carried out a reform benefiting the local producers at the expense of merchant capital. Examples of such items were decrees repudiating samurai debts or abolishing monopolistic control of special products. And, according to this interpretation, measures such as these had led the rich peasant (*gōnō*) class to support the lower samurai cliques, and this coalition, passing through various stages, ultimately took the form of the samurai-peasant mixed militia which overthrew the Bakufu.

Such a theory can be very stimulating, and has in fact given rise to a considerable volume of interesting and valuable research. Yet, I feel that at best it is oversimplified, and at worst it distorts out of all proportion the true picture of the Restoration. Specifically, I do not feel that the Restoration was the result of an alliance of classes as described above. A Marxist, however, cannot view an event of this magnitude as simply a shift in the "superstructure" of society. Rather, such an upheaval must represent a change brought about by shifts in the basic class structure of the society which, in turn, must reflect a breakthrough in the means of production.

As a consequence great attention has been paid to emergent "industrial capital" in all of its forms. Elaborate arguments have been presented concerning the stage and degree of political maturity of such industry in the late Bakufu period. All of the stages discerned by Marx—that of handicraft production, the domestic system, manufacture (the last stage before machine industry)—have been proposed at one time or another, and lengthy debates have occurred concerning what these stages meant. Since the war attention has tended to focus on the class of rural peasant entrepreneurs.

Yet, to argue in terms of changes on this level distorts the real significance of the Tempō Reforms. It overemphasizes the reforms by making them the beginning of Restoration history. It underemphasizes the meaning of the reforms by seeing in them only a certain class coalition, a symbolic augury of things to come. Far more important than the relation between the samurai reformers and the rich peasants, a relation that was tenuous at best, were certain definite consequences of the reforms on certain areas of han life—their effect on samurai morale and on the clique structure of the han government. Only in these areas did they contribute positively to Chōshū's role in Bakumatsu history.

Background

The first point to be noted concerning Chōshū's Tempō Reform is its independence of the Bakufu's Tempō Reform. Chōshū's reform began in 1838 when the new daimyo, Mōri Takachika, on entering the han for the first time, appointed Murata Seifū to the temporary office of Minister for Reform. The Bakufu reform of Mizuno Tadakuni was not to begin until several years later. Nor was the Tempō Reform of Chōshū exceptional in being independent of the Bakufu. The Tempō Reform of Mito and Satsuma occurred more than ten years before that of the Bakufu, and like that of Chōshū, was independent in program as well as in time. On 4/11/43, when the Bakufu reform came to an end, the Accompanying Elder, the highest official of the han, sent the following letter to the han warning the han officials not to be swayed by the failure of the Bakufu reform:

> For several years, reform measures have been promulgated within the han, and now at last, frugal customs have been established, debts are decreasing, and the retainers are spared [the burden] of extra financial levies. Just when these laudable changes are being carried out . . . if the rumors from Edo are heard in the han by those whose nature tends toward conservative and vacillating politics, the morale of the han officials will naturally decline. Those who prefer the abuses of the past will take advantage of this and with devious arguments play on the emotions of others. The daimyo feels that this is likely to occur; therefore, strict orders should be given to counteract this tendency.[2]

Who were those who preferred the abuses of the past to the

[2] Tanaka Akira, "Bakumatsu-ki Chōshū han ni okeru hansei kaikaku to kaikakuha dōmei" (diss. submitted in 1956 to the Tokyo Kyōiku Daigaku Bungakubu Kenkyūka), p. 233.

frugality of the present? Murata, the Chōshū reformer, writes: "Among the four classes, peasants and samurai are pleased with simple and frugal government; they have limited incomes. The artisans and merchants have no limit on their incomes in time of extravagance . . . therefore among the four classes, they alone are pleased by extravagance." [3]

Why is it that the Chōshū reform took place when it did? Viewed in the light of Chōshū's financial history, the Tempō Reform was merely one in a series of retrenchments designed to keep the han debt under control. Murata, at the time of the reform, called the Chōshū debt "the 80,000 *kan* great enemy," and it was this great enemy, I feel, which was the fundamental target of the reform.[4] It may appear that the operations of the *buikukyoku* rendered reform unnecessary but this was not so. On the contrary, only by periodic retrenchments such as the Tempō Reform had the han debt been kept within manageable proportions. And this, in turn, made possible the success of the *buiku* system, and gave to Chōshū the freedom necessary to carry out its more or less successful reform during the Tempō period.

In addition to the budgetary troubles which were basic to the reform, also present in the background were the peasant uprisings which took place in Chōshū in 1831 and 1836. These undoubtedly intensified the reformers' awareness of the han's financial and social afflictions. The great peasant uprising or *ikki* of the Tempō period took place in 1831. Some accounts estimate the participation of 50,000 peasants out of a total of 500,000 in the han; the true figure seems to have been closer to 2000.[5] The spark that set off the *ikki* was a clash between peasants and those of the *eta* class who dealt in the hides of animals. The peasants in the area believed that if hides of newly-slaughtered animals were transported by their fields at the time when the grain was almost ripe, a bad harvest would inevitably ensue. The

[3] Seki Junya, "Bakumatsu ni okeru hansei kaikaku (Chōshū han): Meiji ishin seiritsuki no kiso kōzō," *Yamaguchi keizaigaku zasshi,* 6:68 (May, 1955).

[4] Misaka Keiji, *Hagi han no zaisei to buiku* (Tokyo, 1944), pp. 260–268; Tanaka Akira, "Chōshū han kaikakuha no kiban: shotai no bunseki o tōshite mitaru," *Shichō,* 51:16 (Mar., 1954).

[5] Seki Junya, *Hansei kaikaku to Meiji ishin* (Tokyo, 1956), p. 91.

merchants who dealt in the hides ignored this superstition, and indignant peasants attacked those transporting the hides and subsequently went on to attack merchant establishments in Mitajiri and Yamaguchi. By a sort of chain reaction the uprising flared throughout the han.

The resentments, frustrations, and antagonisms provoking the *ikki* can be seen in the demands of the peasants who joined the rioting. Though these varied considerably from area to area, in general they fell into three categories. The first type of demand asked for the remission of some taxes. The second pressed for the abolition of the han monopolies, to which peasants were forced to sell their various commercial crops at a low fixed price: in the eyes of the peasants this was tantamount to an added tax. The third type requested the reform of local government. Some asked that some particular peasant official be removed, others that certain village records be made public, or that some specific abuse on a local level be corrected.

After putting down the *ikki* and punishing those suspected of being the instigators (actually, it seems to have been spontaneous and without any element of forethought), the han government made very few concessions to the peasants. In those areas where the *ikki* had been most violent, several peasant officials were removed and punished, although as a rule the removal served more as a penalty for having let the *ikki* begin than as a concession to the peasants' grievances.

In 1836 another lighter flurry of *ikki* occurred, but it soon died out. It did not compare with the first either in violence or duration. These two *ikki* may have influenced the han leaders in their plans for reforms: disturbances among the people were, in the Confucian scheme of things, certainly a sign of a need for moral reform. This in turn may have been associated in Chōshū with the familiar specter of financial unsteadiness. Yet the *ikki* cannot be called the primary cause of the reform. Neither in the han's immediate reaction to the *ikki* nor in the items of reform carried out later can one see that the placation of the peasantry was considered to be of cardinal importance. Those who stress the importance of the Tempō *ikki* in Chōshū argue that while *ikki* continued to increase in number throughout the country as a whole up to the time of the Restoration, in Chōshū there were no further disturbances, and this they attribute to the virtues of the reform. The abolition of han monopolies, welfare distributions of rice, exemptions from taxes given to peasants whose lands were ruined by floods, the enactment of judicial reforms—all items of the reform—to a certain extent undoubtedly did mitigate the sufferings of the peasants. But, this was not the chief aim of the reform, and equally important in explaining the subsequent absence of *ikki* was the recovery after 1836 from the floods, typhoons, and unusually severe insect damage that had marked the years from 1825 to 1836.

Some historians who emphasize the role of the peasant in the Restoration movement describe the Tempō Reform as a response to a crisis that threatened the very social order in Japan.[6] They view the Tempō *ikki* as the first in a series of peasant revolts—a form of the "movement from below"—which if left unchecked would soon have overturned the Tokugawa system. For them the Tempō Reform

[6] Those mentioned in footnote 1 would constitute a fair sample of the Marxist approach as applied to the Tempō *ikki* and reform. I have treated in a very summary fashion this question of the significance of the pre-Restoration *ikki* that many see as the most important question of Restoration history. The first real debate on this question began during the 1920's with the writings of Ono Takeo and Kokushō Iwao. Ono held that the *ikki* were revolutionary in character: that they produced tremors in the basic structure of Tokugawa feudalism, revealing the weakened condition of the samurai class. Kokushō, in contrast, felt that the *ikki* were not basically revolutionary; he held, rather, that they were epiphenomenal and did not feed directly into the Restoration. Research since the war has tended to agree, almost entirely, with Ōno; and it has been concerned to show the particular socioeconomic structure (especially the nature and form of capital development) "underlying" any given upheaval. In my explanation I have stressed natural disasters. This is obviously not to deny that the *ikki* occurred within a specific social context, but only to suggest that in Chōshū the degree of disintegration of the peasant village alone had not sufficiently changed from earlier periods to explain the *ikki* solely in terms of the changed social setting. For research on the degree of capitalist development in Chōshū that would tend to confirm this position see: Furushima Toshio, "Kinsei ni okeru shōgyōteki nōgyō no tenkai," *Shakai kōsei shi taikei*, ed. Watanabe Yoshimichi (Tokyo, 1950), VIII, 105–140; Oka Mitsuo, "Chōshū han Setonai nōson ni okeru shōhin seisan no keitai," *Rekishigaku kenkyū*, 159 (Sept., 1952). I have denied the revolutionary character of these *ikki*. This is not to question the value of the postwar research, but only to question its assumption that the presence of certain forms of capital in a given area automatically guarantees that an *ikki* in that area would be revolutionary. Such an assumption is to history what "guilt by association" is to the body politic.

was the first step taken by the feudal class to control and subvert the energies of this new force to its own purposes, and the Restoration, the final consequence of this controlled peasant energy. The "alliance" of peasant producer and lower samurai was not, thus, a free union of interests, but one in which the peasants were successively blindfolded, duped, used, and discarded.

This is too inclusive an argument to disprove simply. In the face of evidence that the Tempō *ikki* in Chōshū was a protest against certain abuses and not a revolutionary attack on the system itself, they explain that "subjectively" the *ikki* was indeed only a protest, yet "objectively," in its class base, it possessed a revolutionary potential. In support of this contention they cite evidence regarding the breakdown of the traditional village structure and statistics concerning the growth of local industry among the farming and fishing villages in that part of Chōshū which bordered on the Inland Sea. The unfolding of this unconscious yet objective potential, they would hold, is the history of the Meiji Restoration.

In this context the question as to whether there are necessary relations between the means of production and the level of political consciousness of those who produce becomes almost a metaphysical one.

The Reform Program

At the time of the Tempō Reforms in Chōshū, two bureaucratic cliques dominated the han government: the clique of Murata Seifū and an opposing one headed by Tsuboi Kuemon. There also seem to have been quite a few officials in the han government who belonged to neither clique. Although the two groups stood for different programs, the men came from similar backgrounds and were of approximately equal rank. They clearly did not represent different social strata. The group headed by Murata might be called the austere reformers or "hards." Murata, while a moralist in the best tradition of his age, was less concerned with rhetoric than with the actual problems of government. The following *waka* by Murata reflects the practical—perhaps it might even be called iconoclastic—vein in his character:

> Mt. Fuji when seen is lower than its reputation,
> May not Confucius and the Buddha also be so? [7]

During his years at the han school Murata had excelled both in his studies and in the military arts. As a youth he had spent some years studying Zen Buddhism. He had risen rapidly in the han bureaucracy, and his memorials were noted for their analysis and unusual scarcity of moralistic exhortation. The purpose of his reform was to root out contemporary corrupt practices (*ryūhei*) and to return to the "purer" social order of an earlier age. To debate whether he was a reactionary in hoping to turn back the clock or a progressive in striking at certain vested interests does not concern us here: we are characterizing his group as "hards" because of their willingness to institute measures against the interests of some in the han in order to achieve the ends of their reform. In contrast to this, Tsuboi's group were moderates; while cognizant of the need for reform, they were unwilling to offend, and as a result, their program proved to be ineffective in solving the problems of the han.

Murata's first act after his appointment as Minister for Reform was to negotiate with the merchant creditors of the han for lower rates of interest over a longer period of time. Fearing repudiation, most creditors were willing to make concessions. His next step, one unprecedented in the history of the han, was to publish the 1840 han budget plan and at the same time to request the lower officials of the han to submit any suggestions they might have for the improvement of the han finances. Both these measures were taken in an effort to convince those responsible for the use of the han funds of the gravity of the reform and to persuade the samurai that the special exhortations to frugality were not to be ignored.

Not by any means was the reform solely an economic one. Consistent with its declaration to restore every aspect of life in the han to an earlier and purer condition, with either exhortation or edict, it

[7] Nohara Yūsaburō, *Bōchō ishin hiroku* (Yamaguchi, 1937), pp. 14–15.

covered every facet of han organization. Like so many other reforms of the Tokugawa period, it produced a flow of sumptuary legislation designating what products were to be used and what luxuries relinquished. The Meirinkan, the han school in Hagi, was opened to many samurai who had hitherto been excluded because of low rank; schools were established in each area of the han for the education of samurai living outside of Hagi; and a school was established in the Edo han compound for the education of young samurai attached to the retinue of the daimyo. Experts in swordsmanship were invited from other han, and Murata stressed actual practice over formal exercises. Some Chōshū students were sent by the han to study in other areas. Most of these went to study swordsmanship, but a few were sent for other subjects. One was assigned to the school of Takashima Shūhan in Edo, another to Nagasaki to investigate the strength of the foreign weapons and to observe the activities of the various han in the field of Western studies, and Yoshida Shōin was sent to Edo for the study of military science. In addition to these educational reforms, the han military was also revamped: the changes followed traditional lines of thought, but the new units did reflect more effectively the realities within the han in 1840.

Yet, in spite of its wide scope, the core of the reform remained economic, since, as we have seen, most of the troubles in other areas of life stemmed from the plight of the samurai under the pressure of the financial difficulties of the han. An attempt could be made to resolve these difficulties and stabilize the han finances by either of two courses: uncovering new sources of income or retrenching han expenses. The Tempō Reform, with one exception, took the latter course. One aspect of this economy policy can be seen in the Tempō budget program.

As was mentioned earlier, according to the 1840 budget estimate published by the reformers, the real income of the han government amounted to 75,780 *koku,* the taxes on 146,161 *koku* of production. This was equal to 3789 *kan* of silver, at the estimated price of one *kan* for twenty *koku*. The Tempō reformers decided that 2295 *kan* of this would be used for han expenses in Edo, Osaka, and Kyoto; the remaining 1494 *kan* to meet expenses in the han.

In addition, 5244 *kan* of "supplementary" revenues were to be obtained by borrowing from the income of the samurai and by levying extra taxes on the peasants. (The latter taxes, known as *chika chisō,* were over and above the ordinary extra taxes.) To this were added 575 *kan* saved by various special frugalities. The total of 5819 *kan* was allotted for the annual payment of interest and principal on the han debt. The annual payment of principal and interest on the han debt had previously amounted to 12,175, but through Murata's negotiations with the Osaka merchants the figure was reduced to 6468 *kan* per year. This was still 649 *kan* in excess of the sum available for the purpose. To this anticipated deficit, Murata added the further expenses: 865 *kan* for special traveling expenses, 100 *kan* for flood relief, and 900 *kan* for general relief and other special expenses. The total deficit in this published budget thus came to 3400 *kan* even in the first year of the reform.[8]

From these figures it would appear that the Chōshū Tempō Reform, even in the estimates of the reformers made before the fact, was doomed to fail. But, it must be remembered that this budget was devised to impress the retainers with the need for reform. Failure was not as inevitable as it would appear. First, the additional expenses were probably estimated somewhat in excess of their reality. Second, the value of rice was greater than that used in the above budget calculations. The 5244 *kan* to be taken from the stipends and fiefs of the samurai was the silver equivalent of 104,880 *koku* of rice at a market value of twenty *koku* per *kan*. But, in fact, in 1840 one *kan* could buy only 15 *koku* of rice (and by 1845 only 11 *koku*) therefore the actual income from 104,880 *koku* of rice, assuming that it was collected in kind, must have been closer to 6900 *kan*. The normal tax revenues of the han must also have been appreciably higher for the same reason. Third, the above figures did not include the income which the han obtained from special products such as paper, wax, and so forth.

One study of the results achieved by the reforms states that by 1842 the han debt had been diminished by 30,000 *kan* and that by

[8] Misaka, *Hagi han no zaisei to buiku,* pp. 269–277.

1846 it had been more than halved.[9] Even taking into account the meliorating circumstances suggested previously, this would seem to have been almost impossible. One authority, Naramoto Tatsuya, has suggested that the profits from the *koshinigata* may have been responsible for the diminution in the debt. Horie Yasuzō has suggested that the reduction was possibly obtained by repudiating a portion of the han debt.[10] If this was the case, the portion repudiated was no doubt the 32,842 *kan* on which interest payments were no longer made.

Whatever the immediate effects of the Tempō Reform, its legacy to the Bakumatsu period from a strictly financial viewpoint was probably not too great. By the late 1840's it had again become necessary to borrow, and the expenditures on fortifications and the dispatch of han troops to Sagami and then to Hyōgo in the years following the arrival of Perry caused a new spiraling of the han debt, which in turn prompted the reforms of the Ansei period (1854–1859), of which we will speak later. Since it was *buiku* money rather than any surpluses from the regular accounts of the han that contributed to Chōshū's military successes later on, one can safely say that the financial contribution of the Tempō Reform was at best an indirect one: by giving a respite to the mounting debt, it had made it possible for the han to relieve the pressure on the samurai and thereby to bolster their morale.

The Tempō Reform and the Han Monopolies

A second aspect of the basic Tempō policy of economic retrenchment was its abolition of the various han-controlled commercial monopolies. Many Western writers have made a great deal of these han monopolies, seeing in them the beginning of samurai entrepreneurial activity which, gaining force, went on to become the mainstream of Meiji capitalism. On the whole, this contention is extremely doubtful. A few examples may be found of samurai entrepreneurs of the Meiji period who received their early training in the adminis-

[9] *Ibid.*, p. 275.

[10] Both of these suggestions were made in the course of conversations with the author in Kyoto in the spring of 1956.

tration of han commerce, but these appear to be exceptional. This contention assumes that han monopolies were a late Tokugawa phenomenon, and this was not the case. It is based primarily on the experience of Satsuma which strengthened its monopolies while Chōshū and many other han were abolishing theirs.

Another thesis concerning the han monopolies is that they were the institutionalized form of the alliance between upper samurai and commercial capitalism, which controlled Japan during most of the Tokugawa period. Thus, their abolition in Tempō times was a victory for early, rural industrial capitalism against the commercial capitalism of the cities. This thesis is in accord with the facts insofar as it recognizes that the Tempō Reforms in general moved away from, rather than toward, official control over monopolies. It also stresses, with some justice, the rise of a class of local producers. But the further assumption that the monopolies were discarded in favor of the local producers remains doubtful.

The Tempō Reforms, on the surface at least, were very little different from those of the Kyōhō and Kansei periods. All were "reactionary" and Confucian in character. Tokugawa economic thought in general was Confucian. Ogyū Sorai, an earlier reformer and Confucian theorist, summed up the attitude of Tokugawa economic thought toward the merchants and commerce in the following words: "The peasant cultivates the fields and so nourishes the people; the artisan makes utensils and has the people use them; the merchant exchanges what one has for what one has not and so helps the people; the samurai rules so that disorders will not arise. Though each performs only his own job, he is helping the other; if even one of the four is lacking, a country cannot be maintained."[11] The samurai may have considered the merchants to be inferior beings, but they also regarded them as necessary. Yet, by the beginning of the eighteenth century, almost all samurai in official positions complained that the organic balance described above had been lost and that the excessive activities of the merchants were damaging the well-being of society. In the face of this, some demanded that commerce be restrained by abolishing the special privileges of

[11] Misaka, *Hagi han no zaisei to buiku,* p. 4.

merchants and encouraging frugality. Others argued that commerce was too important to be left to merchants. It must be controlled by the han which would share or monopolize its profits.[12] Neither of these alternatives was a new development in Confucian theory by Tokugawa thinkers. Both were part of a long tradition of Confucian economic thought going back at least as far as the Iron and Salt Debate of Han dynasty China.

Yet, in spite of the fact that the idea of abolishing state-operated monopolies is old, this does not preclude the possibility that its emphasis by the Tempō reformers was new. It might be a case of new wine in an old bottle. Marxist historians emphasize that even though the Tempō Reform was subjectively Confucian and negative, it was objectively the recognition of a new state of affairs—that is to say, of the power of the local producers. In contrast to this, I feel that the abolition of han monopolies was carried out as one element in a broad program of fiscal retrenchment primarily designed to improve the condition of the han finances.[13]

[12] Nomura Kentarō, *Gaikan Nihon keizai shisō shi* (Tokyo, 1949), pp. 239–250.

[13] Regarding this point, I have used, among other materials, those gathered by Seki Junya who, stressing the reactionary nature of the Tempō Reform in Chōshū, writes that, "while inheriting the basic structure left by the reforms of Kyōhō (1716–1735) and Hōryaku (1751–1763), it attempted to respond to the rate of change in the structure of the peasant village." He also wrote of it as, "an attempt to re-establish the feudal control structure based on just benevolence i.e., on traditional values." See Seki, *Hansei kaikaku to Meiji ishin*, pp. 2, 103. To use materials and specific criticisms gathered by a historian who uses the Marxist framework to refute the general position taken by most Japanese Marxist historians is, perhaps, not quite fair. Yet I feel, in this instance and others, that the new materials uncovered by men such as Seki have almost burst the jacket of theory into which they have been fitted. Seki has been criticized by a more orthodox adherent of the Marxist position as inconsistent: "If we say that the Chōshū Tempō Reform was completely reactionary, if we understand it to have been limited to an attempt to fasten the serfs to the soil and restrict merchant capital, the problem then arises, how was it that the later reformists who followed in Murata's footsteps became the advance force of absolutism"? See Niwa Kunio and Unno Yoshio, "Bakumatsu yūhan no keizai kōzō," *Rekishi hyōron*, 79:89 (Sept., 1956). From the Marxist point of view this criticism seems to be justified: how indeed could Chōshū emerge as the leader of the alliance between industrial capital and lower samurai if less than two decades before its basic structure was that of the early Tokugawa period. The resolution of this contradiction, however, lies not in attacking Seki's estimation of the reform, but rather in discarding the facile concept of absolutism which so beclouds contemporary historical views.

In the course of the Tokugawa period the han monopolies took many forms. Sometimes the actual purchase and sale of commodities was carried out by samurai officials of the han. More often samurai officials would supervise merchants who carried out the actual work. Even the *kabunakama* can be thought of as one mild form of han monopoly dealing with products which the han found unprofitable to control directly. At times the head of such a local *nakama* would be given the status of a low official of the han. Still another method by which revenues could be obtained from commercial production was to sell licenses permitting local merchants and even peasants to deal in local products. In every case considerations of profit for the han largely determined the form chosen.

One of the earliest han monopolies was that on the sale of paper. The following tabulation[14] indicates the rise of paper production in Chōshū:

Year	*koku*
1523	1,008
1589	8,300
1607	19,000
1622	48,000
1686	58,000

The han monopoly on paper was begun sometime in the 1624–1644 period. It was carried out by merchants under the supervision of samurai officials. In 1643 the han revenue from the sale of paper was 700 *kan*. By 1686, excluding the land tax, 74 per cent of all the han revenues were derived from the sale of paper (of the remaining 26 per cent, 15 per cent came from salt fields and 11 per cent from supplementary taxes on land). Production of paper reached a peak during the Genroku period (1688–1704), and by the middle of the eighteenth century it had dwindled to the point where its sale was no longer regarded as an important source of income to the han. The product which the peasants had developed as an extra source of income had been so ruthlessly exploited by the han that it was no longer worth producing.

[14] Naramoto, *Kinsei hōken shakai shiron*, p. 43.

Possibly the administrators had learned a lesson: when the han monopoly on wax was begun, the producers were guaranteed one-half of the profits. In contrast to paper, which was produced only in the Yamashiro region, wax was produced throughout the han, and before long the han found its supervision too demanding and readily turned the monopoly over to a merchant guild in return for a fixed tax. In order that the head of the guild might exercise police powers to punish infractions of its special rights, he was given the status of a low official of the han.

During the 1751–1764 period, oil, linen, raw cotton, cotton cloth, indigo, and other products were added to the list. As in the case of wax, the han soon found it more lucrative to permit guilds to handle monopoly sale of these items, and by about 1772 all of them had been changed to guild monopolies. Accordingly, the guilds paid the han for the monopolies and were given permission to impose punishments themselves. (This contrasted with other areas where merchants were required to petition a local government office to take action against anyone encroaching on their monopoly.)

The case of the salt monopoly reveals another feature of the dynamics behind the rise and fall of han monopolies. Like the monopoly on paper, it had been one of the earliest in the han and because of its importance, production had been encouraged. A similar pattern took place in most of the other han along the Inland Sea until the 1760's when overproduction caused salt to glut the markets. Profits dropped to almost nothing and salt workers were impoverished. Finally, an agreement was reached between the producing han to limit their production. With production fixed, the annual profits were also stabilized and the han soon relinquished its monopoly rights to a merchant guild in return for a fixed yearly payment.

Variations on this system continued throughout the eighteenth century in connection with most of the han's commercial crops. In general, guilds displaced the han in controlling the monopolies. Then in the 1818–1830 period, han monopolies were again established. It is probable that the han cherished hopes of plucking for itself the fruits of merchant activity that had grown vigorously since the middle of the eighteenth century. But combining as they did monopoly with the forced sale of products by the producers at a fixed price, the han monopolies were even more exploitative than those of the guilds. Together with the natural disasters of the 1825–1836 period, this new burden brought great suffering to the peasantry and undoubtedly contributed to the peasant uprisings in 1831 and 1836.[15]

The reaction of the han to these uprisings was by no means immediate or direct. In 1831 the forced sale of indigo at a fixed price lower than the Osaka market price was terminated and free sales were permitted; the han continued to obtain profits by levying a fixed tax on every bale sold. In 1843 at the time of the Tempō Reform, the han office for the purchase of indigo was closed and the purchase was left to a guild for a yearly payment proportional to the volume of sales. It was not, however, as an answer to peasant demands that this action was taken; rather, the han was losing money on indigo because of the competition of indigo producers in Awa han.

At the time of the Tempō Reform, Murata and his faction also abolished the han monopolies on wax, salt, sake, cotton, cotton cloth, and other items. But in almost every case this was done in hope of procuring greater profits by selling monopoly rights to a guild rather than by directly supervising the monopoly. There was, of course, another consideration: all things being equal, the han would prefer to protect its samurai from the corrupting practices of commerce. Murata had warned: "Although some will propose profit-making schemes in the name of reform, in both China and Japan, in ancient and recent times, there have been too many instances where such schemes led a country to ruin. . . ."[16] Notwithstanding, there is nothing in the Chōshū background to indicate that, if profits were at stake, such feelings were really allowed to hinder the use of profit-making schemes.

From the Tempō period until 1858 the guilds continued to handle all the commercial crops of the han with the exception of wax. The

[15] The materials on the han monopoly system are taken largely from the writings of Naramoto and Seki.

[16] Seki, *Hansei kaikaku to Meiji ishin,* p. 105.

Tsuboi clique, which was back in power in 1856, re-established a han monopoly on wax. This monopoly was extended by the reformist "hards," who were reinstated in 1858; when evidence of profits was presented, even the Sufu clique, the heirs of the Murata clique, did not hesitate to forget temporarily their own stated principles. However, on most products, not only was the han monopoly not re-established, but even the existing guild monopolies were dissolved in favor of the sale of licenses to local producers and peasant merchants.

Fortunately, in one administrative area of the han (Yoshida *saiban*), records down to the village level, registering the sales, distribution, and profits of these licenses, have been preserved. This area was not commercially the most important, but it was located in the Inland Sea region, where most of the han commerce centered. In one village of 423 houses, 161 licenses were sold; in another of 211 houses, only 6 were sold: throughout the area about one house in ten bought a license (646 licenses sold to 5375 houses). Each license cost between two and ten *momme* (.002 and .01 *kan*), depending on the number of items it covered. The sales in this single area totaled 3.775 *kan*. There were fourteen such areas in the han.[17]

The willingness of the peasants and local merchants to buy the licenses certainly indicates a desire on their part to free themselves from the constricting fetters of the monopoly guilds. And the fact that the total profit from the sale of licenses was small, the fact that the sale of licenses meant an added burden to the local officials of the han, and the fact that in the minds of the rulers the ideal peasant was one with no head for figures, all can be used to suggest that the han had indeed been forced to recognize the new strength of the local producers. Yet there is no direct evidence that this was the case, and the same facts are equally open to a different interpretation. The reformers may have considered the added administrative burden on the peasant officials a slight thing; they may have felt that it was more important to restrict the activities of the merchants than to curtail the economic rationality of peasants who were in any case

[17] Tanaka Akira, "Tōbakuha no keisei katei: Chōshū han bakumatsu hansei kaikaku to kaikakuha dōmei," *Rekishigaku kenkyū,* 205:6 (Mar., 1957).

accustomed to monetary exchange; and even the small return from the sale of licenses may have been greater than that obtained from the lagging monopolies: the exception made in the case of the wax monopoly suggests that profits were close to the heart of the matter. In the wave of total retrenchment even small profits were not to be ignored.

The most important exception to the antimonopoly policy of the Tempō reformers was their expansion of the *koshinigata*. This had been established at the beginning of the nineteenth century to sell the surplus rice of the han, to provide storage facilities for merchants, and to lend money to the merchants of other han while holding their goods as securities. The operations of this office were successful, and under the direction of the Tempō reformers it was expanded into a full-scale *toiya*. Situated at Shimonoseki, it was located strategically on the route to Osaka for merchants' ships coming both from Kyūshū and from ports on the Sea of Japan. It kept itself informed of the Osaka market; when prices were low, it would offer to store the goods of other han until such a time as the Osaka market improved and would make loans on the security of the stored goods. The *koshinigata* would also buy up certain goods when the Osaka market was low and keep them until the market improved. Like the later monopoly on wax, the *koshinigata* was an example of a profit-making scheme which, because it was shown to be profitable, was endorsed even by the "hards." Not that this signified a *rapprochement* with possible political overtones between the Chōshū merchants and the Chōshū samurai; the primary focus of its operations lay first and last in its dealings with merchants outside the han.[18]

It will be interesting at this point to compare Chōshū's monopoly policy with that of Satsuma where, contrary to the case of Chōshū, the Tempō Reform brought about a strengthening of the anti-merchant government monopoly that from the start was already strong. In its finances as in other aspects of its character Satsuma was more given to extremes, more prodigal, than Chōshū. During the first two centuries of Tokugawa rule its debt grew more slowly than

[18] Misaka, *Hagi han no zaisei to buiku*, pp. 222–230; Suematsu Kenchō, *Bōchō kaiten shi* (Tokyo, 1921), VII, 338–339.

that of Chōshū—largely because of its income from the sale of cash crops other than rice. Then within a relatively short span of time, its debt leaped to 312,000 *kan,* over three times greater than that of Chōshū. The debt was so enormous that, despite Satsuma's great natural riches, the Osaka merchants refused to grant it further loans. This precipitated a financial crisis and in 1827 Zusho Hirosato, a confidant of the Satsuma daimyo and a man who had risen from rather humble origins, was appointed to carry out a reform.

Seeing that there was no way to repay such an enormous debt, Zusho simply repudiated it. Debts to merchants within Satsuma were totally repudiated; debts to the han's creditors in Osaka, Edo, and Kyoto were virtually repudiated by the promulgation of a scheme according to which the creditors would be repaid less than 0.5 per cent a year of the principal of the debt over a 250 year term. Thus, with a single stroke he not only disposed of Satsuma's incubus, but also succeeded in fomenting a climate of crisis that might very well further his purposes. The han was now thrown on its mettle: no future loans could be obtained and the han had no choice but to live within its income.

Zusho provided for this by establishing within Satsuma a rigid state-operated monopoly on sugar. Because of their geographic location certain islands to the south of Satsuma were the only places in Japan where cane sugar could be cultivated. Exploiting this geographic advantage Satsuma came to furnish over one half of the sugar sold in Osaka. Revenues from this source alone came to 6300 *kan* of silver annually. Using these and other revenues, Satsuma was able not only to meet its fiscal requirements but also to accumulate substantial reserves. By 1848, the year of Zusho's death, a surplus of about 62,400 *kan* had accumulated in the Satsuma treasury. This provided the financial base for its activities in the Bakumatsu period. Thus in a sense, the cane sugar monopoly of Satsuma played a role corresponding to that of the *buikukyoku* of Chōshū.[19]

Upon examining closely the inception of this reform we find that

[19] Tsuchiya Takao, *Hōken shakai hōkai katei no kenkyū* (Tokyo, 1953), 386–475; Tōyama, *Meiji ishin,* pp. 35–36.

both the plan for the repudiation of the han debt and the idea of a strengthened state monopoly on sugar were suggested to Zusho by Hamamura Magobei, an Osaka merchant who with four of his fellows subsequently handled the sale of the Satsuma products. The repudiation of so great a debt—a debt which had become a part of the financial structure of many merchant houses—produced an uproar in Osaka. Hamamura was put in prison for a short time in 1836 and then banished from Osaka until 1841 by the Bakufu commissioner of that city. Of course, no action was attempted against the samurai leaders of Satsuma. There existed no court to which the merchants could appeal the Satsuma action.

Viewing the organization of the reform within Satsuma itself one is struck by the establishment on its sugar islands of a despotic plantation type system. No crops could be raised on those islands other than sugar; their paddy lands were all destroyed. All other commodities—literally all other commodities, 457 items ranging from rice, cloth, wood, to pots and pans—had to be obtained within the moneyless economy established on these islands by barter for fixed amounts of sugar. Rice for example cost, in terms of sugar, six times what it was worth at the Osaka market. Land was allocated in a fashion reminiscent of the T'ang system: so much per man, so much per woman, according to their ages. Each stage in the productive process was supervised by samurai. Private sales of sugar were punishable by death. Licking one's fingers in the process of production was punishable by whipping. For the production of poor quality sugar one was put in stocks. Altogether it resembled nothing so much as a penal colony producing delights which its inmates were forbidden to consume.

Satsuma thus presents something of a paradox. Its financial machinations involved a greater contact with merchants than did those of Chōshū; yet at the same time it was more markedly antimerchant in its policies. I feel that one cannot view its sugar monopoly as evidence of a class alliance. That Hamamura was willing to "do in" his fellows merchants by means fair or foul rather testifies to the lack of solidarity in the Osaka merchant community in the midst of its palmiest days in the period before the Bakufu Tempō Reform. Hamamura gained but not as a representative of his class, the interests of which he damaged severely.

Nor can one speak of merchants as having possessed any power within Satsuma. On the contrary, the merchant class within Satsuma was hard hit by the twin blows of the repudiation of han debts and the inauguration of the samurai run monopolies, which covered, in addition to sugar, a variety of other products such as wax, oil, cinnabar, turmeric, paper, medicines, and sesame. Like sugar these were supervised by samurai at every stage of production until the time they were sold.

Some credit has been given to Satsuma for the early introduction of Western industry, and a few have gone so far as to view this as the extension of the han monopoly system and a sign of a proto-capitalistic han. However, to quote Tōyama Shigeki: "The fact that the han was able to realize han-operated manufacture and machine industry earlier than others was due to the complete sacrifice of the free development of the producers. Not for a moment does it indicate the culmination of a normal bourgeois development; rather, it points out feudal strength."[20] The strength which has won for Satsuma the name of "the North Pole of feudalism" or the "Prussia of Japan" was also apparent in its peasant village society, where direct rule by *gōshi* caused it to be the only han in Tokugawa Japan without one *ikki* during the entire Edo period.

Of the two traditional attitudes towards economic reform which were sketched earlier, Chōshū took one, retrenchment coupled with a repudiation of han involvement in commerce, while Satsuma took the other, stronger political controls over commerce. The differences of the two reactions to financial crisis, however, stemmed largely from the different conditions in the two han and not from a basic difference in outlook. In Satsuma monopolies were lucrative and therefore continued; Chōshū's monopolies, no longer profitable, were abolished. In other respects the reforms of the two han were much alike: both used debt repudiation, both were negatively inclined

[20] Tōyama, *Meiji ishin,* p. 36.

towards the merchant class as a whole, and both relied on institutional innovations to re-establish their finances. And in both cases I feel that it was the vitality of their traditional samurai governments that gave them the willingness and strength to innovate and that enabled them to override the interests of other groups within their spheres of control. This, rather than any capitalistic character, is also the connection between the Bakumatsu "progressive" experiments with Western industry and the military-industrial ventures of the early Meiji period: the use of political power to create the means necessary for political ends.

To return for a moment to the views of those who see the Tempō Reforms as the starting point of Restoration history: in the case of Chōshū, they see the Tempō Reforms as the han's response to the Tempō *ikki,* and the change from guild monopolies to the sale of licenses in 1858 as the negation of traditional merchant capital in favor of youthful industrial capital. We have already challenged the first part of this contention. Han monopolies had given way to guild monopolies at the time of the Tempō Reform for the same reason that they did in the 1770's—because they were found to be more profitable arrangements for the han. As yet no figures showing the han incomes from this source before and after the reform have been uncovered (except in the case of indigo, where there is specific evidence that the change was made for financial reasons). It is quite probable that the changes in 1858 were also made to procure more revenues for the han. Until more evidence is gathered, revealing the specific intent of the reformers and comparing the payments made by the guilds with those obtained from the sale of licenses, this supposition for the moment seems the most plausible.

And yet the number of peasant houses buying licenses in the area for which figures are available does indicate a real desire for freedom on the part of the peasant producers and local merchants. But the desire for freedom does not necessarily imply that these peasants and merchants had advanced to the "first stage of industrial capital." Moreover, the sale of licenses did not mean that the monopoly of the former merchant monopolists was actually broken. Purchase of a license gave the peasants the right to bargain for the sale of their products; it did not mean that merchants from other han could freely enter Chōshū, nor did it mean, obviously, that the thousands of peasants who bought licenses could travel to other areas to sell their products. A very interesting study on the relation between cotton producers and cotton merchants in the Osaka region has shown that the loss of monopoly rights by the latter did not cause an increase in the prices they paid to the former for cotton; the producers, who, on first glance, seem to have been favored by the Tempō Reform, actually made no gain; if anything, they were woven even more tightly into the *de facto* monopoly of the great merchants.[21] It would not be at all surprising if the same thing occurred in Chōshū where the producers were far less advanced than those in the Osaka region.

Even if it could be proved that local producers had made some gains in 1858, this still would not constitute evidence of a union between embryonic industrial capital and lower samurai (*sotsu*), simply because neither the Tempō reformers nor the Ansei reformers were of lower samurai origin. Thus "absolutism" as defined by some, can not rightly be considered to have begun either in 1840 or 1858.

The monopoly policy of the reformers reflected the economic contradictions discussed in the last chapter. The reformers wanted an almost natural economy within the han, and yet they needed cash. Their solution was to tax and tax again, and to sell the taxes in kind outside the han in exchange for cash. This gave rise to a mode of financial thought that, passing beyond mercantilism, almost resembled bullionism. Murata wrote: "The first principle of finance is to sell all the goods produced in Chōshū to other han in exchange for gold and silver and not to let any gold or silver produced in Chōshū out of the han."[22] To the extent that taxes were collected in cash and not in kind (either from peasants or merchants), the government was merely making use of merchants and the existing degree of merchant activity as an indirect market. In the minds of the han leaders, the rightful scope of merchant activity consisted in

[21] Furushima Toshio and Nagahara Keiji, *Shōhin seisan to kisei jinushi sei* (Tokyo, 1954).

[22] Naramoto, *Kinsei hōken shakai shiron*, p. 62.

merchants serving the han without procuring great profits for themselves.

The Tempō Reform and the Chōshū Samurai

The contribution of the Chōshū reform to the morale of its samurai was not the least of its achievements. One dilemma faced by the Chōshū government was the contradiction between its desire to preserve its finances and at the same time to maintain a healthy samurai class. Both Chōshū and Satsuma were among the first of the han to supplement their regular revenues with a portion of their samurai's income or stipends. Perhaps it was more than coincidental that two of the han with the heaviest samurai populations should have taken the lead in this matter. Chōshū first resorted to this sort of "borrowing" in the 1644–1648 period and the amount taken gradually increased, until in 1704, for the first time, the han appropriated 50 per cent of the income of its samurai. Reforms took place at various intervals in an effort to lighten this burden, as at the time when the *buikukyoku* was founded, but from about 1700 up to the Restoration, an average of 40 per cent was taken each year.

Murata describes the condition of a samurai with 25 per cent of his income taken:

> When ten *koku* is borrowed by the han from the income of a samurai with a fief of one hundred *koku,* he is left with an actual income of thirty *koku* [assuming a tax rate of 40 per cent]. With seven mouths to feed in his family, and one person requiring five *gō* of rice a day, it will require 12.6 *koku* a year, leaving him with 17.4 *koku.* In recent years the price of rice has been low, one *kan* of silver being worth from eighteen to twenty *koku* of rice, so, taking the average of nineteen *koku,* the above rice will be worth nine hundred and fifteen *momme* [.915 *kan*] with which he must meet the expenses of living in Hagi. Of this, the fixed costs come to 490 *momme:* 120 *momme* to rent a small tenement house, 250 *momme* for a servant, and 120 *momme* for firewood. The remaining four hundred and twenty-five *momme* will not buy food and clothing for [even] half a year. [The situation of] the samurai whose income is under one hundred *koku* goes without saying.[23]

[23] *Ibid.,* p. 41.

When one bears in mind that over 88 per cent of the Chōshu samurai (even excluding *baishin* or rear vassals) had incomes of less than one hundred *koku,* it is not difficult to imagine the plight of the samurai who had their incomes halved for several years in succession. And, higher samurai with many rear vassals to support did not fare much better.

The high cost of living in the castle town of Hagi had led many samurai to request permission to live outside of the castle town on their fief or in a village until their finances had recuperated. In 1669 the debt of the samurai was such that all samurai with fiefs less than two hundred *koku* were given permission to reside temporarily in the country. Nor was this expedient limited to lower-ranking samurai: as early as 1775 several of the highest-ranking retainers (han Elders) had also asked for permission to live on their fiefs. The incidence of this practice at one time lent itself to the theory that these samurai living in the villages constituted a class much like the *gōshi* of Tosa and Satsuma, but later research has shown that although the numbers absent from Hagi remained fairly constant, there was a constant turnover within their numbers. In Bakumatsu times, about 1299 samurai families were living outside of Hagi for this reason. Many laws were enacted to prevent continued country residence, and those who were found guilty of staying longer than the time needed to repay their debts were punished.[24]

The Tempō Reform improved the condition of the samurai in two ways. The first was to reduce the amount taken from their income. During the years immediately before the reform, 50 per cent of a samurai's income had been taken; this was reduced to 30 per cent. And since the price of rice was higher than the nineteen *koku* per *kan* of silver mentioned in the above quotation, even shorn of 30 per cent of his income, the samurai described by Murata would have been able to balance his budget in Hagi. Moreover, most samurai had some sources of income beside their regular stipend or fief; these ranged from household handicraft production to special emoluments for the

[24] A brilliant article has been written on the subject of these village living samurai. Kimura Motoi, "Hagi han zaichi kashin dan ni tsuite," *Shigaku zasshi,* 62:727–750 (Aug., 1953).

upkeep of weapons, grants for travel, or extra pay attached to certain official posts.

The second positive measure taken by the reformers to relieve the samurai was known as the "Thirty-Seven Year Debt Repayment Act." This was certainly the most controversial measure of the Tempō Reform in Chōshū. It was promulgated in the fourth month of 1843 at the height of the han reform; it decreed that all debts incurred by Chōshū samurai, whether to the han or to private parties, were to be paid back over a period of thirty-seven years, during which time the annual interest would not exceed 2 per cent. Any future loans made on terms other than those stipulated above would be illegal and would not have to be repaid. This law, if enforced, would have meant the ruin of many Hagi merchants, who, like the *fudasashi* in Edo, had made a business of lending money to samurai at high interest rates on the security of their future rice stipends; it would also have ruined many samurai by making impossible future loans. In the end it proved too strong, and following some political changes, the measure was replaced by a more moderate act which canceled all samurai debts to the han and made the han responsible for all private debts. So that the act would not seem to vindicate those who had fallen into debt, samurai without debts were given one *kan* of silver for every one hundred *koku* in their fief. Needless to say, the new act pleased samurai and merchants alike, although by now the reform had lost its coercive bite.

The material benefits derived from these measures were obvious. The condition of the samurai in Chōshū was now better than it had ever been in the preceding decades. But one can only speculate what influence this had on their actions at the time. It is also very difficult to compare them with the samurai of other han. Likely, they were much better off than the samurai of most other han, but, nothing final can be said until comparative studies of the finances of individual samurai households are made. Still, one or two comments made by samurai of other han who observed Chōshū, and by Chōshū men, on the eve of the Bakumatsu upheaval do suggest that Chōshū was in better condition, and therefore its samurai in better morale, than most han. When in 5/44 Maki Yasuomi, a samurai of Kurume han, viewed the retinues of several daimyo returning from Edo, he noted in his diary:

Rain in the morning clearing before noon. I rested at Mitsuke where I encountered the retinue of the daimyo of Kumamoto on his way back to the han. Both in accoutrements and in the number of men and horses, it is probably the finest in Japan. I also met with that of the daimyo of Chōshū, and observed it while resting in a house by the side of the Road. His cortege, if one compares it with that of Kumamoto, was only half as large; however, the samurai were extremely well disciplined and the effects of the daimyo's reform is a sight to behold. I also met with the daimyo of Fukui, Ōsu, and Kurushima, none of whom are worth looking at.[25]

Nine years later, Mitsukuri Gembo of Tsuyama han recorded his impressions after passing through Chōshū on his way to meet with the Russian mission at Nagasaki: "The people are simple and pure, the women and young girls are little given to coquetry, the innkeepers attend to customers . . . , it is not frivolous like Aki [han]. Food is plentiful and the houses are practical with little ornamentation. The majority of the peasants' dwellings are thatched and white walls and tile are rarely seen. In this too, the healthy condition of Chōshū can be seen."[26]

The Tempō Reform and Tempō Politics

In Chōshū it was the custom for high officials to tender their resignations every second year on the occasion of the daimyo's return from Edo. The resignations were accepted only when the daimyo wished to make a change in his government. The fiction of changing every two years was probably maintained by the daimyo to demote or relieve officials without causing them to lose face. In 6/44 when the daimyo returned to the han from Edo, he accepted the resignations of Murata and the other leaders of his group, replacing them with the moderate reformers headed by Tsuboi. This marked the end of the austere phase of the reform.

[25] Misaka, *Hagi han no zaisei to buiku*, p. 274.
[26] *Ishin shi* (Tokyo, 1939), III, 5.

At the time many attributed the downfall of Murata to the merchants who resented his severe Thirty-Seven Year Debt Repayment Act.

> Hagi stood alone on the edge of the Sea of Japan, a place through which the world in general did not pass. It [its merchants] had no commercial or other relations with other han, nor did it [its merchants] share in the profits from the han monopoly system. There were gathered in Hagi those who carried out the business of the government offices and of the households of the retainers. There were five retainers with fiefs of over ten thousand *koku,* three or four with fiefs of about seven or eight thousand *koku,* and many with fiefs of three to five thousand *koku.* The finances of all these were in the hands of merchants, and the merchants rejoiced when the samurai became poor. When this happened, money was lent at [a high rate of] interest, and in its place, the retainer's portion of the rice from his fief was received by the merchants who bought it cheaply and sold it dear. . . . This was the primary business of the Hagi merchants who, therefore, disliked this reform . . . and finally came to destroy it. What means they used I do not know . . . but eventually doubts arose even in the mind of the daimyo and it ended in this way. If one asks why [it failed), it was the fault of the merchants.[27]

In addition to merchant pressure, whatever form it may have taken, there might have been others in the han who wearied of the reform. Some samurai may have found it constricting to live within their incomes; their wives may have resented the prohibitions against the luxuries to which they had become accustomed; and even the members of the family of the daimyo may have been dissatisfied with their shrunken allowances. Finally, the members of the han government who traveled to Edo with the daimyo may have been caught up in the reaction against the extremism of Mizuno and decided that even in the han the milder policies of Tsuboi would be more prudent.

Tsuboi's first act on taking office was to replace the Debt Repayment Act with the milder act described earlier. In most ways he tried to carry on the reform spirit of Murata, even if in attenuated form. In spite of his reformist impulses, his policies were more inclined

[27] Tanaka, dissertation, footnotes 94–99.

to curry favor with public opinion than to deal with the unpleasant realities of han finance, and the reform soon began to lag. Samurai again incurred debts and the daughters of the daimyo began to spend more than their allowances permitted. By 4/45 it had become necessary to borrow 2000 *kan* from Osaka, and this situation steadily deteriorated until in 12/47 Tsuboi was relieved of his office and ordered into domiciliary confinement as punishment for the failure of his reform.

The failure of the Tsuboi clique which had directed the second phase of the Chōshū Tempō Reform did not lead to the reinstatement of the opposing Murata clique. Instead, both leaders withdrew from active politics while the pendulum of government swung back to a position midway between the two groups. From 1847 to 1853, both in its policy and personnel, the Chōshū government represented a compromise between these two groups. Neither could completely dominate the other, each resembled the other in social status, and each had the support of certain segments of opinion within the han. And, in spite of the mingling of forces within the han government, the opposition or tension between the two cliques continued, perhaps even more vigorously than ever.

This tension may have been one of those crucial aspects of Chōshū's total character which made possible its activities during the Bakumatsu period. The balance between opposing bureaucratic cliques—each of which claimed that its policies best represented the true interests of the han—seems to have kept alive the possibility of alternative courses of action. The content of the policies evolved with the times as each group jockeyed for power within the han government. The existence of alternative policies, however, gave to the han government a flexibility that later enabled it to switch policies in response to a changed situation. This flexibility—to which, of course, many other factors also contributed—seems to have been lacking in the case of many other han.

Kumamoto, for example, a great han with an assessed gross product of 721,000 *koku,* played almost no role in the politics of the Restoration. It also had no reform of note. Both instances of inaction may in part be due to the domination of the Kumamoto government by a single faction of high-ranking retainers who were deeply committed to their privileges and to the *status quo,* and this may also have been the case in other han as well.

The Murata clique in Chōshū was, in terms of lineage, the direct predecessor of the bureaucratic faction that guided Chōshū onto the central stage of Bakumatsu politics in 1861. The Tsuboi clique, on the other hand, became in Bakumatsu times the opposition, *status quo* party, which fought against involvement in national politics. Possibly, the small degree of difference visible in their respective reform programs had assumed a new dimension and meaning in Bakumatsu times, when shifting circumstances opened up opportunities for change in the political structure. The name *zokurontō* which was applied to the *status quo* conservative forces in the Chōshū civil war in 1865 was first used at the time of the Tempō Reform. The name is a pejorative term meaning "party of the pedestrian view," and was applied to the Tsuboi clique which tried to please everyone at the expense of the reform. It seems probable that, in Chōshū, the type of temperament favoring radical economic reform went on to favor an extremist political position under the changed conditions of the Bakumatsu period.

Such an inference seems justified in the case of Chōshū. Yet in the case of Satsuma the pattern was reversed. Like Chōshū, Satsuma showed a considerable flexibility of policy during the Bakumatsu period. This was probably due more to Hisamitsu's personal leadership than to the presence of contending cliques. But, in Satsuma the radical Tempō reformers headed by Zusho did not go on, as in Chōshū, to become the radical pro-Emperor party during the later Bakumatsu period. Instead, they became the leaders of the conservative opposition.

Some have ascribed this reversal of the Chōshū pattern to the difference in the nature of the reform in the two han: Satsuma's reformers were backward (supporting as they did a union of feudal power with merchant capital), therefore they naturally became the politically unprogressive faction; Chōshū's reformers were advanced (joining with nascent industrial capital against merchant capital), therefore, by the same historic inevitability, they became the enlightened force in Bakumatsu times. I find this sort of explanation unconvincing. First, it is difficult to argue that Satsuma's reform was really more favorable to the merchants than that of Chōshū. Second, the reforms were much broader than the question of han monopolies alone and even the "class relations" of the reform cannot be seen solely in terms of this item. Third, if one accepts my earlier suggestion that, in essence, the reforms of the two han were more alike than different, then later differences perhaps may be explained largely in terms of personal and clique rivalries within the upper echelon of the Satsuma ruling class—rivalries which at the time of the Satsuma Tempō Reform became entangled with a succession dispute.

The Satsuma daimyo at the time of that han's Tempō Reform was Shimazu Narioki. Narioki favored as his heir Hisamitsu, the son of a concubine, over Nariakira, his eldest son, who was generally recognized as the better man of the two. (Although considering the ability shown by Hisamitsu in the pre-Restoration period, this may be historical bias.) Zusho, the Tempō reformer, favored Hisamitsu in this struggle: as the protégé of Narioki he had no option other than to support Narioki's choice, and as his protégé he felt that the succession of Nariakira would signal the end of his reform. This is not to say that Nariakira was antireformistic in character; it was Nariakira who went on to carry out a program of experiments with Western industry in Satsuma. Rather, since Nariakira and his adherents could not directly oppose the will of the daimyo, their opposition took the form of an attack on Zusho and his reform. Zusho supported han monopolies; Nariakira therefore opposed them. The implications of the two positions, however, were not radically different, except on the personal and clique level.[28]

The Mito Tempō Reform resembled the Chōshū reform in some ways, and the Satsuma reform in others. In a sense the reform had come about as the result of a succession dispute. The reformists, headed by Fujita Tōko, had their base in the Mito school, the Shōkōkan; they sent in several petitions asking the daimyo Narinaga

[28] *Ishin shi,* I, 380–383.

for reform, but, as these were turned down, they put their hopes in the succession of Nariaki. The opposition, desiring aid from the Bakufu, tried to have the succession decided in favor of the son of the Shogun Ienari. The reformists won, and, upon Nariaki's succession, all those who had opposed them were relieved of their positions. Here, as in Chōshū, the reformists became the activists in the Bakumatsu period. The aftermath of the succession dispute continued to influence han politics, and, when Nariaki died in 1860, the conservative opposition, with its strong pro-Bakufu orientation, crushed its opponents and ended Mito's role in the Restoration struggles. In comparison, clique rivalry in Chōshū before the 1860's was much less violent, since it was concerned only with questions of reform policy.

To sum up the tentative conclusions of this chapter: the Tempō Reform in Chōshū was carried out in an effort to restore the han to the "purer" condition of an earlier age and to rid the han of the evils of the present (*ryūhei*). This objective was to be reached chiefly through financial reform. With the single exception of the *koshinigata,* its program was negative: it sought to decrease expenditures rather than to increase revenues, since within the Tokugawa framework it had no effective way of increasing revenues. Whether the reform was a failure or a success is hard to say; it certainly was not an unqualified success. Probably its chief contribution to Chōshū's role in the Bakumatsu period was the improvements it made in the condition and morale of the individual samurai. And, although not a consequence of the reform as such, its cardinal feature was the clique opposition in the han government which continued into the Bakumatsu period. One cannot say of Chōshū what can be said of Satsuma: that its Tempō Reform laid the foundation for its political successes in the crucial later years.

The limited success which Chōshū was able to achieve in its reform resulted from the fact that it began the reform in much better financial condition than the Bakufu or most other han. The financial solvency which had been obtained through the *buiku* system gave Chōshū the leeway to carry out an effective reform, and the reform, in turn, contributed to the condition of its morale and to its political flexibility. Thus, Chōshū, with its large production, its relatively large samurai class, and its traditionally anti-Tokugawa bias, entered the Bakumatsu period with a much greater capability for action than most han.

CHAPTER IV

The Early Stage of Bakumatsu Politics, 1853-1861

The National Scene, 1853–1861[1]

The etiology of the Meiji Restoration has been frequently debated in terms of two alternatives: the impact of the West or the weight of cumulative change within Tokugawa Japan. The earliest interpretation of Meiji history, one accepted for a time by Japanese and Westerners alike, was that the Restoration was the consequence of Western pressure. After a time there occurred a reaction against this view in which the primary emphasis came to be placed on factors present within Japan. Today the circle has been closed; most historians agree that both the impact of the West and the particular nature of late Tokugawa Japan were important. But today these are no longer seen pitted one against the other as mutually exclusive categories. Rather, as in the old nature-nurture controversy, the important question is the relation of one to another within a structured context.

Had Japan not possessed the particular character that it did, its

[1] Most of the material in this short survey of the national scene between 1853 and 1861 is available in the standard histories of the period such as the *Ishin shi* (Tokyo, 1939). In particular I am indebted to the following: Tōyama Shigeki, *Meiji ishin* (Tokyo, 1951); Oka Yoshitake, *Kindai Nihon no keisei* (Tokyo, 1947); Osatake Takeshi, *Meiji ishin* (Tokyo, 1942).

reaction to the West might not have differed from that of many other Asian nations. This reaction obviously must be explained in terms of internal factors. On the other hand, had the industrial West not forced open the door to Japan in the mid-nineteenth century, Tokugawa society might have continued in its slow decline for another fifty to one hundred years without appreciable change in its political structure.

The decline was obvious, but, viewed within the total sweep of Japan's feudal history, this decline had been going on for several centuries and, judging by internal factors alone, it might have continued for many more years. Or, viewed within the more limited scope of the Tokugawa system—that was only partly feudal—there were signs of decay and occasional crises, but there was no *crise de système*. The peasants were not on the point of national uprisings or peasant wars, nor were the great han about to overthrow the Tokugawa hegemony. Chōshū, in spite of its size, its tradition, its economy, and so on, was still a very tame and obedient han on the eve of the Bakumatsu period. Therefore it is to the external forces and their initial impact on the Tokugawa state that we must turn in order to comprehend the later developments within Chōshū.

On July 8, 1853 (3/6/53), the four "black ships" of Commodore Perry entered the harbor of Uraga in Japan. After various negotiations, the letter from the President of the United States was given to the Bakufu officials and Perry's ships left Japan with the understanding that they would return again the following spring with a stronger force, at which time, if necessary, they would use arms to obtain a treaty of "friendship." Having realized their own incompetence in face of the strength of the foreign ships, the Bakufu officials knew that they must agree to the demands of the foreigners. They feared, however, that once relations were established with foreign countries, they would no longer be able to preserve the network of political and economic controls by which they had maintained their hegemony during the years of seclusion. They also distrusted the daimyo who, they thought, would use their new contacts with the foreign powers as *points d'appui* from which to seek new privileges and rights; they anticipated that in this process new political forces inimical to themselves would emerge. Abe Masahiro, the head of the council of *rōjū,* who, owing to the weakness of the shogun, was the effective head of the Bakufu, hoped to prevent the budding of such movements by forming a united national front which would function as a basis for concerted action against the foreign powers. It was with this intention that he broke with the 250-year-old tradition of Bakufu autocracy by ordering the daimyo to present their opinions concerning the American demands.

The reponses of the daimyo were diverse, naturally, but the majority recommended the continuance of seclusion. This tendency was particularly pronounced in the opinions presented by the larger han. The response of Chōshū was typical:

> Although there are various requests [in the letter from the President], the essence of the letter is that they are seeking friendly trade with Japan, and that they intend to send warships if they are not accommodated. . . . If, for the present, trade is permitted according to their request, then for the time being, the affair will be peacefully settled. However, if we permit the Americans to trade, then the other barbarians will make similar demands until, finally, Japan's national strength will be weakened by trade. Recently, in nearby China, a dispute arising from trade has turned into war, and one hears of the suffering and misery of the people. Moreover, trade also paved the way for the downfall of the Sung and the Ming dynasties. . . . Therefore, refuse their [demands] in such a way as to strike terror into the hearts of the foreign pirates. . . .[2]

The dominant note in the official responses of the han was antiforeign. Some of the more realistic han suggested that Japan adopt a lenient policy toward the foreign nations until adequate military preparations had been achieved, whereupon Japan could forcefully sever its relations with the foreign powers and return to seclusion. The han in favor of opening Japan were usually driven by the desperate hope that foreign trade could be used to improve the condition of their finances. Foreigners were barbarians, albeit barbarians with a superior technology, and though many Japanese as individuals were curious about them, the official reaction was negative. Many more han than official opinions would suggest may have

[2] Suematsu Kenchō, *Bōchō kaiten shi* (Tokyo, 1921), II, 27–28.

secretly felt that trade with the foreigners was not altogether a bad idea, but pride vis-à-vis other han demanded a strong answer properly reflecting the military spirit of their samurai.

Perry returned to Japan in March 1854, and despite the opposition by the majority of the han the Kanagawa Treaty of Friendship (Kanagawa *Washin Jōyaku*) was concluded some weeks later. The Bakufu, however, was disconcerted over the responses of the daimyo in favor of seclusion; it feared that the negative responses of the daimyo would lead them to criticize the treaty and, in turn, the Bakufu for having signed it in spite of their disapproval. Anticipating this the Bakufu broke with another precedent by asking the Court for Imperial approval of the treaty; it hoped that this would quiet any opposition from the han. The Court replied that it was satisfied with the treaty and, heartened by the action of the Bakufu, itself broke with precedent and sent a note to the Bakufu exhorting it to take effective measures toward national defense. The Bakufu leaders debated at some length what attitude they should take toward the note; many were in favor of punishing the Court nobles for interfering in what was not their concern, but Abe, who was bent on unifying the country under the Bakufu, astutely decided against any punitive action and forwarded the note to the daimyo.

Abe also tried to strengthen the national front under the leadership of the Bakufu by allowing some of the daimyo from the leading *shimpan* and *tozama* han to enter into its councils. Nariaki of Mito, a brilliant reformer, the spiritual head of the Mito school of nationalistic studies, and the spokesman for most of the Tokugawa branch houses, was appointed the Councillor for Maritime Defense. Abe also established close relations with Matsudaira Keiei, the daimyo of Echizen and leader of the *ōrōkanomazume kamon* daimyo, and with Shimazu Nariakira, the daimyo of Satsuma and leader of the *ōbiromazume* daimyo.[3] The marriage of the foster daughter of

[3] In characterizing the clique struggles of the time in terms of *ōrōkanomazume kamon*, *ōbiromazume*, and *tamarinomazume* daimyo, I am following Tōyama. See Tōyama, *Meiji ishin*, pp. 82–83. The *ōrōkanomazume kamon* daimyo were the "Three Houses," the three *shimpan* daimyo closest to the Bakufu. Next most favored in terms of seating order at the Edo castle were the *tamarinomazume* consisting of the next most

Nariakira to the Shogun Iesada in 11/56 was a component link of the latter relation. This national front proved to be mutually satisfactory to both the Bakufu and the outside powers: the Bakufu had earned the cooperation of the most powerful among their potential opposition and the daimyo were able to get a foothold on grounds hitherto forbidden to them.

The national front policy was attacked by the *fudai* daimyo and especially by the *tamarinomazume fudai* daimyo, who were the chief officials of the Bakufu. Their leader was Ii Naosuke, the daimyo of Hikone. Most of these han were so small and ineffectual that their influence was hardly felt; it was only as the wielders of Bakufu power that they were important. To them, the continuation of Abe's policy would mean that the voices of the great daimyo would eventually drown out their own in Bakufu councils. The tension between these *fudai* han and the national front powers was further complicated by their respective stands on the foreign question. Nariaki and some of the other non-Bakufu powers were fervent supporters of seclusion. The *fudai* daimyo, perhaps because of their intimate knowledge of the weaknesses of the Bakufu, favored opening the country. This called forth further attacks on the "jingoistic" Nariaki.

In the years immediately following the signing of the treaty, Abe was faced with the formidable task of keeping the support of the national front powers, without letting them grow too strong, while executing the *kaikoku* (open the country) policy advocated by the *fudai* han. The haughty Bakufu officials from the *fudai* han were disgruntled and sought to remove Abe from his position as head of the council of *rōjū*. These officials headed by Ii gradually gained power. But when they were on the point of securing Abe's removal,

---

important *shimpan* daimyo (such as Aizu, Kuwana, etc.) and the six most important *fudai* daimyo (such as Ii, Honda, Sakai, etc.). These were followed by the *ōbiromazume:* eleven *shimpan,* three *fudai,* and twenty-four *tozama* daimyo, in general, daimyo with domains of over 100,000 *koku.* During these early political struggles, the *tamarinomazume,* who were the controlling faction in the Bakufu, joined with the *teikannoma,* the fourth ranking tier of smaller *fudai* powers, against a coalition of the *ōrōkanomazume* and *ōbiromazume* daimyo. The cleavage was of course far from perfect.

he foiled their plan by stepping down in favor of Hotta Masayoshi, an ardent supporter of *kaikoku* policy and a moderate in relation to the internal politics of the Bakufu. This was a concession to the group headed by Ii, but one which enabled Abe to preserve the ties between the Bakufu and the national front powers—his main objective at this time.

Abe's departure from traditional Bakufu autocracy in consulting the opinions of the daimyo and entering into relations with han outside the traditional ruling clique of *fudai* daimyo, has often been criticized, and perhaps justly, as the first step leading to the downfall of the Bakufu. But one must acknowledge that at least up until 1857 this policy achieved a remarkable degree of success: for three years the han did work together under the leadership of the Bakufu. Following Bakufu orders, the han strengthened their defenses, intensified the military training of their samurai, and introduced and developed Western studies within their domains.

In 1858 the American consul, Harris, advised the Bakufu leaders that it would be more prudent to enter voluntarily into a lenient treaty of trade with the United States than to be forced into a harsher treaty by the gunboats of other nations at some subsequent date. Faced with a decision it could not make alone and aware of the precedent it had established barely three years earlier, the Bakufu for the second time ordered the han to present their opinions on the matter. This time there were more han in favor of opening the country than there had been in 1853. In general, the opinions followed the lines drawn by the power struggle within the country and the Bakufu; the *fudai* han were in favor of a treaty and the *tozama* han and some *shimpan* were against it. Again, the opinion of Chōshū is fairly representative of the *tozama* han, although Chōshū was still inactive in national politics:

If the English are coming next spring, as the American mission says, it is inevitable that the other Western countries will follow in succession. Our Imperial country faces the sea on all four sides and we cannot predict from which side they will approach. The feelings and conditions of the barbarians are difficult to surmise. It is important to take measures so that our country will not be disgraced. The customs of peace have naturally led to a neglect of readiness; therefore, let preparations be quickly and strictly ordered so that whenever war begins, our military glory will be established. . . .[4]

Abe died in 6/57, and about the same time the problem arose of choosing an heir to the Shogun Iesada who was weak, sick, and childless. Two candidates were proposed. The Ii faction supported as the heir apparent Iesada's closest blood relation, Tokugawa Yoshitomi (later Iemochi), the daimyo of Kii. The opposing national front faction contended that Yoshitomi was much too young and inexperienced for the position in this time of stress and proposed instead Keiki, the daimyo of the Hitotsubashi Tokugawa branch house and the seventh son of Nariaki. The delicate balance between the two factions that had been achieved by Abe was now due to collapse. The candidate who won would give his faction the decisive voice in Bakufu politics. Each group, therefore, exerted itself anew to advance its own candidate.

If the dispute were decided in favor of the Ii faction, then Ii and his group of *fudai* officials would enforce a return to the Bakufu autocracy of former days; there was no doubt that as the ranking ministers of a youthful shogun, they could dominate the Bakufu. On the other hand, if the opposing *shimpan* and *tozama* faction won, then Nariaki of Mito stood to gain the dominant voice in the Bakufu, and the daimyo of Tosa, Satsuma, and Uwajima, all of whom were active at this time, could expect a larger role in the formation of national policy. Some of the national front powers even hoped to institutionalize their new positions. Matsudaira Keiei, for example, favored reinstituting a system of rule by five *tairō* under the shogun, such as had existed in the last years of Hideyoshi's rule, and a retainer of Keiei, Hashimoto Sanai, wanted to establish a council of all those daimyo with domains of over 100,000 *koku*.

Both sides tried to further their cause through intrigues at the Court and the Bakufu. After conferring with Tosa, Keiei delegated Hashimoto Sanai to agitate at the Court. Nariakira of Satsuma hastily despatched letters to important Court officials in support of Keiki. On the other side, Ii sent a retainer to persuade the *kampaku* Kujō Hisatada, one of the most influential of the Court nobles, to

[4] Suematsu, *Bōchō kaiten shi*, II, 194–195.

support his candidate at Court. Until this time, in spite of its earlier approval of the first treaty and its subsequent brave note to the Bakufu, the Court had not been drawn into the factional struggle within the Bakufu. Now, however, courted by both sides, the Court suddenly became an important factor in Bakufu politics, and factions among the nobles aligned themselves with the contending groups of daimyo. Meanwhile, within the Bakufu the Keiki group was directing its energies toward the council of *rōjū,* while the opposition was trying to win over the high Bakufu officials closest to the shogun.

The Ii group won this contest when on 23/4/58 Ii was appointed *tairō*. This automatically made him the head of the council of *rōjū*. Using the authority of his new position, Ii privately settled the question of succession in favor of his own candidate, Tokugawa Yoshitomi, or Iemochi as he was known when he became shogun on 25/10/58. Ii's opponents, as a last-ditch move, now tried to obtain from the Court an Imperial order to replace Yoshitomi by their own candidate. But they were fighting a losing battle, and in the sixth month Ii publicly proclaimed Yoshitomi as the shogunal heir.
heir.

Even supported by the national front powers, the Court had been unwilling to issue an edict directly opposing a Bakufu decision. But, encouraged by the national front powers, it did adopt a negative position concerning the proposed Treaty of Commerce (*Tsūshō Jōyaku*). Moved by the arguments of Harris and the news of Western victory in China, Ii decided to sign the treaty first and to obtain Imperial approval later. He signed a week before publicly announcing that Yoshitomi would be the heir. At this point the struggle for power that had first focused on the foreign problem and had afterward come to bear on the internal problem of succession, once again returned to the external problem of the new treaty. The national front powers attacked the rule of Ii on the grounds of a lack of proper respect for the Court. It was from this time that the slogan *jōi* (expel the barbarian) became joined with *sonnō* (honor the Emperor), and on the other side, *kaikoku* (open the country) with *sabaku* (support the Bakufu).

At the same time that the Court began to emerge as a political force, the currents of political activity within the country deepened. The political activities of which we have been speaking had been confined mainly to the top echelon of the political hierarchy, that is, to the daimyo and their chief retainers. Now the *sonnō* thought, latent in many of the han, began to merge with a consciousness of the han's position in the power struggle, and small groups of samurai intellectuals began to enter into the politics of the day. For the first time one can rightly describe the *sonnō jōi* agitation as a movement.[5]

Ii rightfully blamed the opposition party for the Court's reluctance to approve of the new treaty, and, in the autocratic fashion he thought fitting, he moved to punish them. On 5/7/58 he ordered Nariaki of Mito into domiciliary confinement (earlier in 1844 Nariaki had been forced to abdicate as daimyo of Mito as punishment for interfering in Bakufu affairs). He also ordered Matsudaira Keiei of Echizen and Tokugawa Keishō of Owari to abdicate as daimyo. This provoked considerable enmity in these han, and *sonnō jōi* intellectuals, partly motivated by a sense of loyalty to their daimyo, conspired to assassinate Ii. Rumors of these plots soon reached the ears of Ii; they prompted him to carry out what in later times has been known as the Purge of Ansei. Arrests were made of *sonnō jōi* ideologues from all the han and especially from Mito, the heartland of the most dangerous sort of *sonnō* teachings. They were seized, interrogated, imprisoned, and some, such as Yoshida Shōin of Chōshū, were executed.

As one might expect, these methods were very effective. The daimyo and their chief retainers were silenced; even most of the lower-ranking ideologues were temporarily hushed. In order to avoid the wrath of the Bakufu, the han governments took it upon themselves to curb their own samurai. The most fanatic among the *sonnō jōi* radicals, however, became even more inflamed. On 3/3/60 a Mito samurai, one of the participants in an ineffectual plot by samurai of Mito and Satsuma to re-establish the Emperor, assassinated Ii. Andō Nobuyuki, Ii's successor, presumably hoping to escape a similar fate,

[5] Tōyama, *Meiji ishin,* pp. 87–88.

reverted to the policies of Abe, but events had reached a point of no return. The way was now clear for the great han to emerge as mediators between the Bakufu and the Court.

Chōshū Politics, 1853–1861: Cliques and Reforms[6]

Through most of the early phase of Bakumatsu politics Chōshū slumbered. The coming of Perry, for example, had little direct effect on the course of Chōshū's internal politics. On the contrary, in most ways the political developments in Chōshū between 1853 and 1858 were the continuations of the policies and clique politics of the earlier Tempō era. The year 1858, not 1853, was the year of Chōshū's awakening. Nevertheless, the structure and forces underlying the 1858 change were the cumulative result of many small changes during the preceding years, therefore it is to these that we must first turn.

Chart 1 illustrates in graphic form the periods during which the Chōshū government was controlled by each of its major cliques. We can see that there occurred three changes in the clique control of the Chōshū government during the years 1853–1858. The first was the resumption of power in 1853 by the reformist "hards" who were now led by Sufu Masanosuke, a protégé of the Tempō leader, Murata Seifū. The second was the rise to power in 8/55 of the Tsuboi clique. During the period from 1847 to 1853 Tsuboi himself had been out of office, and within the han officialdom Tsuboi's lieutenant, Mukunashi Tōta, had acted as the spokesman for this group against the more extreme program of Sufu. In 1855 Tsuboi once again took up the reins of government. The third change was the fall of the Tsuboi clique and the rise of the Sufu clique which was to hold power from 1857 until 1864. Let us first examine the period of the first Ansei reform.

Why after six years of indecisive action, six years during which

[6] Most of the materials in this section on the development of cliques in Chōshū during the 1853–1861 period were taken from the writings of Tanaka Akira and Seki Junya. The materials on the structure of the han government and on the political developments within the han were taken from the *Bōchō kaiten shi*.

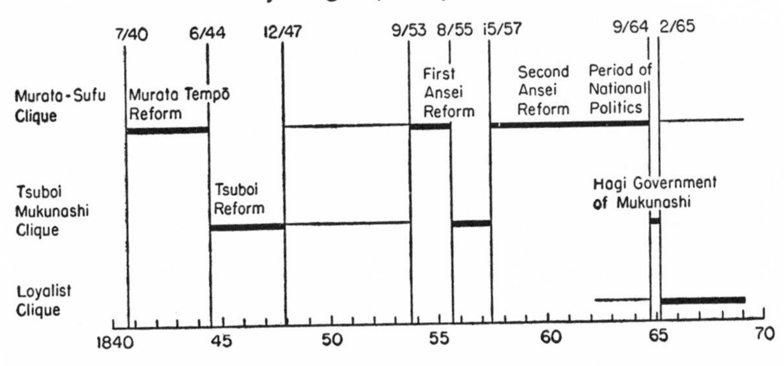

CHART 1. The position of the three dominant cliques in the Chōshū government from 1840 to 1868

no single clique controlled the han government, did the Murata-Sufu clique regain power in 1853? In part at least this was an indirect consequence of events on the national scene: the arrival of Perry caused the Bakufu to place new levies on the han, thus increasing their financial difficulties. New arms were needed; fortifications had to be built. Moreover, Chōshū forces were assigned to defend certain coastal areas within the Tokugawa domains. (Was this a consequence of the Bakufu's "weakness of numbers?") The expenses of these Chōshū forces had to be borne by the han. Five days after Perry's arrival the Bakufu had ordered the various han to send their Edo troops to defensive positions along the perimeter of Tokyo Bay. This order threw most of the han into confusion, for almost none had an adequate supply of weapons in their Edo residences. Chōshū, however, was able to send out five hundred men fully equipped: one item of Murata's Tempō Reform had provided for the storage of weapons in Edo. Then in 8/53 Chōshū samurai were ordered to take up defensive positions at Sagami Bay. These were held for five years after which the Chōshū forces were reassigned to other positions in Hyōgo.

These new expenses, together with the familiar specter of the han debt, called for a new reform. Because of its affinity with the earlier Tempō Reform, the Council for Reform which met on 21/6/54 called its reform the Second Financial Reform. Tempō was the explicit model. Like its prototype, it made no attempt to uncover new sources of income, probably because there were so few available, but rather concentrated on curtailing expenses. In the months that followed Sufu's appointment, over fifty measures of this sort were enacted. One of the most important among these dealt with administrative retrenchment. Those holding office in the han government had always been given certain emoluments in addition to their hereditary stipends. By reducing the number of officials in an office, or by combining the functions of several offices, the number of these emoluments could be decreased and substantial savings realized. During the course of the reforms, over fifty offices were eliminated. In addition the allowances of the daimyo's family were cut, many of the han ceremonies simplified, the cortege of the daimyo reduced, and many other comparable frugalities enforced. The reformers followed further in the footsteps of the Tempō Reform by enacting in 4/55 a law concerning samurai debt. It stipulated that, during the years when the han government had borrowed one-half of the samurai's income, no repayment of principal and only 5 per cent interest need be paid on any debt, private or public.

As during the Tempō period, these measures provoked a reaction. And again, it was the law concerning samurai debts that appeared to be the chief target of the criticism. On 11/8/55 Sufu was relieved of his office, and the Tsuboi-Mukunashi group again rose to power. They kept power until 1857. During the two years of their rule there was virtually nothing of note except the very lack of activity itself. They did try to increase production by buying up some of the production of the peasants at fairly high prices for resale outside the han. The twofold purpose behind this was to bring prosperity to the han by encouraging production and to increase the supply of money in the han in "exchange for worthless roots and grasses."[7] As one might expect, the han government lost money and soon abandoned the plan. Its only important consequence was that the Tsuboi

[7] Tanaka Akira, "Bakumatsu-ki Chōshū han ni okeru hansei kaikaku to kaikakuha dōmei" (diss. Tokyo Kyōiku Daigaku Bungakubu Kenkyūka, 1956), pp. 275–276.

faction alienated the local peasant officials, who had been bypassed in the administration of the plan. Some military and educational reforms were also put into effect, but even these were halfhearted and inadequate: in any case they were not controversial. In general, the moderates maintained their power by abstaining from anything that might offend; the Sufu opposition could only attack them for their do-nothing policy.

But the times were against the moderates. As tensions mounted within the country, even within Chōshū an awareness grew of impending change, and the demand grew that Chōshū, as a great *tozama* han, take a strong position. The moderates opposed action on the national scene, but the Sufu faction gradually came to feel that Chōshū must act if an opportunity occurred.

On 25/6/58, having returned from Edo a few days earlier, the daimyo of Chōshū began the traditional transfer of officials. The Tsuboi-Mukunashi group was swept out of office; only the two leaders remained in official positions, and they were both removed from the center of government. Tsuboi was relegated to one section of the *buikukyoku* and Mukunashi to the han school. The Sufu group again gained control of the government.

A third reform was announced. Remembering the antagonism aroused in 1843 and 1855, they issued no act concerning samurai debt, and yet in many ways this reform proved to be more extensive than its predecessors. New military units were formed, provisions were made for a new supply of traditional weapons, and proposals (on which no action was taken) were put forth concerning the formation of peasant troops. The entrance requirements of the han school were again relaxed to admit the most able from among the lower-ranking samurai. It was also decided that, since a school by its very nature was a specialized institution, students should be seated according to ability and achievement rather than by rank as had hitherto been the case. But, the students were cautioned that this rule would apply only within the school and that in other situations the distinctions and etiquette due to different gradations in rank would continue to be observed.

Following the example of earlier reforms, the 1858 reformers also ordered a tightening up of morals, the renunciation of extravagance, a regime of austerity, and in all things a return to the good customs of olden times. Those who disobeyed would be chastised and those who set the best examples would be rewarded. Peasant officials were instructed to observe benevolent paternalism, granaries were to be maintained, production increased, and extra taxes collected from "wealthy" peasants were to be allocated to public works such as irrigation ponds, waterways, and so on. In addition, as we have seen earlier, monopolies were discarded in favor of a system of licenses for trade within the han. None of these measures merits further discussion; they are mentioned primarily to illustrate the fundamentally old-fashioned nature of the reform. When Western innovations are mentioned in the next chapter, it should be kept in mind that they comprised only one small facet of the reform activities.

The real significance of Sufu's program in 1858 was that, for the first time since the arrival of Perry, reform politics within the han had been eclipsed in importance by national politics. The rise of the Sufu clique in 1858 can be said to have been the rise of a clique that would put national politics, that is to say, han participation in national affairs, before reform within the han. Before treating Chōshū's program for national action, however, we must first explain how the shift from one clique to another within the han government came about. What was the relation between the cliques and the daimyo and Elders (*karō*)? What sort of men were Sufu, Tsuboi, and Murata, and how did they rise to power? Who dictated han policy? Only by exploring the dynamics of government and the power structure within the han can one adequately explain Chōshū's inactivity during the early years of the Bakumatsu period and the subsequent changes in policy that took place in 1858.

Chōshū Politics: the Dynamics of the Han Government

The structure of the samurai class in Chōshū during the Tokugawa period was a smaller replica of the Mōri hierarchy which had existed during the period of Warring States. Under the daimyo were

5675 vassals and over 5000 rear vassals.[8] These figures actually indicate the number of families of samurai rank rather than the total number of persons in the class as a whole. Consequently, the total samurai population would be much larger, probably very close to 50,000 persons in all. Since the total peasant population of the han was about 500,000, the samurai class made up roughly one-tenth of the population of the han. When we remember that at least one-third of the peasants' production was taken as taxes, we could reasonably say that one-tenth of the han population monopolized one-third of the wealth for its own consumption. Also, as it will be made evident later in this chapter, the distribution of wealth within the samurai class was as uneven as it was for the han as a whole.

Viewing the structure of the han from a distance, the strongest impression would be of one great pyramid—the han—containing many lesser pyramids. Of these lesser structures, the largest and most autonomous were the four branch han (*shihan*). Three of these branch han were ruled by daimyo who, like the daimyo of Chōshū, were recognized as daimyo by the Bakufu and in consequence were required to spend alternate years in Edo. The fourth branch house had an indeterminate status: it was not recognized as an independent han by the Bakufu but its government was separate from, and independent of, the government of Chōshū proper. That their domains were exempt from the land surveys which took place from time to time in the main han (*shuhan*) shows the extent to which all four branch han were financially autonomous of the main han.

Yet, in spite of their administrative independence in both the political and financial spheres, the branch han were subordinate to the main han. In part this subordination derived from their inferior position in the kinship structure of the house of Mōri and was ceremonial in nature. In part it was customary and stemmed from the fact that they had been founded by an ancestor of the main house and consequently were obligated to it. Therefore like the Tokugawa *shimpan* vis-à-vis the Bakufu they were expected to act in the best interests of the han. Their advice was accepted—when

[8] Umetani Noburu, "Meiji ishin shi ni okeru kiheitai no mondai," *Jimbun gakuhō,* 3:44–45 (Mar., 1953).

requested—on all matters affecting the total han, yet with rare exception they concurred in all decisions made by the main han. Therefore, when treating such decisions in this book we omit the mention of the messengers who were sent each time to obtain the opinions of the branch han daimyo. We have already seen that, of the han's total production of 895,158 *koku,* 183,022 *koku* or about one fifth went to the branch han. This figure would be considerably augmented—probably it was well over one fourth—if the domains of the branch han had been surveyed as recently as had those of the main han.[9]

After the branch han, the next largest pyramids within the han were those of the eight Elder (*karō*) houses, the highest-ranking vassals of the daimyo. Their incomes ranged from 16,000 *koku* to 5000 *koku*. The most striking feature of these large Elder houses is that they were almost exactly like the han structurally, though put together on a smaller scale.

We will take as a median example of this type the house of Masuda, with fiefs amounting to 12,063 *koku*. Masuda had 538 vassals or a little over 8 per cent of the total number of rear vassals in the han. Of these vassals 263 were classed as *shi* (knights) and 275 as *sotsu* (soldiers). Most of these vassals lived at the administrative center of the Masuda fief, just as most of the Mōri vassals resided at Hagi. Masuda and his chief retainers, however, spent most of their time at Hagi, in much the same way that the Chōshū daimyo spent part of his time in Edo. It is also significant that, while most of Masuda's vassals received their incomes in stipends (*kirimai* or *fuchimai*), there were a few with fiefs within the Masuda fief. The house of Masuda also possessed a house government that administered its fief. It distributed stipends, settled disputes between vassals, and operated a school to train Masuda's vassals in the military arts and to teach them the rudiments of

[9] Actually, what I have referred to as the fourth branch house was officially considered as a fief, and was raised to the status of a branch han only after the Restoration. Yet, since it was by blood a branch house of the house of Mōri, and different from all other fiefs, I have included it with the other three, which it resembled on most points. This "han" of Iwakuni is further discussed in Chapter IX.

Confucian learning; in short, it ruled in much the same way that the government of the han ruled Chōshū, although, of course, it was functionally less specialized. The loyalty of the vassals of Masuda was directed to the house of Masuda as that of Masuda and the other vassals of Mōri was to the daimyo of Chōshū. Moreover, this was not merely a formal relation. Even during the Chōshū civil war, although the Masuda retainers were split into two camps reflecting the situation in the larger unit of the han, the struggle between the camps did not merge with the larger struggle but took place within the Masuda fief, physically and structurally apart from the larger unit.[10]

The 5675 direct vassals of the house of Mōri were divided into two groups: *shi* (knights) and *sotsu* (soldiers). There were seventeen ranks of *shi* and twenty-three of *sotsu*. It is interesting to note that in the status system of the han, a *shi* rear vassal was lower than a *shi* vassal, but higher than a *sotsu* vassal. A *sotsu* vassal was higher than a *sotsu* rear vassal. Of the seventeen ranks of *shi* vassals, the highest ranking, of course, were the Elders. The next highest group was known as the *yorigumi;* it was composed of sixty-two families whose incomes ranged from 5000 *koku* to 250 *koku*. Like the Elder houses; these too, had vassals of their own. In the 1624-1644 period, possibly a time when many rear vassals were being discarded due to economic pressure, laws were enacted requiring that 2.3 vassals be maintained for every one hundred *koku* of a samurai's stipend (*roku*). At the time of the Tempō Reform this was reduced to two rear vassals for every one hundred *koku* of income. Actually, some of the great houses, like that of Masuda, had almost twice the number of vassals required by han law, and most small houses with one hundred or less *koku* of income had none at all. If one were maintained, it was usually as a hereditary servant. Rear vassalage was hereditary, as was vassalage.[11]

The third stratum of *shi* was the *ōgumi* class. With 1378 members this was the largest single group of *shi*. The incomes of this group ranged from about 250 *koku* to 40 *koku;* 569 of the group

[10] See Chapter VIII on the Chōshū civil war.

[11] Suematsu, *Bōchō kaiten shi,* I, 35–48.

had incomes over 100 *koku*. And, below the *ōgumi* class, in a descending scale of rank and income, were the remaining fourteen strata of *shi* and twenty-three of *sotsu*. In general, as the scale descended, the number of the smaller pyramids increased. The vast majority of these little pyramids were made up solely of the head of the family and his household. Yet in such a society, as one writer has pointed out, even "the family is the polity writ small." [12]

Up until this point our discussion of the general organization of the han and of the composition of the samurai class suggests that there existed a considerable degree of feudal autonomy within Chōshū. Was this true only of the branch houses (*shihan*) and *karō* houses or was this general among the ranking samurai of the han? On the whole it appears that by Bakumatsu times this type of autonomy existed only among those possessing fiefs, and then only among the top few of these. Where it existed it stemmed more from the control exercised by a samurai lord over his vassals (*baishin*, rear vassals vis-à-vis the Chōshū daimyo) than from his control of the land, and even this limited autonomy only appeared in moments of crisis. In general, far more significant on the sub-han level than this truly feudal autonomy was the autonomy possessed by bureaucratic cliques and by certain units of the han military.

Yet, it is nevertheless important to consider for a moment the pattern and extent of fief holding within Chōshū—as a clue to the process of bureaucratic rationalization if not as a simple index to the effective separation of samurai from the land. Two Japanese scholars, Kimura Motoi and Seki Junya, have uncovered very interesting materials dealing with this question.[13] Kimura has put together a table showing the pattern of fief holding in Chōshū and its changes over time. (See Table 2).

This table makes clear that, absolutely, the decline of fiefs was very slight over the course of the Tokugawa period; yet, relatively it was very significant: a 50 per cent decline in fief holdings between the

[12] Robert N. Bellah, *Tokugawa Religion* (Glencoe, Illinois, 1957), p. 18.

[13] Kimura Motoi, "Hagi han zaichi kashin dan ni tsuite," *Shigaku zasshi*, 62:727–740 (Aug., 1953); and Seki Junya, *Hansei kaikaku to Meiji ishin*, pp. 15–20.

TABLE 2. Fief Tenure within Chōshū

Year	Total product of Chōshū	The products of samurai fiefs	Percentage of total in samurai fiefs
1625	473,326 *koku*	270,662 *koku*	57.2
1763	. . .	217,389 *koku*	. . .
1821	712,136 *koku*	205,558 *koku*	28.9
1841	712,325 *koku*	202,448 *koku*	28.4
1854	712,060 *koku*	200,038 *koku*	28.1

Source: Kimura Motoi, "Hagi han zaichi kashindan ni tsuite," *Shigaku zasshi,* 62:728 (Aug., 1953).

early Tokugawa period and the time that Perry arrived over 200 years later. The gradual shift from fiefs to stipends was of course encouraged by the han government since it increased the lands under its direct control. The decline in the number of fief-holders was less extreme: in 1652, 884 samurai held fiefs; by the time of the Tempō Reform this number had declined to 505 (holding 810 fiefs). Thus in Bakumatsu times in Chōshū, which was famous as a han possessing many fiefs, only 9 per cent of the samurai had fiefs and only 19 per cent of the class of *shi* had fiefs. This 19 per cent, however, held in fiefs the land producing over one fourth of the han's total production (excluding the *shihan*). The average size of the fiefs was about 400 *koku*. Excluding the eight *karō* houses whose lands constituted 80,000 of the 200,000 *koku* of lands in fiefs, the average of the remaining fiefs drops to 240 *koku*.

Given, then, this distribution, the significant question is how they were managed. To what extent was the fief a "fief," to what extent was it a hollow form the reality of which was securely gripped by the government of the han? The extension of controls by a daimyo over the fiefs of his vassals began to a very limited extent during the Ashikaga period when the "true" feudal fief first appeared. From the time that vassals began to reside in the castle town of their lord, and particularly from the time that castle towns were removed from secluded mountain sites to centers of trade and communications, it became necessary to establish some controls over the lands of absent samurai. For the most part this was handled by the hereditary retainers of these samurai, but, gradually, some powers came into the hands of those vassals who administered on the local level those lands directly controlled by the daimyo. The pre-Tokugawa history of the development of such local bureaucratic tendencies is beyond the scope of this work; yet we can say that by the beginning of the Tokugawa era these had become formalized into a system of district offices possessing an authority not limited to strictly legal and police matters (although not omnipotent in these spheres either).

Yet in spite of the existence of such offices, at the inception of the Tokugawa period most samurai fiefs in Chōshū were in fact controlled by their samurai lords. At this time legal control was reinforced by the fact that, in addition to peasants, the land in these fiefs was worked by *baishin* and occasionally even by the family of its samurai lord. This afforded a high degree of direct samurai control.

As the han began to emerge as a distinct unit within its Tokugawa setting there appeared in the 1615–1644 period a number of formal controls designed to limit the degree of exploitation of peasants by their samurai masters: the latter must not levy undue taxes, they must not arbitrarily deprive peasants of their land, they must not switch the peasants' fields for their own profits, they must not force them to labor except at certain stipulated tasks, and so on. The remaining powers of the samurai fief-holders were, however, still considerable: they controlled the "mountains, rivers, and fields"; [14] they levied taxes and collected the tax rice; and they still held various special powers over the peasants. And, in fact, they did change peasants about almost at will, they assigned new peasants to cultivate fields, and they collected extra taxes. It was generally recognized that the life of a peasant on a samurai fief was harder than one on lands directly controlled by the han government.

[14] Seki Junya, *Hansei kaikaku to Meiji ishin*, p. 17.

The sale of persons as hereditary cultivators or servants (*fudai hōkōnin*), a lower class of peasants, was prohibited within peasant society in general in Chōshū by the statutes of the 1658–1660 period, yet in the case of the samurai fief-holder, the possession of such persons was recognized in the case of those persons in the service of the samurai house before 1651. Even after this date, however, the second and third sons of peasant families were from time to time accepted in lieu of tax rice; this was especially frequent when a peasant house went bankrupt. In the 1660's and 1670's such hereditary cultivators began to be transformed into *nago,* representing a slight increase in their legal rights, and, in Chōshū, from time to time into *baishin* as well. Yet even at this time the displacement of the officially listed peasants by such hereditary servants still constituted a problem. The infringement from above of peasants' legal rights also occurred when samurai, impoverished by the expenses of life in the castle town, returned for a time to their fiefs and took over and worked land by themselves. Such samurai and their *baishin* caused considerable trouble at the time of the cadastral survey of 1686–1687 by interfering with the work of the surveyors and registering complaints at the local han offices. Finally it became necessary to issue an edict prohibiting such interference.

The survey of 1686–1687 was a watershed as regards samurai fiefs. Before that time they were by and large under the control of their samurai lords and only loosely supervised by the intendant of fiefs (*kyūryō daikan*). After that time the supervision passed into the hands of the regular district offices and little by little direct control by samurai was lost. It is hard to detail the process by which this took place. Enforcement of laws requiring samurai to reside in the castle town was one factor. The strengthening of peasant tenure on the lands directly controlled by the han government was another. The extension of controls by the district offices was yet another (for example, they began in time to appoint peasants to supervise the collection of the tax rice on samurai fiefs).

By the middle of the nineteenth century very little remained of the earlier power of the fief-holder except on the largest fiefs. The form which this rationalization took has been described by Kimura Motoi in the analyses of three samurai fiefs.[15] In general, in smaller fiefs a fief official was appointed by the district office to handle its affairs. If a number of such small fiefs existed in a village, a higher fief official was appointed to oversee the fief officials. In larger fiefs both types of officials (with slightly different titles) were present within a single fief. For example, 8500 *koku* of the *karō* Fukuhara's 11,300 *koku* were located in three official villages (forty-eight small villages and hamlets). The peasant administration of these were for the most part managed by two higher fief officials and eleven fief officials. As in the case of the smaller fiefs these were responsible not to the samurai lord, but to the district office. The extent to which custom paralleled formal administrative organization is not known. The power of the *karō* houses in the central government of the han certainly precluded (almost in Chinese fashion) any thoroughgoing bureaucratic control of the *karō* fiefs by the samurai officials of the district offices. Yet in all but the *karō* fiefs such control was probably almost complete.

What then was the organization of the han government which by the early nineteenth century had gained such a formidable power? In general, the government of the han was the house government of the Mōri family, under the personal control of the daimyo. In principle it was exactly like the governments of the *shihan* or that of the house of Masuda, except that, since the han was a larger unit, its government was more specialized. As a house government in Edo times, the han government was a feudal bureaucracy. The words "feudal" and "bureaucracy" are sometimes taken as antithetical. In a very pure sense of the terms this may be true. Ideally, the feudal ranks in themselves indicate the hierarchy, and government is properly a function of the personal relations between ascending tiers of vassals and their lords. But in fact, as early as the period of the Warring States, when feudalism reached its fullest development, a bureaucracy of sorts had already formed. The unit fiefs had become so large that personal government could no longer handle their

[15] Kimura Motoi, "Hagi han zaichi kashindan ni tsuite," *Shigaku zasshi*, 62:732–733 (Aug., 1953).

multifarious affairs. Even among the "direct" retainers of the daimyo, only certain ones—the so-called *jikisanshi*—had the right of audience. Feudal ranks, however, had not yet become fixed. As long as a state of war remained the normal relation among the territories of different lords, it was still necessary to reward merit with rank and income, and, as such, rank was more likely to indicate position than was the case during the Tokugawa period.

With the commencement of the Tokugawa period, or perhaps even a little earlier, rank became frozen, and with a few exceptions it was determined by hereditary succession. Yet even in times of peace there was a limit to the degree to which ability could be ignored in favor of rank. As a result, it is not surprising that over the Tokugawa period there occurred a gradual widening of the rank base for eligibility in the key positions of the han government. What is more surprising is that even at the end of the Edo period the governments of the han should still have been controlled by so select a group from the top strata of the samurai class. Since this was the case, even during the early part of the Bakumatsu period it is legitimate to call the han government a feudal bureaucracy. But at the same time, one must remember that the nature of the feudal ties altered in several important respects during the transition from the period of Warring States to early Tokugawa Japan and then to the Bakumatsu period.

During the early years of the Tokugawa period the structure of the government was as indicated in the accompanying diagram.

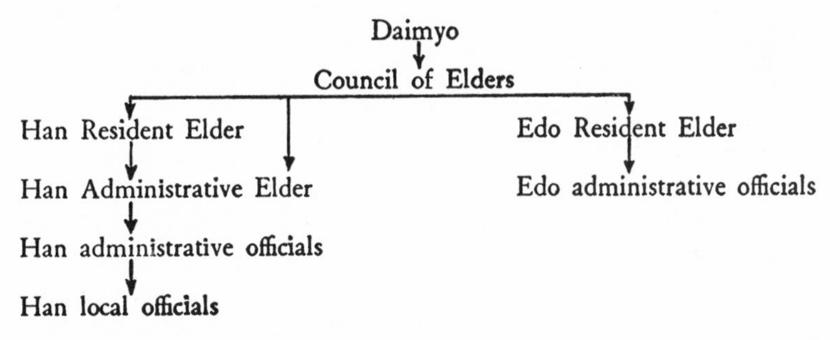

The daimyo stood at the apex of the government. When he was of age and of strong character, he was able to dominate the han government. When he was weak, power in the han came to rest in the Council of Elders (*kahanyaku* or *karōshū*). This council was made up of the heads of the eight Elder houses plus a few Single Generation Elders (*ichidai karō*) appointed from the *yorigumi* stratum of *shi*. It met three times a month and dealt with all problems concerning the han; it was the seat of both legislative and executive power. The latter, however, was exercised through two resident Elders, one for Edo and one for the han, and one Administrative Elder for the han. When the daimyo was in Edo, the Han Resident Elder was left in charge of the operation of the various han offices. One other important official not shown on the scheme was the Direct Inspector (*jikimetsuke*). This office was outside the hierarchy pictured above, and, as the name discloses, the duty of this official was to investigate whether orders given by the daimyo were being carried out and also to keep an eye on the other officials in the han. He reported his findings directly to the daimyo.

During the course of the Tokugawa period several important changes took place, so that by the time of the Tempō Reform the actual structure of the government had changed as shown in the accompanying diagram.

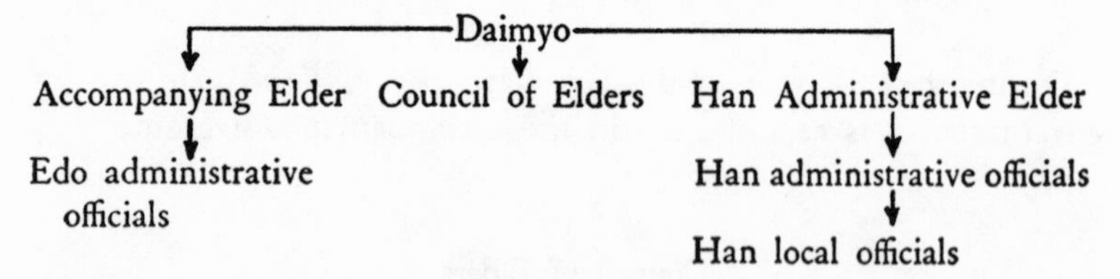

We can see, first, that the Council of Elders had been reduced to an advisory body. It still continued to meet three times a month, but it had very little influence except insofar as it formed the climate of opinion about the daimyo. The office of Han Resident Elder (*kunimoto rusui karō*) lost more and more of its power to the Han Administrative Elder (*tōshoku* or *kokushō*), who was close to the officials actually administering the han, until it became nothing more than an empty title. While this was taking place, another office that had been of negligible importance earlier, that of the Accompanying Elder (*tōyaku* or *gyōshō*), gained in prominence. The Accompanying Elder, as the name would indicate, accompanied the daimyo, whether in Edo or in the han. During the middle of the Tokugawa period, the Han Administrative Elder was still more powerful than the Accompanying Elder, but by the end of the period the Accompanying Elder had outstripped the other. Since both of these powerful Elders would be in the han at the same time when the daimyo was in the han, much rivalry existed between them. When the daimyo was weak, each tended to act as a check on the other. From now on, when speaking of them together, we shall refer to them as the executive Elders.

As the position of the executive Elders acquired importance, a "cabinet" grew up under each. By the end of the Edo period the cabinet of the Han Administrative Elder came to include virtually all of the officials continuously in residence in the han. The cabinet of the Accompanying Elder, which was known as the "traveling cabinet," since it too traveled with the daimyo, became, as it were, an executive cabinet surrounding the person of the daimyo. Why such a group became dominant is obvious enough. It is more difficult to explain why it never completely overshadowed the office of the Han Administrative Elder. Most likely, when the daimyo was in the han, the Han Administrative Elder was able to preserve the power of his position by virtue of his personal contact with the daimyo. And, in the absence of the daimyo, even if the Han Administrative Elder should be completely controlled by the Accompanying Elder, the latter still had to rely on the Han Administrative Elder to carry out his orders from Edo. Also, the close contact between the Han Administrative Elder and the administrative offices of the han served to sustain his power.[16]

Possibly the most important consequence of these changes in the structure of the han government was an increase in the status of certain key officials under the two executive Elders. Where they had previously been two or three steps removed from the daimyo, now they were only a single step away. Where decisions had previously

[16] Suematsu, *Bōchō kaiten shi,* I, 49–68.

required the discussion and approval of the entire Council of Elders and then had to pass from the Resident to the Administrative Elder, now they were made by the executive Elders in conference with those in key positions in their cabinets. When the daimyo and the executive Elders were strong, these changes undoubtedly meant more efficient government. When they were weak, however, it had the effect of transferring the entire weight of the government to the men in the key cabinet positions. Since these key positions had originally been purely administrative and of minor consequence, there were not attached to them the same prerequisites of rank as there were to the positions which had formerly been dominant. As a result, men who were not in the very highest group in rank came to assume leading roles in the han government. This process was one of the modes by which the broadening of the base of bureaucratic power earlier mentioned had occurred.

Murata, Tsuboi, Sufu, and Mukunashi were examples of such key officials. The following tabulations[17] will offer some idea of the class background of the leading figures in the different reformist cliques. They will also enable us to compare the Tempō reformers of the early 1840's with those of the Ansei period (1854–1858), and the Ansei "hards" with the Ansei "softs."

Tempō Reformers	*Hereditary Stipend Expressed in* Koku
Murata Seifū	91.000
Tsuboi Kuemon	100.000
Akagawa Chūemon	180.000
Kihara Genemon	93.500
Nagaya Tōbei	114.000
Niho Yaemon	100.000
Ogawa Zenzaemon	46.939
Fukuhara Echigo	48.800
Nagoya Noboru	83.000
Nakaya Ichizaemon	113.500
Akagawa Kihei	133.447

[17] Tanaka Akira, "Chōshū han no Tempō kaikaku," *Historia,* 18:28 (June, 1957), and Tanaka Akira, "Tōbakuha no keisei katei: Chōshū han bakumatsu hansei kaikaku to kaikakuha dōmei." *Rekishigaku kenkyū,* 205:4 (Mar., 1957).

Ansei Reformers	*Hereditary Stipend Expressed in* Koku
Ansei "hards"	
Sufu Masanosuke	68.275
Maeda Magoemon	173.553
Inoue Yoshirō	202.336
Kaneshige Jōzō	60.000
Naitō Marisuke	110.000
Shishido Kurobei	120.000
Ansei "softs"	
Tsuboi Kuemon	47.500 (demoted in 1847)
Mukunashi Tōta	46.213
Akagawa Tarōemon	133.447
Miyake Chūzō	. . .
Naitō Hyōe	170.500

From these tabulations we can see that the Tempō reformers were of approximately the same rank as those of Ansei. It is also clear that Sufu's group was generally of the same composition as that of Tsuboi and Mukunashi; if anything, the average of the "hards" is slightly higher than that of those "softs" whose incomes are known. In terms of rank, most of the reformers were members of the *ōgumi,* the third stratum of *shi*. During the early part of the Edo period, dominant positions such as those held by these reformists could not have been held by members of the *ōgumi* class. Only a member of the *yorigumi* could have exercised such power, and this only by becoming a "Single Generation Elder." In short, positions that had been open to only seventy persons were now accessible to 1448 persons, although admittedly, this was still a small portion of the entire class.

The most important single figure in Chōshū during the Bakumatsu period was undoubtedly Sufu Masanosuke. His career illustrates admirably the rise of an *ōgumi* samurai with a hereditary income of 68.275 *koku* to a position of power within the han. He was born in 1823 in the castle town of Hagi. His mother's maiden name had been Murata; she was a close relation of Murata Seifū. While still young, his father died, and he was raised by his mother's family, receiving part of his education at the hands of Murata Seifū. As a youth Sufu twice won prizes for filiality. As a samurai of the *ōgumi* class, he was permitted to enter the han school, the Meirinkan, at the age of eighteen. He excelled in his studies and also became the leader of a student group. Several members of this group were to become members of his clique in the han government. Sufu began his official life in 1847 as an inspector in the Office of Revenues. Subsequently he taught for two years at the han school, and in 1850 he became one of the minor officials under the Han Administrative Elder. Then in 1853 he became the chief assistant to the Accompanying Elder, the highest position in the han bureaucracy a person of his rank could hope to attain, a position which because of the changes described earlier gave him a dominant voice in han affairs. It was from this eminence that he carried out his program of reform.

Having established the position of the leading bureaucratic personnel, such as Sufu, in terms of their rank and income, we may ask whether these men should be called upper, middle, or lower samurai. This is extremely difficult to ascertain since "upperness" was a relative and not an absolute quality even in the minds of the samurai themselves. The samurai bureaucrats, for example, were upper, vis-à-vis the *baishin, sotsu,* and *shi* in strata lower than their own, but they were lower than the top strata of *karō* and that of the *yorigumi.* Elsewhere I have criticized as misleading the usual indiscriminate application of the term "lower samurai." [18] Here I will only suggest that the term "lower samurai" should be restricted to *sotsu* and possibly *baishin,* and that of "upper samurai" to *shi.* Finer distinctions could then be made, on the one hand, within the category of *sotsu* by the usage of rank or group names such as *ashigaru, chūgen,* etc., and on the other hand, within the category of *shi* by giving specific ranks or by speaking of high- and low-ranking *shi.* This work will make these distinctions whenever possible.

The following tabulation,[19] illustrating the pattern of samurai income distribution within Chōshū, will enable us to see the relative position of Sufu and his fellow bureaucrats:

[18] Albert Craig, "The Restoration Movement in Chōshū," *Journal of Asian Studies,* 18:187–197 (Feb., 1959).

[19] Suematsu, *Bōchō kaiten shi,* I, 41–46.

Income	*Number of Families in the Group*
Over 100 *koku*	661
Over 70 and less than 100	202
Over 50 and less than 70	339
Over 40 and less than 50	472
Under 40	4001
Total	5675

From this it is clear that, using the criterion of income, the samurai bureaucrats were "upper" in a statistical sense. In fact, anyone with a fief or stipend of over forty *koku* was in the upper third of the samurai class (or including *baishin,* in the upper sixth of the class). Yet, Sufu with his stipend of 68 *koku* was obviously not an upper samurai or upper *shi* in the same sense as a samurai with a fief of 500 or 2000 *koku*. Nevertheless, his position in the han was considerably higher than that of samurai with the same income who lacked official position.

Finally, we must note that "lowness" or "upperness" alone do not determine the historical character of a person or group. By almost any definition, at least 75 per cent of the Chōshū samurai were "lower samurai." Therefore, any movement of samurai not originating from and constituted by the strata of samurai accustomed to rule would be a movement of lower samurai. The movements of the Bakumatsu period were not pure lower samurai movements even by this sort of definition, yet some writers who speak of them as such have suggested that it was the lower samurai character of the participants that determined the nature of the movements. Since the number of participants in these movements was always relatively small, the cardinal question is why some joined in and some did not. The answer cannot possibly be deduced solely from their "lowerness."

Having viewed the changes in governmental structure which permitted upper middle (or lower upper?) samurai bureaucrats such as Sufu to rise to positions of legislative and executive power, let us now examine the "mechanism" by which power was transmitted from one group to the other. First, it must be noted that for the Sufu or the Tsuboi group to control the han government, they had only to secure a very small number of key positions under the two executive Elders. A change from one clique to the other could be effected by switching about a small number of men. Members of the opposing bureaucratic clique might still be in the government, but, once removed from the key positions, their function would be purely administrative. Also, as we have said before, not all of the officials in the han government belonged to the two cliques. There remained a strong uncommitted center of gravity made up of neutralist officials.

Because of this neutral middle group and the small number of key positions, a small change was at times sufficient to transfer power from one clique to the other. The daimyo and the Elders were in the position to make this change. Yet, throughout the Bakumatsu period, the daimyo and Elders of Chōshū were incompetent. This is one aspect of Chōshū's Bakumatsu politics which contrasts strongly with those of Satsuma, Aizu, or Tosa. Mōri Takachika, the daimyo of Chōshū from 1838 to 1871, is almost universally recognized as having been very weak; one historian has referred to him as a robot.[20] The Elders also seem to have been at best mediocre. At a crucial juncture in Bakumatsu history, Sufu sharply called them a "bunch of good-for-nothings."[21] This remark was made after the execution of the three Elders who had supported the Sufu program, but even in the actions of these three there was little to indicate that they were not completely dominated by samurai of the Sufu clique. However, in spite of their ineptitude, the daimyo and the Elders remained the ultimate source of power within the han; rule by either clique was rule in the name of the daimyo and with the approval of most of the Elders.

This combination of weakness and authority accounts for the frequent change from one clique to the other. Since the daimyo and the Elders were weak they were easily swayed by "public opinion."

[20] Tōyama Shigeki, personal communication. Naramoto Tatsuya has also written that "the special nature of Chōshū was that in comparison with other han, the personal influence and popularity of its daimyo were weak." See Naramoto Tatsuya, *Yoshida Shōin* (Tokyo, 1955), p. 119.

[21] Suematsu, *Bōchō kaiten shi,* VI, 58.

When as in 1839 or 1853 or 1858 the dominant feeling in the han was one of the need for reform, they would put the reformists in power. When a reaction arose against the reform, as in 1844 or 1855, they reverted to the moderate Tsuboi.

But in spite of their personal shortcomings, the daimyo and Elders could not be completely dominated by either clique in power, because socially and politically their positions were so far above those of the bureaucratic cliques. This was especially true of the daimyo. We said earlier that by Tempō times the Council of Elders had become a mere advisory group, its power gone to the executive Elders. This was true concerning the ordinary functioning of the government. The power to appoint officials, however, remained in the hands of the daimyo who, it seems, was influenced mainly by the "public opinion" of the collectivity of Elders and some of his highest-ranking retainers. These in turn were sensitive to a much wider compass of opinion that probably included their own retainers, merchants, samurai bureaucrats, and others.

The structure of government also helps to explain Sufu's words in 1864: "As my rank is low, no matter what I think, it is of no use." [22] This may sound strange coming from the man who more than anyone else in the han had chartered its Bakumatsu policies. What he meant was that no matter how much power he might gather to himself, it was contingent upon the continuing approval of the Elders and the daimyo; like an elected official in a democracy, he might exercise great power in his position but the maintenance of the position itself rested ultimately on outside approval.

The importance of rank and the nature of the power held by the daimyo and Elders can also be seen in their frequent punishments of important samurai officials for minor infractions of the laws of the han. For example, even while his group was in power, Sufu was twice ordered into domiciliary confinement for short periods. Once he was ordered into domiciliary confinement for sixty days for having left his post without the permission of his superiors, in spite of the fact that he had done it to consult with the Elders on han policy. Both times there were some complicating factors, but even then

[22] *Ibid.*, pp. 57–58.

such punishments seem excessively arbitrary. The Elders, unable to formulate policy, gave rein to their desire to exercise authority by taking a strict and petty disciplinary stand on the rules governing the conduct of the lesser officials.

Had there been a man of Sufu's stature among the Elders, he would probably have dominated the government. Had the daimyo been such a man, the existence of cliques might have been relatively unimportant, for they would never have been given an opportunity to formulate policy. And, had this been the case, Chōshū might have played just as important a role as it did in the Restoration struggles but certainly in a different manner. This is not to say, corrupting Carlyle, that Chōshū's history was the history of weak men. Rather, it is to emphasize the importance of the highest vassals even in such a late feudal hierarchy.

The same structure also accounts for Chōshū's inactivity on the national scene during the 1853–1858 period. Some historians have attributed this inactivity to the han's preoccupation with internal reforms. Many reforms did take place, but they were no stronger or more extensive than those of Tempō and were certainly far less extensive than those taking place after 1858. Rather, both the inactivity of the han in national politics and the weakness of its reforms in this period were a consequence of its governmental structure. During 1853–1855 when very scant possibilities for action were open, the Sufu group was in power, though its tenure was extremely insecure; many of the Tsuboi-Mukunashi group held only slightly less important positions, and when reaction rose against the reform the government changed hands. Then, during the 1855–1858 period, when more opportunities were available, the do-nothing government of the moderates was in power. Another change in "public opinion" was needed to dislodge them, and this change in favor of a strong and active policy did not take place until 1858.

Moreover, national politics in the 1853–1858 period meant Bakufu politics. Most of the participants were *shimpan* and *fudai* han, and the few active *tozama* daimyo could only participate by aligning themselves with factions within the Bakufu. Politics was on the daimyo level, and the daimyo of Chōshū was not of adequate caliber.

It was not until 1858 that Chōshū public opinion was finally aroused, and by that time the possibility had appeared for a new type of political action.

The Formation of a National Policy

We previously said that reform politics remained the chief preoccupation of the Chōshū government until the policy change of 1858. There were, however, premonitions of the 1858 change during the preceding years. As we have seen, Chōshū, in its advice to the Bakufu in 1853, referred to Japan as the "honorable country" (*mikuni*) and advocated a firm refusal of the Western demands. During the same year, when Chōshū was assigned defensive positions along Sagami Bay, the daimyo sent the following message to his retainers: "The military glory of the Imperial country depends upon [our] success or failure. This is a critical matter, and if both the higher and lower classes do not join with loyalty and bravery, it will be difficult to fulfill this task. . . ."[23] Here Japan is called the "Imperial country" (*kōkoku*).

In 1855 the daimyo expounded the duties of a Chōshū samurai:

> In recent years, foreign ships have come to Japan and frequently reconnoitre to the east and west. Their activities, evil plots, and alien hearts are difficult to fathom; it is truly a time of difficulty for the Imperial country. Because of this situation the Emperor has ordered that Buddhist bells be melted and cast into cannon. From this we can infer the anxiety of the Court. . . . The house of Mōri is descended from the Imperial line. When the Lord Dōshun donated funds for the enthronement ceremonies of the Emperor Ōgimachi, he put at ease the Imperial mind and in doing so, was both loyal [to the Emperor] and filial [to his ancestors]. Unworthy as we are, we are descended from their blood; therefore, in this troubled time, we must surpass other han in putting up military defenses against the barbarians and so, in our small way, put at ease the mind of the Emperor.[24]

This is only one section of a very lengthy document in which the

[23] *Ibid.*, II, 29.
[24] *Ibid.*, pp. 85–86.

logical consequences of the obligations to one's ancestors are fully developed. The general line of argument was that by fulfilling their obligations to the present daimyo the vassals were being both loyal and filial: loyal since they were doing their duty and filial since they were also paying back the debt of their ancestors to the house of Mōri. It is highly interesting and indicative of things to come that, as early as 1855, the essentially feudal loyalty obligation that, on the retainer level, focused on the daimyo, should on the daimyo level be focused on the Emperor, the symbol of "monarchy," and not on the shogun, the feudal hegemon.

We may also remember that Chōshū in presenting its views on a commercial treaty in 1857 spoke of Japan as the "Imperial country." Even in addressing the Bakufu, Japan in contrast to foreign countries was the Imperial country. The shogun may have been the *de facto* hegemon over the numerous han but the Emperor was the symbol for the country taken as a whole or as a single unit. Until this time, both the Sufu government and the Tsuboi-Mukunashi government had more or less taken the same position on national affairs. When the question of succession to the shogunate arose in 1857 the moderates of Chōshū took the position that this matter concerned the house of Tokugawa alone and refused to become entangled in national politics. One may reasonably conjecture—though without evidence—that by this time the Sufu group was already in favor of some action nationally.

The Bakufu subsequently asked the Court for its approval of the Treaty of Commerce. This the Court flatly refused. The Bakufu then asked the han their opinions on the Court's refusal. The daimyo of Chōshū was in Edo at the time; he had been on the point of returning to the han. Pressed for an answer to the Bakufu's query and without sufficient time to consult with the officials in the han, the officials under the Accompanying Elder drafted an answer. At this time these officials represented the main power of the Tsuboi-Mukunashi clique, and their opinion can be taken as representing the position of the clique vis-à-vis the Bakufu. It was submittted on 3/5/58. "Even if a peace is arranged with the foreign countries, internal complications will arise if the hearts of the people are not united within the country. . . . Therefore, the only course is to carry out the Imperial wishes and, by uniting the hearts of the people, to take measures against the various foreign countries."[25]

In the meantime, the power of the Sufu clique had been rising in the han. This was partly due to the support of Masuda Danjō, the Han Resident Elder. Sufu himself, dismissed from office in 1855, was reappointed to office as the intendant (*daikan*) of Ōtsu *saiban* (one of the fourteen administrative districts of the han) early in 1856. On 1/9/57 he was appointed to head the office of finances, one of the important offices of the han under the Han Administrative Elder. When the news reached the han that the Bakufu had asked the daimyo of Chōshū for his opinion concerning the earlier refusal of the Court to approve the Treaty of Commerce, a conference was called in the han. At the conference it was decided to draft an opinion on the question representing the position of the officials in the han and to send it to the daimyo in Edo. The Sufu clique, which now controlled the offices under the Han Administrative Elder, feared that the Tsuboi-Mukunashi clique under the Accompanying Elder would submit too weak an answer to the Bakufu. This action proved to be the first clear indication of a disagreement between the two cliques on national policy. One can safely assume that this difference had been developing during the years 1855 to 1858 without appearing in documents, since official documents would only record the position of the group in power. The opinion drafted by the Sufu group was longer and stronger than that of the Tsuboi group, but the sentiment remained essentially the same. The following is part of the opinion drafted by the han conference:

Although the words of the barbarians appear to be those of kind advice, their evil intentions have been visible in our various contacts with them. In spite of there being a temporary expedience in the magnanimity of the Bakufu [i.e., in wanting to grant them a treaty], if the ports are opened and if the barbarians live among the Japanese and heathen churches are established, then people's hearts will be deceived, disorders will arise, and the national polity cannot be maintained. Therefore, following the Imperial will, we must refuse [the demands of] the

[25] *Ibid.*, p. 231.

barbarians, and though this may cause a great upheaval, the nation will be without humiliation. If we defend ourselves with unswerving determination, then the hearts of the people will change, the evil customs arising from [250 years of] peace will be reformed and we will rapidly accomplish the training of war horses and the construction of warships. Other than the above, there is no good policy for the maintenance of the Imperial country.[26]

In addition, the memorial stated that the official Chōshū policy should be one of "loyalty to the Court, trust to the Bakufu, and filial duty to the ancestors [of the house of Mōri]." With the rise to power of the Sufu clique in 1858, this became the official position of the han, which was maintained until the start of the *tōbaku* (overthrow the Bakufu) movement in 1865. We see again that loyalty is accorded to the Court and not to the Bakufu. Sufu took the message to the daimyo but the shorter opinion had already been submitted, and unsuccessful, he had to return to the han.

A great change in officials took place when the daimyo returned to the han in 1858. Masuda, the Elder favorably disposed to the reformists, became the Accompanying Elder, and Ura Yukie, another Elder sympathetic towards the reformists, became the Han Resident Elder. Too much should not be made of the significance of their role, for Ura had previously been the Accompanying Elder under whom the Tsuboi-Mukunashi faction had been grouped. But now all the key positions under the two executive Elders were given to the Sufu clique, and all the members of the Tsuboi-Mukunashi clique, with the exception of the two leaders, were dismissed from office. The reformists now had as fast a hold on the government as was possible at this time; they had not only garnered all the key positions for themselves, but they had excluded their opponents from every office of any note. They directed the han under two Elders favorable to their policies and also had the strong support of public opinion in the han for their program. This strength plus certain factors that would only emerge afterwards were to keep them in power until 1864 (when the Mukunashi faction again came into

[26] *Ibid.*, pp. 232–233.

control of the government for a brief span). On 18/7/58 they began their second Ansei reform in which, as we have seen, they were careful not to pass the sort of debt legislation that had proved a stumbling block in 1844 and 1855.

An open split between the Court and the Bakufu occurred on 19/6/58, when the Bakufu signed the Treaty of Commerce against the wishes of the Court. Two months after this, the Court sent its first message to Chōshū; it arrived on 21/8/58. It was couched in extremely cryptic language, not unlike that used by astrologers. The substance of the message was that the Court was waiting for someone to champion its cause and that this "someone" had not as yet come forth. Three days later the Kyoto Resident Elder returned to the han with a second message, an official "secret edict" dated 11/8/58.[27] This was similar to the edicts sent to Mito, Owari, Echizen, Kaga, Satsuma, Higo, Chikuzen, Aki, Inaba, Bizen, Awa, Tōdō and Tosa. It predicted that the Treaty of Commerce would cause dissension within the country, and it asked Chōshū to work for the cause of the Court.[28]

A conference was called in the han and the following reply was composed:

Though we are unworthy, we have venerated the Imperial Court for many years. When the Bakufu questioned us concerning the Imperial answer, we replied that the Imperial will should be followed; that if the foreigners are dealt with by a country where hearts are united, then the Imperial prestige would be established. [Our answer in full] can be seen in the enclosed papers. . . . Unite in support of the Bakufu, deal with the foreigners, then even if things are more difficult for a time, in

[27] The office of the Kyoto Resident Elder was unimportant within the structure of the han government, therefore it was not mentioned in the earlier description of the government. It was established because of Chōshū's special relation to the Court. See Chapter I.

[28] All communications from the Court had to be countersigned by the *kampaku*, the highest official at the Court. Any communication sent out without first passing through the hands of the *kampaku* was unofficial and considered a *naichoku*. The fourteen orders sent out at this time belong to this category. The *naichoku* should be distinguished from the *mitchoku* which was both unofficial and secret. See Osatake Takeshi, *Meiji ishin*, pp. 206–207.

the end a time of peace will arrive to allay the anxiety of the Emperor. We have been ordered [by the Bakufu] to guard the Hyōgo coast and are resolved to protect the Imperial residence with all the strength of our han.[29]

It is plainly an innocuous answer; it assumes as a basis of national unity what in fact was impossible, namely, that the Bakufu would follow the Imperial wishes and abrogate the Treaty of Commerce. Chōshū would then have been able to fulfill its duties to both Court and Bakufu without contradiction. Sufu carried this message to the Court, although his ostensible purpose in going to Kyoto was to perform certain duties concerning Chōshū's defense of the Hyōgo coast. On 24/9/58 he returned to Hagi.

Now for the first time in han politics, national affairs took precedence over questions of reform. This situation had been brought about by the sense of impending change within the country and the quickening in the han of a desire to act on the national scene. At this time it was decided that Chōshū would act on the national scene when an opportunity presented itself. One cannot stress enough the significance of this decision. Once the decision to act had been made, Chōshū was committed. And with the first step taken Chōshū became more and more deeply involved, until it could no longer withdraw. Why the decision was made when it was, and why opinion in the han tended to support this decision is difficult to explain, since we lack documentary information concerning the immediate circumstances affecting the decision. In general we can say that it was the result of the totality of background factors discussed in the earlier chapters. Yet the contribution of some of these factors to the decision seems to have been purely negative. Size, numbers, solvency, and such factors seem to have contributed to the decision only in the sense that they did not become obstacles to action. The positive stimulus to action on the national scene seems to have been the awareness on the part of the Chōshū samurai of the pre-Tokugawa glory of the house of Mōri, and the desire to revive its past greatness. In the case of the han officials (or at least of the Sufu clique) who made this decision, loyalty to the han was identified with active

[29] Suematsu, *Bōchō kaiten shi,* II, 253–254.

achievement for its sake. Successful action on the national scene, they felt, would on the one hand satisfy their loyalty and on the other confirm their own bureaucratic power.

Whatever the reasons for this decision, it immediately provoked a reaction within the han and drew upon the Sufu government the criticism of two very dissimilar groups. On one side the Tsuboi-Mukunashi faction argued that entrance into national politics was dangerous and would spell certain disaster to the han. On the other side, a group of younger samurai centering on Yoshida Shōin argued in favor of openly supporting the Court against the Bakufu; for them the intention to act at some future opportunity was too vague a policy when the moral duty of the han was so apparent.

One happy consequence of this criticism from the historian's point of view is that it led Sufu to obtain the daimyo's approval for his policy. He did this by submitting the drafts for three alternative policies: the one chosen by the daimyo would become han policy, and with the official approval of the daimyo it would no longer be vulnerable to criticism. The first two alternatives presented in this vitally important document were the following:

> Policy One: Shall we go one step further with the policy which we have been presenting in our statements of opinion to the Bakufu and, standing between the Court and the Bakufu, carry out the Imperial will, establish a union of Court and Bakufu, and bring together the various han to aid the Tokugawa house? With honorable intentions such as these, the noble work of honoring the Emperor and expelling the barbarians will certainly be achieved, and the repute of the house of Mōri will shine forth for ten thousand generations.
>
> Policy Two: During the last few years we have presented our opinions to the Bakufu several times, but to date it has taken no action. Shall we, then, according to the movements of the han, carry out that which is right, and when an opportunity occurs, act in response to changes, so that as much as it is possible the intentions of honoring the Emperor and expelling the barbarian can be carried out?

The third alternative, given immediately below, needs a few words of explanation, since it is written in a very confused fashion. It is composed of four parts: a defense of the first two alternatives, an attack on the conservative policy, the statement of the conservative position, and a second attack on the conservative policy. For the sake of clarity I have italicized those parts stating the gist of the conservative position.

Policy Three: When the two previous alternatives are seen with vulgar eyes [*zokugan*], it can be said that they will bring misfortune to the han and the daimyo. We are naturally taking care day and night to distinguish between what is safe and what is dangerous. The truth of the matter is that those who act first will control others, and those who procrastinate [in acting] will be controlled. If the two previous policies are chosen, then the han will become stronger and stronger. However, if we act according to the vulgar view [*zokuron*] and *make the patching up of the han [i.e. internal reforms] our primary objective, and if we must silence those within the han who possess a will to action, and put a stop to trade with the Kyoto region and all other items that might incur the displeasure of the Bakufu, and plan so that the effect of all of our proposals of the past years will be obliterated,* then, contrary to our expectations, the Bakufu will observe the weakness of the Chōshū samurai and it is difficult to say what sort of thing will be done to us. And, even if this sort of thing does not come to pass, can [the samurai of Chōshū] be at ease knowing that the reproach of the intelligent will be difficult to avoid? Can our government stoop to such a lowly policy as this? [30]

It is obvious from the wording of the last proposal that Sufu did not intend to have serious consideration given to it—this, even the most incompetent daimyo could not fail to perceive. Rather, it was intended as a slap at the Tsuboi-Mukunashi faction, or the *zokurontō,* as it had been disparagingly called by its opponents ever since Tempō times. What then, of the first two alternatives? The first carried with it a commitment for immediate action on the national scene; the second, containing the qualification, "when an opportunity occurs," was less definite. Therefore, in spite of the fact that the second alternative appears to contrast invidiously the Bakufu's action with the morally correct course of action, the first policy is actually the stronger of the two.

[30] All three policies, *ibid.,* pp. 241–242.

The reply of the daimyo, after consultation, one suspects, with his Elders and the leaders of the Sufu clique, was as follows:

> The first policy may appear appropriate for the times; however, in various ways, it is difficult to take action in the present situation. Therefore, accept the second policy as the correct one. . . . However, if from the start, the second policy is set up as our goal, then it will be hard not to slip into the third one. Since this is the case, it is necessary to investigate carefully possibilities [for action under] the first policy, so that when the time comes to act, it may be done immediately. As it was stated before, it is not easy to act, so while setting up the first policy as our goal, we must resolve to act according to the second policy.[31]

As the reply would indicate, in spite of its decision to act, Japan in 1858 offered Chōshū no opportunities for mediation. These were the years of Ii's ascendancy; most of the daimyo who had been active on the national scene had been ordered to retire, and the Purge of Ansei was soon to wreak its vengeance on the enemies of the Bakufu. Late in 1858, Masuda Danjō, the Accompanying Elder, wrote to Ura Yukie, the Edo Resident Elder: "Now though we wish to go one step further with the policy which we have been presenting in our statements of opinion to the Bakufu and to mediate between the Court and the Bakufu, at the present time there is no opportunity for action."[32] Like the other great *tozama* han, Chōshū sat quietly, watching and waiting.

[31] *Ibid.*, p. 243.
[32] *Ibid.*, pp. 244–245.

CHAPTER V

The Background of Ideas in Chōshū

The Problem

One of the most interesting problems concerning Tokugawa society is the manifold nature of its intellectual life. It saw the rise of a great number of systems of thought, in many cases competing systems, which treated a wide range of problems. Why did such ideas arise, what did they mean? One is tempted to compare it with nineteenth-century Europe, the "age of ideology," except that in Europe such systems of thought often represented the conflicting interests of different groups or classes, while those in Tokugawa Japan seem, insofar as they deal with political topics, to reflect the variety of interests of a single class. Tokugawa political thought is, in the Marxist sense in which the term is frequently used, a single ideology. This undoubtedly reflects the fact that most of the Tokugawa systems of thought were the work of an officially subsidized, occasionally idle, ruling stratum.

In this sense, the intellectual production of the Edo period resembles more that of Italy in the early Renaissance. The parallels that can be drawn are extensive. In each case a powerful wave of humanism led to the refinement of an aristocratic class: the Italian ideal of the courtier was paralleled in Japan by the transformation to a certain degree, of the samurai into *chün tzu*. In each case, out of a background of monkish thought, a rising secularism semiurban in nature led to the decline of the established religion of the previous age. In each case there arose a form of "local nationalism" different from the personal loyalty that had existed earlier. In both societies there arose a cult of antiquity: in Italy humanists, both Christian and Greek, sought renewal in the past; in Japan both the Shinto-oriented School of National Studies (*kokugaku*) and the Confucian School of Ancient Learning (*kogaku-ha*) attempted to reconstruct the ideas of an earlier, purer age. Both Italy and Japan saw the emergence of schools of history, each had its bibliophiles, and in each there arose schools of textual criticism: compare de Valla with the School of Textual Analysts (*kōshōgaku ha*) or the methodologically related School of National Studies. In both cultures there appeared new forms of art and poetry and a popular literature ranging from the mildly religious to the profane. Science was also introduced in each. Astronomy, botany, medicine (even anatomical experiments), and military science advanced, and parallel to these there arose the study of foreign languages and a new interest in geography.

In the literal sense of the term, the Tokugawa peace brought with it a renaissance of the civil arts. Yet in spite of the similarities noted above the comparison breaks down when we examine the vastly different content of the two cultures. The Japanese aristocracy remained a military one. Vague words such as secularization or humanism mean radically different things in different contexts. And more important, the ideas reborn in Japan were so different from those of Greece that implications for modernity were largely lacking. But the comparison does serve to point up the surprising range of Tokugawa intellectual activity which is, as others have noted, one reason why the samurai did not become a decadent class during two hundred and fifty years of peace.

What then were the roles which these various systems of thought played in Bakumatsu politics? In particular, what were the roles in Chōshū of Western learning, *kokugaku* (National Studies), and Confucianism? Can any of these be said to have predisposed Chōshū to take action in Bakumatsu politics in the same sense as did the *buikukyoku* or Chōshū's position as an anti-Bakufu *tozama* han?

My answer, at least as far as Chōshū is concerned, is that only Western learning and Confucianism were of consequence; *kokugaku* was of negligible significance to the rise of the *sonnō* (honor the Emperor) movement in Chōshū, and its importance within Japan as a whole has probably been greatly overstressed.

Western Learning

The rise of Western learning in Tokugawa Japan was the direct consequence of official Bakufu policy. Like Christianity, Western learning had been proscribed shortly after the decisive battle at Sekigahara, and during the first one hundred years of the Tokugawa period it all but disappeared from the Japanese scene. Its subsequent revival in the eighteenth century stemmed from the support given to it by Arai Hakuseki in his *Seiyō kibun,* and by other Tokugawa thinkers. From the "positivist" position of the orthodox Chu Hsi philosophy, Hakuseki argued that Western science was not allied with Christianity and that, since it was superior to Japanese science, it ought to be adopted to benefit Japan. As a result, European texts were again permitted to enter Japan, Western learning was again encouraged by the authorities, and most of the important developments in the field took place under the patronage of the Bakufu. Many of those who first took up Western learning were Confucian scholars who saw it as supplementary to the learning of the sages. Hashimoto Sanai, who acted, as we have seen, for Tokugawa Nariaki in Kyoto, wrote: "We shall take the machines and techniques from them, but we have our own ethics and morals." [1]

Western writers on this subject, intent on explaining the rapid post-Meiji Westernization of Japan, have too often dwelt on Western learning as an anti-Tokugawa force. They have emphasized that the samurai students of Western learning took up these studies because they felt that their native Sino-Japanese culture had become exhausted and sterile. They have stressed the alienative aspects of Western learning. Viewed in this manner, the rise of Western

[1] Numata Jirō, "Acceptance and Rejection of Elements of European Culture in Japan," *Journal of World History,* 3:241 (Part I, 1956).

learning or *rangaku* (Dutch Studies), as it was called during the Tokugawa period, seems to furnish an element of continuity with the post-Restoration destruction of the old feudal society.

Actually, to speak of Western studies in this manner misses the cardinal fact that they were developed under the aegis of the Bakufu, to be applied as needed to certain areas of Tokugawa technology. Official support, of course, was limited to the support of Western learning as a means and instrument for the execution of officially approved plans. When, as it happened from time to time a student of *rangaku* criticized some aspect of official policy, the Bakufu would clamp down; but this occurred only rarely and to single out such cases distorts the nature of the phenomenon. Watanabe Kazan (Noboru), for example, is often spoken of as one who had become alienated from the society of his day; yet we must remember that he was of Elder rank in his han and that when he committed suicide so that his actions would not bring punishment to his daimyo, he left as his last words: "Disloyal and unfilial, Watanabe Noboru." [2] This is not exceptional,; most of these early Westernizers were motivated by a stronger than average "feudal" consciousness.

There were always a few in the Bakufu and elsewhere who felt that the learning of the sages was not in need of supplement, but these were almost always overruled by the majority who favored the judicious use of Western learning. The attitude of Matsudaira Sadanobu is perhaps typical:

> The countries of the barbarians are well-versed in science. Particularly in astronomy and geography, armaments, internal medicine and surgery, we have much to gain from barbarian works. Sometimes, however, they but excite idle curiosity and spread views that are quite inappropriate. And yet we cannot ban them for that would only make them spread afresh. . . . We should therefore do all that is possible to see that they do not pass into the hands of those who are not prepared for them.[3]

Here we see the official ambivalence toward Western learning, which led the Bakufu to establish in 1811 an office for the translation of

[2] "Watanabe Kazan," in *Dai jimmei jiten* (1956), VI, 607.

[3] Numata, "Acceptance and Rejection of Elements of European Culture in Japan," *Journal of World History*, 3:241 (Part I, 1956).

foreign works and in 1857 schools to teach foreign languages, foreign science, and the Western military arts, while on the other hand it banned the Barbarian Society (*Bansha*) when one of its members criticized the Bakufu policy toward the U.S.S. *Morrison,* which had come to Japan to return shipwrecked Japanese sailors.

Moreover, even such occasional criticism did not represent a fundamental alienation of the critics from their own society. The shelling of the *Morrison* was a mistake in terms of national defense; it was not in any sense a transgression of the teachings of *rangaku*. The question of alienation is an extremely complex one since the awareness of being different, which the students of *rangaku* naturally felt, in itself constitutes a certain measure of alienation. The only real differences, however, were concerned with empirical or cognitive matters in certain narrow fields; their values and their fundamental orientations by which their values were organized were no different from those of their fellows.

First, for reasons given earlier, *rangaku* was directed chiefly towards the sciences, and for this reason there were few areas in which it could encroach upon the central sociopolitical doctrines of Confucianism. Secondly, even in its scientific studies it tended to be practical rather than theoretical. It is questionable whether the view of science as a body of systematic theory arrived at by generalization from experience had been reached even by Bakumatsu times. Thirdly, even though Western science was very different methodologically from the Confucian mode or argument by examples, it was not necessarily irreconcilable with it. One has only to think of the great variety of metaphysical underpinnings which Western thinkers have from time to time given to science.

The Chu Hsi school, for example, taught that principles were intuitable, but must also be sought as they are embodied in things. The vocabulary of *rangaku* physics was largely derived from this school: "natural philosophy" being equated with *kyūri,* the search for principles, and "physics" with *rigaku,* the study of principles. The same sort of compatibility also seems likely in the case of the School of Ancient Learning with its emphasis on practice. Associated with and influenced by this school was the school (literally, method) of Ancient Medicine (*koihō*) which, attacking the Yin Yang metaphysics of the Chu Hsi school of medicine, argued that only what one had personally experienced was of value.[4] This was coupled, characteristically, with a demand for higher medical ethics. It is significant that Sugita Gempaku had been an adherent of this school before beginning his studies of Western medicine. Finally, the existence of men such as Sakuma Shōzan suggests that it was not impossible even to combine the metaphysics of the Wang Yang-ming school with an interest in Western learning. Perhaps, in the same sense in which Zen in Japan is often interpreted as a philosophy in which the self is conquered for secular ends, the intuitive Wang Yang-ming school led to a new freedom which, in achievement-oriented Japan, could be applied to Western science.

What we have said above is of course tentative: studies illustrating in detail the way in which Western science and Japanese metaphysics met and became fused have not yet been made—although the materials are easily available. It does suggest, however, that one cannot easily dismiss as absurd programs such as that of Sakuma Shōzan, "Eastern ethics and Western science." Over the very long run this is of course a fallacy, but over the short and even medium term it embodies a considerable truth. This can be seen in the development of Western studies in the han even as late as the Bakumatsu period.

Western learning in the han followed very closely the Bakufu pattern and example. In almost every case it developed under official patronage and was carefully maintained within officially prescribed limits. Among the han, Saga, Mito, and Satsuma were the leaders. Saga, by reason of its proximity to Nagasaki, the only point of contact with the West during the Edo period, had early developed schools of Western studies and, importing equipment from the West, it had even begun to experiment with Western type industrial processes. Mito was also one of the first to develop a school for the promotion of Western learning; its primary concern was with Western military arts, which it hoped to employ to strengthen Japan's defenses. Satsuma, as always, was exceptional. In spite of its relative isolation, it was one of the most ardent supporters of Western studies. Besides promoting the study of foreign language and

[4] *Meiji-zen Nihon igaku shi* (Tokyo, 1955), I, 27.

science, it embarked upon a series of industrial experiments based entirely on research in Western books. Without the advantages of Saga or the Bakufu, it progressed the furthest in the utilization of Western techniques in the period before the coming of Perry.

Chōshū was a late arrival on the scene, but like the others its actions closely followed the example set by the Bakufu. The first action was taken by Murata Seifū, who, at the time of the Tempō Reform (1840), set up a small group for the study of Western medicine. The second step was taken in 1847 with the establishment of the position of Official in Charge of Western Books. The two samurai appointed to the post were charged with the translation of Western works on medicine, geography, and the military arts. Several years later the Kōseikan, a school for the study of Western medicine, was established in Chōshū with an initial enrollment of sixteen students.[5]

The arrival of Perry and the end of seclusion gave increased impetus to the development of Western studies in Chōshū. On 1/9/55 new Western studies posts *(Seiyōgaku yōgakari)* were added to the infant medical school, and eleven new posts created: five instructorships and six research assistantships. Several months later three of these instructors were transferred to the han school, the Meirinkan, and from this time Western studies became a part of the official han curriculum, although they were not required of all students. History, military science, geography, astronomy, and various other subjects were taught, but the majority of students were interested solely in military science. Consequently, a samurai was sent to Nagasaki to study the latest information on this subject; he returned to Chōshū in 1857 and was appointed to head the Western studies staff at the han school. The following year this section was again reformed by Tōjō Eian, a student just returned from the Bakufu school in Edo. By 1859 there were over twenty students of Western military science at the han school, and in the same year a section was added for the study of Western naval science.

In addition to these developments in Chōshū, younger samurai

[5] Most of the material concerning the development of Western science and learning in Chōshū is taken from Suematsu Kenchō, *Bōchō kaiten shi* (Tokyo, 1921), II, 376–510.

were also given "scholarships" to study in other han. In 8/54 several were sent to study Western ships at Nagasaki, and the following year seven more were sent to join them. Since no one student could master everything, each was instructed to learn a particular field: one, cannonry, one, marine carpentry, another, military formations, and so on. Students were also sent for training aboard the *Kankō-maru,* a ship which had been presented to the Bakufu by the Dutch government. In 1858 instruction in foreign languages was begun among the young samurai on duty at the Chōshū Edo residence, and in the spring of the following year others were sent to the school of Takashima Shūhan and Egawa Tarōzaemon to study the more advanced military techniques of the West. At one time more than half of the students in their school were either from Chōshū or Kumamoto.[6] Also in the same year, Kusaka Genzui and others from Chōshū entered the Bakufu school for Western studies. Yet, one must emphasize that Western learning did not replace traditional studies; it merely supplemented, and the number of those studying the new learning was a mere trickle beside the mainstream of those engrossed in swordsmanship and the Confucian classics.

Chōshū also lagged behind the other han in its experiments with Western industry. Its first enterprise, a bureau to produce "modern" rifles, was established only in 1854. Its products, however, were relatively primitive, nonstandardized weapons, and five years later it was found necessary to send gunsmiths to Edo to acquire more up-to-date methods. Chōshū also set to work to build a reverberatory furnace, but this was abandoned in 1856, when it was learned that the products of a similar furnace in Saga had proved unsatisfactory. Plans for the construction of steamships were advanced further, but these came to the same end. A shipyard, a lumber yard, a copper works, and a steam engine workshop were established in which over one hundred persons were employed. Two schooners were finally produced, but the original plan for the construction of a steamship was never realized.

But Chōshū lost very little by its late start. It had only begun to experiment when the more advanced han, discouraged with the

[6] *Ibid.*, p. 452.

results of their early attempts to duplicate Western technology, began to turn to Western merchants for more satisfactory products. Satsuma, Saga, and Mito had actually built steamships, all of which had been planned even before the arrival of Perry, but these were sub sequently discarded in favor of Western ships purchased in the late 1850's. Chōshū had already made inquiries regarding the buying of a Western ship early in 1855, but it gave up the idea when it was discovered that the cheapest ship would cost 20,000 gold *ryō* (about 1360 silver *kan*). In 1860, following the examples of Satsuma and Saga, Chōshū began to purchase foreign ships, and no sooner had the Bakufu repealed its ban on the purchase of foreign weapons than it bought a thousand rifles. Chōshū's *buiku* savings had more than made up for its late start; from 1860 on, the military potential of the han gradually gathered momentum. Yet all these measures were at best fringe activities which did not encroach seriously upon any sphere of vested interest within the han. Only in 1858 did the Western reform begin to intrude upon one of the strongholds of Chōshū's feudal society, the han military.

The first step toward the modernization of the han military was taken under the direction of Yamada Uemon. Yamada was not only an advocate of Western learning but also a prominent member of the Sufu clique. In 1847 he had imprudently sent a memorial to Tsuboi recommending the formation of rifle units and had been ordered into domiciliary confinement for his rashness. But his appointment eleven years later to carry out the reform was to vindicate his earlier judgement. He began his reform by sending thirty students to Nagasaki, twenty to study military tactics and ten for the study of naval technique. These returned to Hagi the following year, and together with other samurai interested in the new type of military they formed a rifle unit which practiced daily on a field outside of the castle town. Yamada's views on the purpose of the training can be seen in the following letter that he sent to the han government, recommending the organization of additional rifle units:

In time of battle the existence of the han will be at stake. Those in a life or death struggle ought not to rely on futile forms like those of children at play. In the movements of a Western rifle company there is a system for advancing or retreating, for grouping and deploying, and in general adaptability one can use troops as one moves hands and feet. Seeing this, one must say that the art of manipulating infantry has been achieved. In the traditional system of Chōshū, there are no fixed formations for the ranks of *ashigaru* and below. And since the skills required for the use of the bow, rifle, pike, and sword are each different, weapons are not standard, and their formations are without control, so that the three thousand lower soldiers (*keisotsu*) cannot help being like a gathering of crows.[7]

Yamada went on to propose that the entire three thousand lower soldiers be organized into rifle units. He argued that, at a time when the formation of peasant troops to cope with the foreign threat was being seriously propounded, it would be absurd to have three thousand lower soldiers "sitting about eating and doing nothing." [8]

Most of the Sufu clique welcomed these reforms, and even the Elders, possibly persuaded by their powerful subordinates, were convinced of the need for Western style rifle units. One of them, Ura, the Han Administrative Elder, wrote to Masuda, the Accompanying Elder in Edo:

The art of war lies in opportunities. To see an opportunity and to respond has been a constant principle from ancient to modern times. The house of Mōri is of famous lineage. . . . If we do not know how to watch the situation and respond to change, we shall, in the end, be unable to escape the ridicule of society. Therefore, let us discard all weapons, military techniques, and formations that are not suitable to the present and replace them with things of true worth.[9]

The path to military reform was not without its obstacles. A review of the new rifle unit, at which all the important officials of the han were to be present, was scheduled for 25/6/59. Opposition proved so vehement that it had to be cancelled, and the high officials conveniently excused themselves on grounds of illness. Yamada's rifle unit seems to have included many middle- or upper-ranking *shi*. Records

[7] *Ibid.*, p. 462.
[8] *Ibid.*, p. 462.
[9] *Ibid.*, pp. 464–465.

inform us that these were often jeered at by others of the same rank for "drilling in the field like raw recruits" in spite of being "officers" (*shikan*) of the han.[10] They were also mocked by the students of swordsmanship, who claimed that the sword was a weapon of honor but the rifle, a utilitarian device to be toted by those of inferior rank. In promoting the latter was the han not debasing its military spirit, they asked. But the han continued to carry out its reform in spite of the attacks from conservative quarters.

Sentiment against the reform was further aggravated by the impudence of some of the students just returned from Nagasaki. Not only did they pride themselves on their skill in giving commands in Dutch, but they also affronted han opinion by dressing in a composite manner, which they deemed the most suitable for military exercises. A shirt with Western style buttons and tight-fitting sleeves was worn; the hat was an ingenious combination of the headgear of East and West; and the costume was completed with a short *hakama,* a Japanese kilt of sorts. The conservatives were outraged: "They are already aping the barbarian in their garb; their hearts must also be barbarian. Are we, the sons of the Divine Land, to imitate even the grimaces of the barbarians?"[11]

Giving no heed to the criticism, two of the students, Kuruhara Ryōzō and Awaya Hikotarō went so far as to wear their "Western clothes" to town. Criticism mounted until finally news of it reached Sufu in Edo. His advice to Maeda Magoemon, one of the chief members of the clique and a high official in the han, was as follows:

We have heard that Kuruhara Ryōzō has walked about wearing Western clothing with sleeves and that Awaya Hikotarō has gone about dressed in a buttoned shirt and outlandish hat. [What they wear] when going to and from the drill-field shall be left to their discrimination. However, though it may seem harmless to walk about [elsewhere] without bothering to change, it might interfere with the proper etiquette, and furthermore, just when we are particularly anxious about [setting up] rifle units, this will antagonize people's hearts until finally the intentions of the daimyo will not be carried out. Therefore, if the two persons

[10] *Ibid.*, p. 386.
[11] *Ibid.*, p. 442.

mentioned above have in fact acted in this manner, it should be brought to their attention and they should be ordered to desist in the future. Although we urgently need to adopt the good techniques of the foreigners, intoxication with them to the point of adopting their clothes and copying their way of speech will only lead to the deterioration of the samurai spirit.[12]

The preface of the plan to expand the Western studies section of the han school reflects a similar spirit: "Study diligently so as to be of use to the han. To yearn foolishly after the customs of foreign lands is to miss the true purpose for which the school was founded. Therefore, clearly distinguishing between ends and means, it is important to act for the establishment of the national polity (*kokutai*)."[13] The frivolous among the samurai may have been captivated by the lure of the exotic West, but they were quickly suppressed. The han government had encouraged their studies in order to increase the military power of Chōshū, and it would not tolerate the antics of the irresponsible few.

Thus, Western learning in Chōshū was to be used for traditional ends. It is important to note that the Sufu clique which gave the most vigorous support to it was the same group that had carried out the reforms of 1854 and 1858, the same group that would launch Chōshū into the uncertain seas of national politics in 1861. Underlying each of these actions was the same strong "feudal" bureaucratic consciousness. That this should have led to action and not conservation was in no small measure due to the flexibility present in the structure of the Chōshū government.

National Studies

Kokugaku (National Studies) offers us an excellent example of a school of thought whose contribution to the Meiji Restoration has been given far more attention than the contribution merits, simply because it was new and unorthodox. And not only was it new, but like Western learning, it seemed to provide a link between the sub-

[12] *Ibid.*, pp. 442–443.
[13] *Ibid.*, p. 422.

surface ferment of late Tokugawa Japan and the nascent forms of the early Meiji period. *Kokugaku* stressed that which was peculiarly Japanese, early literature and the Imperial house; the Restoration movement centered on the Imperial house and gave rise to an Imperial government. Can it be that the two were not connected?

Our answer is that while *kokugaku* was a conspicuous signpost in the history of Japanese nationalism, a literary movement of note, and a scholarly form of religious atavism, it had little if any influence on the rise of the *sonnō jōi* movement within the great han which led in the Restoration movement. That is to say, insofar as we can treat *kokugaku* as an institutionalized movement, its importance was negligible. The chief areas affected by its teachings were the smaller *shimpan* and *fudai* han in those regions of central Honshū close to the heartland of the Bakufu itself. The great outlying *tozama* han were almost exclusively Confucian in their make-up. (Of course, in many cases, their "Confucianism" would not have been recognized as such in China, molded as it was by the late Tokugawa climate of opinion.)

Scholars have applied the name *kokugaku* to a wide range of thought, but its central concern was the study of ancient Japanese literary and religious texts. For the most part it arose during the Tokugawa period. Some studies on this order were attempted by Confucian scholars but the main body of "national scholars" were in opposition to the Confucian schools. Even the teachings of Confucius, they maintained, were foreign to Japan and alien to its true spirit. One must not, however, stress too much the polemical nature of these studies. In almost every case they dealt exclusively with poetry, literature, linguistics, or Shinto, and consequently tended to repress rather than to awaken the political awareness of those who pursued these studies.

The single exception to this rule was the school of Hirata Atsutane, the last of the "four great scholars" in the *kokugaku* tradition. Hirata was political with a vengeance. In his writings and "sermons" he lamented the decline and low estate of the Imperial house, and he argued that the Emperor was the sole, rightful source of all political authority. Hirata was far more eclectic than his predecessors: the political content of his system was undoubtedly "Confucian" in origin, and in spite of a sweeping xenophobia, he was willing to utilize Western techniques.

Since the Hirata school of *kokugaku* was the only school in the tradition that concerned itself with politics, the direct effect or influence of *kokugaku* on the Bakumatsu political scene can be determined, with a few possible exceptions, by ascertaining the achievements of that school. The Hirata school, we find, is particularly amenable to study, since a register was kept, both during Hirata's lifetime and after his death, of all who enlisted as disciples. The register, containing 4283 names, has been carefully analyzed by Itō Tasaburō, an eminent Japanese historian, who has traced the rise of this school from 1813 to 1874 and the geographical distribution and class background of its members.[14]

The life cycle, the rise and fall of the Hirata school, can be roughly broken down into five periods: 1) During the first period from 1813 until the late 1820's Hirata won relatively few disciples. The fourth son of a samurai of Satake han, Hirata attempted during these early years to propagate his ideas among his fellow samurai but he met with very little success. Even those samurai who sympathized with him personally were unwilling to accept his teachings inasmuch as he advocated a nation united under a divine Emperor, a concept at variance with the hierarchical structure of Tokugawa Japan. At one time Hirata journeyed to Mito, hoping to persuade the leaders of the Mito school of the worth of his ideas. Fujita Tōko and Aizawa Yasushi, who received him, both recognized the superiority of Hirata as a person over any of their own teachers, yet they denied him permission to spread his ideas in Mito, preferring their own Confucian system of thought. 2) This sort of rebuff was typical, and Hirata soon realized that little could be done with samurai. Therefore, from the late 1820's until his death in 1843 he directed his primary efforts not toward samurai but toward the well-to-do

[14] Professor Itō very kindly lent me the proofs of his work on *kokugaku, Sōmō no kokugaku,* together with a chart showing the distribution of Hirata's disciples by han and by period. The example of Hirata's experience in Mito was also given to me by Professor Itō.

peasants and the merchant community; during this period he gained about fifty disciples a year. Shimazaki Tōson writes in "Before the Dawn": "Of those who followed Hirata's teachings, there were only a few who belonged to the samurai class; the majority were village headmen (*shōya*), merchants (*toiya*), keepers of official inns, doctors, peasants, or townsmen." [15] Several writers have interpreted the early participation in the *kokugaku* school by members of this group as the first sign of the political consciousness that was to appear in full force after the Restoration in the form of the *jiyū minken* (free people's rights) movement. Denied a voice in politics within the Tokugawa framework, they accepted *kokugaku* as the promise of a political order in which all men would be equal under the Emperor. It furnished them with a rationale by which they could consider themselves the direct subjects of the Emperor, rather than as commoners below samurai who were below a daimyo below the shogun, who was only nominally below the Emperor. Moreover, membership in the school afforded to the wealthy commoner peasant a definite if limited field for political activity. 3) In the decade from 1843 to 1853, following Hirata's death, the number of new disciples began to decline, falling to an average of about twenty each year. The decline was undoubtedly due to the death of the prophet himself. 4) The Hirata school made its greatest gains in membership in the period between 1853 and 1868, under the influence of the disorder caused by the political struggles, the intrusion of foreigners, and the rise in the prestige of the Court. Some 40 new members joined in 1857, 50 in 1859, almost 80 in 1861, 130 in 1864, 243 in 1867, and 981 in 1868. This marked the high point of the Hirata school.[16] 5) In the period following the Restoration the movement lost its force and soon died out. As early as 1874 new disciples had almost ceased to enter the now decrepit organization. Again quoting from "Before the Dawn": "Viewing it from the present, one could say that the year 1868, when in the entire country the membership reached 4000, was the peak of the Hirata school. That year in Ina, where the study of [Hirata] Atsutane's [teaching] had always flourished, there were

[15] Shimazaki Tōson, *Yoakemae* (Tokyo, 1936), I, 581.

[16] See footnote 14.

120 new members. But in 1870 the number of new members dropped to 19, and in the following year, there were only 4 new members."[17] In "Before the Dawn" Shimazaki Tōson has told the story of Aoyama Hanzō, a resident of a small country village, who took great pride in his membership in the *kokugaku* school and looked upon himself as a member of an intellectual *avant garde*. After the Restoration, when the rumor circulated that these teachings were already old-fashioned, Aoyama refused to believe it and clung even more tightly to his beliefs. And when the story ended: "The Hirata disciples . . . were treated as stubbornly blind obstructionists. And as for foolish fellows like Hanzō, not only were his deeds and accomplishments misinterpreted by those about him, but it had come to the point where he could hear the sound of voices, jeering, 'Get rid of that lunatic'."[18]

For the purposes of this book, even more important than the class composition of the Hirata school or the pattern of its rise and fall, was the distribution of its membership among the various han. Table 3 shows clearly the geographical distribution of the membership of the Hirata school.

The significance of the distribution is that there are relatively few adherents among the great outlying han. One explanation for this unequal distribution is that, in the case of the large outlying han, the resistance of the han authorities, as in the case of Mito, served to thwart the spread of the Hirata school. Another factor is that, with few exceptions, the second generation of teachers was more active in central Japan than in the peripheral areas. Yet another reason is that there were more potential members in the Kantō and Kinki areas of central Honshū, since these were the areas where, with the commercialization of agriculture, the breakdown of peasant society had progressed the most, and more rich peasants or well-to-do merchants had appeared. Satsuma, with ninety-three disciples, seems to be a partial exception to the rule. Why should Satsuma, the most distant of the han, have more disciples than Kaga or Tosa or Chōshū? The answer seems to be that many of the disciples in

[17] Shimazaki, *Yoakemae,* II, 415.
[18] *Ibid.,* p. 672.

TABLE 3. The geographical distribution of the Hirata school.

Han	Number of disciples: Before Hirata's death	Number of disciples: After Hirata's death and before Meiji	Number of disciples: After Meiji	Total
Central han:				
Shimōsa	110	27	23	160
Edo	61	34	118	213
Shinano	4	196	433	633
Ōmi	5	74	113	192
Mino	1	72	289	362
Outlying han:				
Echizen	4	4	29	37
Kaga	5	3	14	22
Tosa	6	17	10	33
Chikuzen	4	26	3	33
Saga	1	1	21	23
Chōshū	1	8	7	16
Satsuma	11	58	24	93

Source: The figures appearing in this table were given to me by Professor Itō Tasaburō.

Satsuma were *gōshi,* and some have suggested that the trammeled, circumscribed world of the *gōshi* resembled in many ways the cramped position of the rich peasant. Yet even the Satsuma disciples were relatively few compared with those in the central han.

The nine pre-Meiji Chōshū disciples of the Hirata school were almost all rich peasants. Most had been converted by Suzuki Shigetane, one of the most renowned of Hirata's pupils. Several of these rich peasants played fairly important roles during the Chōshū civil war in 1865, but that was all. Apart from this, Hirata *kokugaku* had almost no influence on the Chōshū *sonnō* movement. There was a teacher of *kokugaku* at the Meirinkan, the han school, but he only serves to corroborate our earlier contention regarding the apolitical character of most schools of *kokugaku;* he taught the Japanese classics in the style of the Motoori school and had no concern with politics. There were also one or two Shinto priests in country shrines who lectured to their constituents on the principles of *kokugaku,* but these were not concerned with the samurai of Chōshū and were insignificant in the total picture of the han. One writer on the peasant loyalist movement in Chōshū has reached the conclusion that even many of the peasants who supported the *sonnō* position were motivated by the Confucian ideology that had been drawn "from above." [19]

What then was the influence of *kokugaku* thought as diffused, rather than the thought as formally propagated within the various schools of the Hirata sect? Even if the members of the Hirata school were few in number within the great han, may not the great han still have come under its influence as it contributed to the intellectual climate of the age? May it not be, as historians say of Christianity in Asian countries, that it was more influential than the number of its adherents would indicate? In this sense it undoubtedly did contribute something, even to the great han, yet its contribution was probably smaller than that from several other sources. In Chōshū, for example, the Confucian emphasis on origins, the influence of the Mito school, or even the awareness of Chōshū's special relation to the Court, were more likely to have influenced samurai opinion than the Hirata school with its anti-Confucian bias and the stigma attached to it of being a doctrine for commoners. Yoshida Shōin, like his counterparts in Mito, was a student of Confucianism, and it is to this philosophy that we must turn to understand the background of the Chōshū *sonnō* movement.

Confucian Sonnō-ism

The most important formal source of *sonnō* thought in Chōshū was Confucianism. Unlike Hirata *kokugaku* with its register of names and ascertainable geographical distribution, Confucianism

[19] Ogawa Gorō, "Chōshū han ni okeru shomin kinnō undō no tenkai to sono shisōteki haikei," *Onoda kōtōgakkō kenkyū ronsō,* 8:16 (Dec., 1953). There is no question about this. Time and again, both in memorials from peasant officials to the han government and in exhortations from the han government to the peasants, the "manorial" relation between peasants and the han was conceptualized within the same framework of duty and obligation as was the feudal relation between lord and vassal.

was distinguished by its ubiquity. It would be pointless to say that one area was more Confucian than another, for, almost every school in every han was Confucian, and scholarship, if not otherwise designated, meant Confucian scholarship. Since this is the case, our problem here is not to speak of the relative strength or weakness of Confucianism, but rather to show how *sonnō* thought emerged from the Confucian background.

Sonnō means "honor the emperor," but in the context of Bakumatsu society, it may also be interpreted as "loyalty to the emperor." Indeed, the first point of Chōshū's 1858 national policy statement was "loyalty to the Court" (*chōtei e wa chūsetsu*); [20] loyalty to the Court was thus expressed in the same Confucian language as was feudal loyalty to the daimyo. It is the contention of this book that much of the emperor-centered Meiji nationalism as well as Bakumatsu *sonnō* thought was essentially a form of transmuted Tokugawa loyalty. It is therefore necessary to examine the latter very closely, asking how it could so quickly be transformed into something as modern as nationalism. This can best be done by contrasting the function of the loyalty bond within the Tokugawa polity with the somewhat different loyalty of the preceding period of decentralized feudalism.

During the period of Warring States, loyalty was conditional, power was private, and the most important fact was the unit-fief centering on the castle town of the feudal lord. (Barony might be a better term than fief, since the authority of the Ashikaga Shogun was purely nominal.) Potentially, at least, the normal state of relations between any two such fiefs was that of war, and this external condition largely determined the nature of relations within the various fiefs. The lord of the fief was dependent on the collectivity of his vassals; without them his fief could not be maintained. The vassals, in turn, were organized under the lord into many ranks, some having vassals of their own. They received fiefs or stipends from the lord in return for which they owed him their allegiance.

[20] Seki Junya, *Hansei kaikaku to Meiji ishin: han taisei no kiki to nōmin bunka* (Tokyo, 1956), p. 128.

However unilateral the vassal's oath to the lord (it seems to have been far more so than in European feudalism), the bilaterality of the relation was apparent in the balance of forces within the fief.

Loyalty was the ideal statement of the vassal's relation to his lord; if necessary, the vassal had to be willing to die for his lord. And, since the lord was dependent on his vassals, their loyalty was functional. Those who proved their loyalty were not only praised but rewarded for their virtue. They were given larger fiefs, special titles, preferential treatment in the court of their lord, and so on, while those who were found to be disloyal were punished by death.[21]

Since loyalty was functional and power, private, the lord had no court of appeal beyond his own strength should a vassal be disloyal. The equilibrium within each unit-fief was one of tension within reciprocity. The goal of the system was power and this was communicated to each of the vassals who constantly attempted to better their position vis-à-vis other vassals. During the intermittent wars of the period the vassals had ample opportunity to increase their own personal power. In most circumstances this could only be achieved by contributing to the power of the fief itself. At times, however, the strongest general of a lord would find himself in a situation where loyalty would result in a decline in his personal power. Even in the face of temptation many remained loyal, but others broke away either by switching their allegiance or by becoming independent lords. Thus arose the paradox that an age in which loyalty was the highest value was also an age in which disloyalty was commonplace. From Ashikaga Takauji until Tokugawa Ieyasu, *gekokujō,* the overthrow of lords by their vassals, was one of the most salient features of political life.

In contrast to this Janus-faced and vitally pivotal loyalty of the period of Warring States, that of the Tokugawa was substantially different, reflecting the changed relations between fiefs. Joseph Strayer has written of feudal Europe: "Later feudalism is more like a series of holding corporations; the local lord still performs important functions but he can be directed and controlled by higher author-

[21] Ishii Ryōsuke, *Nihon hōsei shi gaisetsu* (Tokyo, 1948), pp. 398–399.

ity."[22] This statement can almost be taken to characterize the Tokugawa polity. In some ways the control and direction exercised by the Bakufu was more circumscribed than that ventured by the emerging Western European monarchies, in other ways it was stronger; but on the whole, they were remarkably similar.

In Tokugawa Japan, as in Japan during the period of Warring States, relations within the fiefs (now called han) were largely determined by the positions of the fiefs within the polity as a whole. The changes which had occurred in the nature of relations between fiefs brought about certain shifts in the loyalty bond between lord and vassal within the fiefs. The Tokugawa control system guaranteed the security of the Bakufu; but it also guaranteed the existence of the fiefs, to the extent that they deported themselves properly. Protected by the Tokugawa system against subversion from within and attack from without, the lord or daimyo was no longer dependent on his vassals in the sense that he had been during the earlier period. As a result, the *de facto* bilateral dependence of the earlier period now changed into one which in practice was far more unilateral: the vassals were dependent on the lord but the lord was no longer dependent on his vassals. The increased strength of the lord under this system was apparent in new land surveys and in the practice of borrowing a portion of the vassal's stipend, measures which no daimyo would have dared to take during the period of Warring States.

Yet in spite of the fact that reciprocity was lost, and that the daimyo increased their power at the expense of the vassals, the loyalty of the vassals did not weaken. It could not; within the Tokugawa framework vassals could not be disloyal. As their relation to the daimyo became unilateral, their loyalty of necessity became unconditional. This unconditional loyalty of the Tokugawa period was not functional in the same sense in which loyalty had been functional in the earlier period; it was, nevertheless, genuine loyalty. The children of samurai were raised as members of a unique military class. They lived in a castle town whose society reflected a hierarchical military order. They were "educated for death"; from an early age they were

[22] Joseph R. Strayer, "Feudalism in Western Europe," in *Feudalism in History,* ed. Rushton Coulborn (Princeton, 1956), p. 19.

taught the use of the sword and other weapons and instructed in the samurai code of honor. They policed and guarded the han, they escorted the daimyo to and from Edo, and they participated in the house government of the daimyo. The nature of the Tokugawa economy reinforced the frugality ethic of the earlier period, and the emphasis on sacrificing oneself for the lord was transferred in part from the political to the economic sphere.

Not only did the vassals' loyalty not weaken, but reinforced by the above practices and by the fact that it was no longer possible for a samurai to gain by disloyalty to his lord, it was internalized far more completely than during the earlier period. It is not without significance that the episode of the Forty-seven Rōnin was a Tokugawa and not a Warring States phenomenon. Of course, 150 years elapsed between this incident and the coming of Perry, during which time some feudal vigor had been lost and some earlier practices had become vestigial. Still, even within the narrow purview of Chōshū's Bakumatsu history, there are more than enough examples of suicide or self-sacrifice in the name of personal honor or that of the daimyo to show that, in the main, the force of the peculiar, unconditional, Tokugawa loyalty continued unabated.

Another change in the nature of Tokugawa loyalty was that relatively speaking it was more impersonal than the loyalty of the earlier period. This is not to say that the samurai now had less contact with their lord. During both periods the taking of the oath of fealty was more a ritual than a personal contact. Only the upper strata of samurai had the right of audience with their lord, and only the retainers in his house government or those of the highest rank were in frequent contact with the daimyo. A somewhat greater range of contact may have existed during the earlier period because of the presence of the lord in camp with his warriors, but even here effective control of vassals was exercised largely through personal contact with the top few. Yet we must stress that during the early period this personal contact with the top few was necessary. When this contact was lost through the succession of an inept lord or a minor, not infrequently, either through usurpation or by internal fragmentation in the face of external attack, the fief would fall.

In contrast to this it mattered little during the Tokugawa period whether a daimyo was strong or weak, for in either case the fief would not fall; from the late seventeenth century on the fiefs were guaranteed by the system. Owing to the working of the *sankin kōtai* system, most daimyo were born and raised in Edo and spent half of their lives there even in their maturity. Thus, in much the same way that the court of Louis XIV at Versailles weakened the ties between the nobility and their lands, the daimyo, knowing little of what went on in the han, could not rule even when personally able. As a result, during the course of the Edo period the role of the daimyo changed from the actual head of the government to the titular head of the government. The daimyo became the symbol of the han, the government of which was actually run by samurai officials. And, insofar as the daimyo remained nonfunctional and isolated as the weak center of government, loyalty to him became loyalty to a status rather than personal loyalty to an individual. In some ways this impersonal loyalty could almost, if not quite, be described as "han nationalism." I say not quite because, on the one hand, this tight in-group feeling was limited to the samurai class and perhaps to the upper stratum of peasant officials and, on the other hand, because it was much more structured than the diffuse feeling usually suggested by the word "nationalism." A personal identification with the han and some sense of participation in its affairs, however, did exist.

We may conjecture that the samurai's relation to the land also reinforced this han loyalty. As we have seen, a considerable number of the highest strata of *shi* possessed fiefs (*chigyō*) within the han. Yet the majority with fiefs as well as those with stipends possessed few if any local ties. The system of fief administration changed throughout the period, yet it appears that by 1700 most fiefs were managed by lower-grade peasant officials under the control of the local han office (*saiban*) and that the samurai received their share of their fiefs' produce from the han. This, together with the fact that they lived, for the most part, in a castle town, was one reason why the samurai class as a whole could not effectively resist the post-Restoration reforms. It also suggests one reason why the samurai could not act as a class, a gentry class, with common class interests in the Bakumatsu period. It reinforced the vertical nature of the han loyalty structure, adding another dimension to samurai dependence on the han.

To conclude this brief survey of loyalty as it was institutionalized within the structure of Tokugawa society, we have seen that it was unconditional, impersonal, vertical, and directed towards a status which symbolized the han. Within the rigid Tokugawa framework such loyalty was fixed—its application was narrowly delimited—yet, since it was directed to a status and not a person, it was potentially, at least, what may be termed free floating loyalty. It was certainly this character that permitted the symbolic shift from the daimyo to the emperor and the shift in content from han nationalism to nationalism proper.[23]

Up to this point we have analyzed the meaning of Tokugawa loyalty as it was embodied in Tokugawa social structure. We have found that the changes in society which occurred between the period of Warring States and the Tokugawa period profoundly conditioned both its function and its potentialities. It is equally true, however, that the implications of the meaning of loyalty as an element in Tokugawa Confucianism and, more broadly, in the Tokugawa ethic, had a definite influence on social structure. Ideas and values are obviously independent of society in that they lead ahead, lag behind, or are at times irrelevant to it. They also control institutions in that they define their meaning and the limits within which they function. Consequently, for the structure of society to change significantly, there must also be a correlative change in its guiding ideas and values. Before illustrating the way in which Confucianism served as the chief vehicle of this change in the Restoration period, we must see what loyalty meant in Tokugawa Japan.

Tokugawa loyalty was partly defined in a Confucian framework. It was one of several virtues possessed by a man of good character.

[23] This discussion points up the dangers of placing too heavy a burden on the word "feudal." It is a useful term, suggesting as it does a hierarchical society, lords and vassals, fiefs and peasants, and so on. Too often, however, it glosses over the profound differences existing between different "feudal" societies.

For this reason the cultivation of other associated virtues was viewed as essential if loyalty was to be developed to its fullest extent. Good character was formed by education; the Confucian emphasis on the close relation between knowledge and virtue was certainly one reason why so many schools were founded during the Tokugawa period. Confucianism also stressed the idea of a graded social order in which superiors and inferiors follow certain patterned rules of conduct. In Tokugawa Japan, as in Chou China, the demand that reality correspond to names (*meibun*), and that men act according to their station (*bungen*), served to bulwark modes of proper action among which proper loyalty was central. Moreover, underlying such particular observations and prescriptions there existed the notion of a moral and rational universe. The samurai (especially in their role as bureaucrats) must act as exemplars for others. If the samurai are dutiful, then others will imitate them, society will be stable and Nature, benevolent.

Yet, as Robert Bellah has demonstrated, in the basic Tokugawa ethic, loyalty was not justified primarily in these Confucian terms.[24] Rather, it was based on the duty of a samurai to fulfill his obligations (*hōon*), a duty that was to be performed selflessly (*muga ni*). Thus, the same idea of loyalty, that on a scholarly level was sanctioned by the moral Confucian universe, was justified on a deeper level by a Buddhist concept of the annihilation of self and the release from a universe of morality and immorality. In an everyday secular setting the annihilation of self became the denial of self. What one ought to do contrasted sharply with what one might desire to do. Virtue consisted in putting aside selfish desire in order to act according to the requirements of one's station. This sort of ethic is neutral in that it can be used to support almost any system of defined duties; yet in any given case its force is conservative. During the Tokugawa it was used to bulwark the pyramid of feudal loyalties.

Equally central as a sanction for loyalty was the concept of *hōon,* the repayment of obligations. Like *muga* this was a term which

[24] Robert N. Bellah, *Tokugawa Religion* (Glencoe, Illinois, 1956).

both in China and Japan had definite Buddhist overtones. However, as Ruth Benedict has shown,[25] it was not limited to Buddhist contexts, but was used to regulate a wide range of behavior. In Tokugawa Japan loyalty was such an obligation, one which bound the daimyo to the shogun, the samurai to the daimyo, and which, in a certain sense, extended even to the lowliest peasant. The samurai were obligated for their fiefs and the protection which they received, the peasants for the land they tilled and the benevolence with which they were governed. It bound not only the living to the living, but was also inherited by the living from the dead. Loyalty to the lord constituted a repayment of the *on* owed by one's ancestors to those of the lord, and so by transposition, even filial piety became dependent on loyalty.

At this point, having contrasted loyalty as a philosophical-ethical concept with loyalty in its institutionalized setting, and having seen the mutual dependence and independence of these two levels, we are now able to assay the contribution of Confucianist thought to the Restoration movement. Confucian doctrine was the form in which certain values present in Tokugawa Japan were explicitly restructured, making possible new patterns of action. The Mito school seems to best represent the typical form in which this restructuring or reordering of values took place.

The growth of the Mito school can be broken down into two periods. The first extends from the second Mito daimyo, Tokugawa Mitsukuni, to the sixth daimyo. It was during this period that the monumental *Dai Nihon shi,* the Emperor-oriented Confucian history of Japan, was written. The second phase extends from the sixth daimyo to the ninth, Tokugawa Nariaki; during this time the earlier historical trend became espoused to current internal and external political problems and gave birth to an early form of *sonnō jōi* thought. It is difficult to say whether the early emphasis on the Emperor in the *Dai Nihon shi* merely embodied traditional feeling concerning Japan's polity that had survived the period of Warring States, or whether it stemmed from the Confucian doc-

[25] Ruth Benedict, *The Chrysanthemum and the Sword* (Cambridge, 1946), pp. 98–177.

trine of "one sun in the sky, one ruler in the land."[26] It may be that both feeling and doctrine were joined by the general search for origins that animated so much of Tokugawa thought. Whatever the answer, one must remember that this early emphasis on the Emperor was not calculated to weaken the Bakufu's claim to legitimacy.

On the contrary, the Mito doctrines accepted completely the many-tiered hierarchy of Tokugawa society, placing the emperor at the top. The Bakufu rule, the rule, in effect, of one feudal lord over others, was thus sanctioned by its relation to the Court in somewhat the same manner that the rising kings of medieval Europe had used the backing of the Church to strengthen their positions against other feudal lords. The shogun, as the term implicitly suggests, was thus openly recognized as the hegemon appointed by the Court to administer the affairs of the nation; the perfected hierarchy under the Emperor was known as the national polity (*kokutai*). Whether the later course of development of the Mito school was consciously produced to give needed support to a flagging Bakufu, or whether it was simply the only line of development which a *shimpan* fief could follow with an Emperor-oriented philosophy, it is difficult to say. What we must note here are the great competitive advantages which this doctrine possessed over the Hirata school.

First, the Mito concept of *kokutai* (national polity) affirmed the existing social order. The loyalty of a samurai to his daimyo was at the same time the measure of his loyalty to the Emperor. (How unrealistic in comparison was the Hirata school with its concern for the institutions of ancient Japan.) However, it should be noted that it was this hierarchical form of the concept of *kokutai* which made the Mito doctrines largely inapplicable after 1868.

Second, and more important, in the affirmation of an ideal hierarchy, the Mito school never made clear which loyalty was primary or what should be done if the different claims were to conflict. This lack of clarity was of crucial importance. Had the shogun's claim to loyalty been primary, then the struggles of, and within, the han,

[26] *Ishin shi* (Tokyo, 1939), I, 137.

which constituted the *sonnō* movement, would have lacked a legitimating principle, and their outcome would not have been accepted by the nation at large. On the other hand, had loyalty to the Emperor been primary, the Mito synthesis would have been rejected outright.

This ambiguity in the Mito philosophy concerning the central value of the Tokugawa ethic enabled groups with radically different orientations to accept variations of the same ideology. In the course of the Bakumatsu period this proved to be very useful to the anti-Bakufu han. Yet at the same time it imposed a definite limit on the statement of objectives of the *sonnō* movement. Its leaders, no matter how impossible in fact, attempted at every point to reconcile their loyalty to their han with their loyalty to the Court. In doing so, they were prevented from seeing beyond the han and the Court even until the eve of the Restoration.

As a result, one must be very wary of speaking of a loyalty that transcended the han; one must carefully distinguish between the objective consequences of an action and what that action meant to those involved in it. The difficulties of interpreting this ambiguous Bakumatsu loyalty are further compounded by the nature of clique struggles within the han. In Mito, for example, the existence of the Mito school, coupled with the political machinations of Tokugawa Nariaki, gave rise to a strong *sonnō* clique earlier than in any other han. This clique, however, rose only to stumble and fall. The failure of Nariaki in the Bakufu political struggle and his subsequent confinement opened the way for the victory of the opposing conservative clique, which promptly suppressed the *sonnō* enthusiasts. Within this context, did the resistance of the *sonnō* clique to the conservative government indicate a loyalty to the emperor transcending the han, or was it merely the continuation of clique struggles within the han? A similar process took place in Owari and Fukui; as long as the early *sonnō* movement served to buttress han policy, it was given free play, but, as soon as Ii rose to power, the han authorities themselves silenced the movement which they had earlier favored.

Very much the same thing occurred in Satsuma and Tosa. The

early participation of Satsuma in the national power struggles of 1854 to 1857 laid the ground for the rapid emergence of a strong *sonnō* faction. One section of this faction, acting against the orders of Shimazu Hisamitsu (who in this case, as the father of the daimyo, would appear to represent his will), decided to stage a military coup on behalf of the *sonnō* cause. Incensed at their disobedience, Hisamitsu sent against them other samurai, who crushed the group in pitched battle at Teradaya. On the surface this would seem to be a clear-cut case of a loyalty transcending han limits. But if it is seen in terms of han clique politics, one suspects that the situation may have been much more complex. Since in many cases the daimyo were nonfunctional and policies were merely issued in their names by one clique or another, struggle between different cliques was not viewed as disloyalty, even if at times it did involve disobedience to the expressed will of the daimyo.

In Tosa the *sonnō* party arose relatively early and for a period in 1862 was even able to dominate han policy. But late in 1862 when Yamanouchi Yōdō regained control of the han government, he reversed the han policy, stamped out the *sonnō* faction, and forced its leader to commit suicide. In these circumstances Sakamoto Ryūma and Nakaoka Shintarō left the han to continue their *sonnō* efforts. The complexity of such a situation does not permit us to speak simply of a loyalty which transcended the boundaries of the han—although in one sense it obviously did.

We said earlier that the Mito doctrine was representative of the restructured Confucianism that provided a pattern for action in the Bakumatsu period. To say that it was representative does not mean that *sonnō* thought in other areas developed out of the Mito teachings. It may have, but it equally well may not have. In the case of Chōshū, han samurai traveled to Edo in the cortege of the daimyo, others were assigned to permanent duty at the Edo offices of the han, and still others traveled to Edo and even to Mito itself to study. The Mito ideas were known in Chōshū. Yet there is very little mention of them before 1853—by which time Japan was already being referred to as the Imperial country (*kōkoku*) in Chōshū memorials. It may well have been that *sonnō* thought in

Chōshū emerged directly from the orthodox climate of opinion in the han in response to the changing political situation.

A great many streams of thought converged to form this climate of opinion. The main stream was undoubtedly the orthodox Chu Hsi Confucianism which was taught at the han school. In addition, the Yamaga Sokō (*kogaku ha*) Confucianism of Yoshida Shōin; the search for an identity which, cutting across many schools of thought, stressed the imperial foundation of the Japanese polity; and the awareness of the special relation of the house of Mōri to the Court, all seem to have contributed, if unequally, to Chōshū *sonnō* thought. Because these currents of thought were present in Chōshū, the symbol of Japan was the Emperor, not the shogun. As long as seclusion was enforced, this meant very little; the primary reality in the minds of the samurai was the Bakufu-han structure. The arrival of Perry, however, forced attention on the country as a whole, and the previously innocuous emphasis on the Emperor suddenly became significant.

Yet, even if the Chōshū *sonnō* movement did not rise under the influence of the Mito philosophy, in most particulars it seems to have duplicated in its development the salient features of that school. The same emphasis on hierarchy, the same legalistic stress on the position of the shogun as hegemon appointed by the Emperor, the same lack of clarity regarding the priority of the various loyalties, even the use of the same concepts such as *kokutai,* characterize the Chōshū movement.

The principal difference between the *sonnō* movement in Chōshū and that in other han was one of composition and not of doctrine. In Chōshū, unlike most han, there were two groups supporting the *sonnō* movement. One was the *sonnō* clique, the other the Sufu bureaucratic clique, a clique which in its early years did not wholly accept the *sonnō* doctrines, a clique which clearly placed the interests of the han above those of the Court. The importance of the support given by this latter group cannot be stressed too strongly. At a time when traditional bureaucratic cliques in other han were turning against the *sonnō* movement, it was the continued support by this group which enabled the movement to develop in Chōshū.

However, since we have already seen the origins of this group as far back as the Tempō period, we will turn here to the second group contributing to the development of Chōshū *sonnō*-ism, the loyalist party, which emerged after 1858 around the school of Yoshida Shōin.[27]

Shōin was born in 1830. His future occupation was set when four years later he was adopted as the heir of the Yoshida family. The Yoshida house had for generations supplied teachers of the Yamaga school of military tactics; within a military class Shōin was to become a specialist in military affairs. His early years were marked by precocity and a fondness for the scholarly life. He entered the han school at the age of eight to study the classics of his calling. At the age of ten he won the praise of the daimyo for his recitation of the military classics. He also studied under his uncle, Tamaki Bunnoshin, an important official in the han government. Absorbing all that the han had to offer in his particular field, he then traveled to Edo, where he studied under various famous teachers, among them Sakuma Shōzan.[28] In 1850, at the

[27] The account of Yoshida Shōin is taken from the following: Suematsu, *Bōchō kaiten shi,* II, 222–301; Kumura Toshio, *Yoshida Shōin* (Tokyo, 1944); Naramoto Tatsuya, *Yoshida Shōin* (Tokyo, 1955); Kano Masanao, *Nihon kindai shisō no keisei* (Tokyo, 1956); *Yoshida Shōin zenshū* (Tokyo, 1940).

[28] One of the most interesting questions concerning Confucianism in action is the way in which different schools of Confucianism, at times, have different implications for the actions of their adherents. It is easy to argue abstractly, as I have done in the section on Western learning, concerning what implications such and such a school should have had, but were these implications developed in practice? The case of Shōin, who was exposed to all three schools of Confucianism, indicates how difficult it will be to answer this question. As a boy Shōin studied the writings of Yamaga Sokō, an outstanding proponent of the School of Ancient Learning (*kogakuha*). This branch of Confucianism was directly opposed to the Chu Hsi interpretation and argued for a return to an earlier, simpler Confucian society. Shōin also studied at the Chōshū college, the Meirinkan, where he imbibed the teachings of the official Chu Hsi philosophy. Then, during his later travels, he studied the Wang Yang-ming school of Confucianism with Sakuma Shōzan. The Wang Yang-ming school differs from the other two schools in that it is intuitive: the Confucian principles of reason are to be sought in the heart through meditation, and not in the world. Virtuous action is produced by the cultivation of a pure mind. Probably, the influence on Shōin of the Wang Yang-ming school was not as significant as the other two. Shōin writes: "Although I have not made a special study of the [Wang] Yang-ming teachings, yet, more often than not, I find them truly agreeable." Inoue Tetsujirō, *Nihon yōmeigaku-ha no tetsugaku* (Tokyo, 1903), p. 606.

age of twenty, Shōin realized the importance of Western learning for the defense of Japan. (This sort of idea was relatively widespread in Edo at the time under the name of *kaibōron* [maritime defense] or of *fukoku kyōhei* [rich nation and a strong military].) The following year he set out on a tour to survey conditions in northeastern Japan, without permission from the proper Chōshū authorities. Consequently, when he returned to Edo the following year, he was first confined and then sent back to the han. In spite of this, however, he continued with unflagging zeal to devote himself to the question of Japan's defenses, and, when Perry came in 1853, he determined to travel abroad to study for himself the military techniques of the Western nations. He arrived too late to board the ships in 1853, and, when he attempted to board them the following year, he was arrested. Following his arrest, he was put in the Bakufu jail in Edo, and then, after a time, given into the custody of Chōshū. He remained in the Noyama jail in Chōshū until 1855, when his sentence was reduced from imprisonment to domiciliary confinement. The following year, while still in confinement, he was permitted by the han to open a school and assume the title of teacher of military studies; his role in Bakumatsu history had now begun.

The pupils who came to this school were to form a clique which would play a role in Chōshū politics almost as prominent as that of the Sufu clique. Itō Hirobumi, Kido Kōin, Yamagata Aritomo, Shinagawa Yajirō, Yamada Akiyoshi, Nomura Wasaku, and many others who were destined to become the leaders of Japan after the Meiji Restoration passed through its gates. How these men rose to power will be dealt with later; now we must ask why they chose to study in the school of Shōin.

Undoubtedly the main attraction at the school was the teacher himself. Small, emaciated, quiet-spoken, and always in control of himself, Shōin burned brightly with the intensity of his ideals. His charisma was that of one who somehow embodied successfully an impossible goal. Driving himself mercilessly, he was a paragon of scholarship; his students have recounted that he slept only rarely and would stand or walk in the snow in order to stay awake at his studies. Yet in spite of his ascetic devotion to scholarship, he could admonish his students: "It is no good to become a scholar; what is most important is virtuous action. Anyone who wishes to can read books in the intervals between his work."[29]

Curriculum, fashion, or ambition may have attracted others to his school. The han school, the Meirinkan, was famous throughout Japan as a center of learning, but its methods were antiquated and it employed as its basic texts the Chinese classics. Shōin, on the other hand, used contemporary, controversial texts such as the "Unofficial History of Japan" (*Nihon gaishi*) by Rai Sanyō, or even his own writings. In his teachings Shōin dealt chiefly with the problems of contemporary Japan. One of his students wrote of the school: "Many parents say that Shōin is a criminal [having been under arrest and in jail], and they do not want their sons to attend his school. If [their sons] attend his school, they caution them saying that it is not objectionable to practice reading books, but that it is unpardonable to discuss the political affairs of the han."[30] The same samurai spoke of his own motives for entering Shōin's school as follows: "At the time the reputation of Master Shōin was high, and everyone was going to his school; it was the fashion. Besides, I thought that I might be able to find official employment if I attended the Shōka Sonjuku."[31] At the time such a hope was not unrealistic; Shōin's uncle was Tamaki Bunnoshin, a high-ranking official of the government, and Shōin himself had been restored to the good graces of the han government. Yamagata Aritomo, for example, was sent to Kyoto by the han authorities on the recommendation of Shōin, who at the time only knew of him through another student.[32]

What then did Shōin teach at his school? During its first two

[29] *Yoshida Shōin zenshū*, XII, p. 202.

[30] *Ibid.*, p. 201.

[31] *Ibid.*, p. 206.

[32] Along with Yamagata, Itō Hirobumi and four others were sent on this mission to serve as the eyes and ears of the han. Roger Hackett suggests that "this was probably the first meeting of importance of the two men whose careers were to be so closely interwoven for almost fifty years." Roger Hackett, "Yamagata Aritomo 1838–1922: A Political Biography" (diss. Harvard University, 1955), pp. 28–29.

years, Shōin's political thought was very much like that of the Mito school, except that he placed even more emphasis on the authority of the Emperor. He felt that the Imperial house was the basis of the Japanese national polity, but within this polity he recognized the hegemony of the Bakufu; as late as 5/58 he warned his students not to engage in wild plots against the Bakufu that would only cause disturbances within the country. And although Shōin personally favored opening Japan to foreign intercourse, he supported the *jōi* (expel the barbarians) demands of the Court, since this was the will of the Emperor and since he felt that Japan had been humiliated in being forced open by the threat of Western guns.

But when the Bakufu signed the Treaty of Commerce against the wishes of the Court, the ideal polity of Shōin was rudely shattered. Almost overnight the sequestered scholar became the *sonnō jōi* radical. It was from this time that there emerged in Chōshū the clique of young *sonnō jōi* intellectuals. In comparison with the Sufu wing of the Chōshū *sonnō jōi* movement which was tempered by the responsibility of bureaucratic office, Shōin's group was composed of those outside of the ranks of the bureaucracy, and in some cases, of those too young for the bureaucracy. Consequently, they formed the radical wing of the movement. Shōin sent some of his pupils to Kyoto to survey the situation there and to make contacts with like-minded samurai of other han, while within the han he presented a series of highly impractical memorials to the han government.[33] When these were rejected, he became increasingly importunate in his demands that immediate action be taken against the perfidious Bakufu. Curiously enough, he never attacked the institution of the Bakufu, but limited his attack to the government of Ii Naosuke which had overridden the will of the Emperor.

At this time one of his students reported to Shōin that the combined *sonnō jōi* forces of Owari, Mito, Echizen, and Satsuma were conspiring to assassinate Ii and were seeking the support of their Chōshū counterparts. A meeting was hastily called to decide whether

[33] *Ibid.*, pp. 30–31.

or not the Chōshū clique should join with the others. Shōin advised his students: "It would be a good thing for our han to join with the other four to punish the evil one. However, if the other han are the leaders and our han the follower, would this not be shameful to the patriots [of Chōshū]?"[34] He therefore suggested that the Chōshū patriots first assassinate Manabe Akikatsu, the Bakufu emissary sent to the Court to obtain the Emperor's approval of the Treaty of Commerce, and then, after having proved their worth, enter into negotiations with the others on an equal footing. A plan was drawn up, and seventeen of the samurai present signed a blood pact to this effect. It is a telling indication of the strength of han nationalism that it would thwart concerted action even in matters such as assassination plots.

Shōin also tried at this time to dissuade the Chōshū daimyo from setting off for his required biennial residence in Edo; he argued that a loyal daimyo ought not to comply with the laws of a traitorous Bakufu. Just then, Nagai Uta, an inspector directly responsible to the daimyo, returned from Edo bearing disquieting news concerning the first phase of Ii's Ansei Purge: the daimyo of Tosa and Uwajima had been ordered to retire because of *sonnō jōi* activities on the part of certain samurai in their han. In the light of this news, Sufu, the head of the dominant clique within the Chōshū government, became more and more alarmed by the indiscretions of Shōin and his disciples, and he determined to act swiftly so that a like fate should not befall the Chōshū daimyo. Even before the beginning of the purge, Kusaka Genzui and Akagawa Naojirō had been ordered back to the han from Kyoto and Edo. Now, Sufu instructed Irie Sugizō, one of the students closest to Shōin, to inform Shōin that the han had its own plan for *sonnō,* which it would carry out when an opportunity occurred. Sufu warned Shōin that, should he continue "to act like a reckless student," he would again be imprisoned.[35] Sufu also sought to restrain the zeal of Shōin through the mediation of Sugi Umetarō, Shōin's older brother. Considering the situation prevailing in Japan

[34] Naramoto, *Yoshida Shōin,* p. 132.
[35] Suematsu, *Bōchō kaiten shi,* II, 263.

at the time, one cannot but think that Sufu's actions were moderate to the point of showing sympathy for Shōin. Shōin, however, only became more obdurate under pressure. He began to feel that he must act alone even though the han government opposed his action; and he began to label the inactivity of Sufu's government as "traitorous." [36]

Consequently, on 29/11/58 Shōin was ordered into confinement, and a week later he was put in the han jail. The charge against him was "impure teachings." [37] This charge aroused a furor among his teachers as well as among his disciples. If his teachings were impure, they maintained, so were theirs and they should also be punished. Tamaki Bunnoshin began to talk of resigning from the han government; therefore, Sufu made explicit what had been apparent all along, that Shōin was in prison so that his activities would no longer cause trouble to the han. Subsequently, eight of Shōin's students who continued to criticize the han government were ordered into domiciliary confinement for a short period.

In jail Shōin's views became more and more extreme, and he worked feverishly on schemes to achieve his ends. Not only did he plan other assassinations, and this during the period of the Ansei Purge, but, realizing that other samurai would not support his plans, he staked his hopes on uprisings of peasants guided by a mystic loyalty to the Court. In the face of views such as these, even his closest disciples turned against him. On 5/12/58 Shōin received the following letter from five of his disciples, explaining why they could not accept the program of their teacher:

We are deeply moved by the brilliance of your just views which you have so painstakingly expressed. However, the situation in Japan today has changed markedly. It is extremely regrettable that the han merely collect their weapons and look on; nevertheless, since the shogunal succession has been determined and the temper of the people quietened, it would not only be truly difficult to rise in just rebellion [at the present], but it would inevitably bring harm to the house of Mōri. Yet, though this is so, the Bakufu officials behave as madmen. Not only [do they persecute] the

[36] *Ibid.*, p. 265.

[37] Naramoto, *Yoshida Shōin*, p. 137.

loyalists, but they also force daimyo to retire and open [the country] to trade [against the wishes of the Emperor]. Certainly a reaction [against this] will arise, and at that time we must truly cooperate on behalf of the country. Until then we must restrain ourselves and, collecting our weapons, do nothing that might bring harm to the house of Mōri.[38]

Other pupils wrote that action at that time might serve to gain a "valorous name" but that nothing practical could be accomplished.[39] To this Shōin indignantly replied, "Is this not an ugly way of speaking?" [40]

By 1/59 Shōin had severed all ties with his former students, with the exception of the Irie brothers (Irie Sugizō and Nomura Wasaku) who were also in confinement. To them he wrote with some sadness and perhaps a note of bitterness: "If we go first and die, then will not the others become aroused? My friends in Edo, Kusaka, Nakaya, Takasugi, and others, all have views differing from my own. The point of divergence is this: it is my intention to be loyal and theirs to perform meritorious deeds." [41] By this time he had resolved to die for his beliefs. On 5/59 Chōshū was ordered by the Bakufu to send Shōin to Edo. Questioned there by officials of the Bakufu, he not only answered questions concerning his political beliefs, but he went on, to the amazement of the officials concerned, to recount the details of his plot to assassinate the Bakufu emissary Manabe. This sealed his doom; he was sentenced to die. Before his death he was reconciled with his disciples, although one must stress that this reconciliation was personal and not one of ideas. On 27/10/59 he was beheaded in the Demmachō prison in Edo.

How are we to assess the role of Shōin in Chōshū history? He, more than any other figure, was responsible for the form in which *sonnō* thought developed. He provided the philosophy that linked traditional hostilities, political tensions, and personal ambitions with certain objectives of Bakumatsu politics in such a way that

[38] *Ibid.*, p. 139.
[39] Kano, *Nihon kindai shisō no keisei*, p. 56.
[40] *Ibid.*, p. 56.
[41] Naramoto, *Yoshida Shōin*, pp. 140–141.

channels were opened for political action to samurai who were not members of the han bureaucracy. Moreover, the school which Shōin founded provided a focus about which the young men imbued with the *sonnō* doctrines could organize as a political force. The degree to which Shōin's disciples maintained a formal organization after his death has not been studied, yet there is no doubt that they constituted the nucleus of the later *sonnō* clique. Had he not lived there may not have been another Yoshida Shōin, and the clique which arose about him in the months following the signing of the Treaty of Commerce in 1858 might otherwise have arisen only gradually in the course of several years.

And yet when Shōin died, his disciples were willing to wait. However great his influence, one must stress the fundamental differences between Shōin and his disciples to understand the position of either in Chōshū history. Seen in the context of the society of his day, Shōin's intense preoccupation with certain goals to the exclusion of others, the fixity or rigidity of his views, his will to die as a martyr for his cause, and finally his willingness to invoke even a peasant uprising as the instrument of his ideals all suggest an obsession passing beyond mere eccentricity. He seemed willing to risk the destruction of samurai society, or even that of the han itself, in order that the Imperial cause be served. Viewed in this perspective he seems an almost prophetic figure: what the Restoration reformers arrived at only after long years of struggles, he seems to have gained with a single leap. That he never attacked the institution of the Bakufu itself was characteristic of the period in which he taught and must not detract from the advanced nature of his views. His loyalty was abstract and on this abstract level his duty to the han and that to the Court did not conflict—however much they may have conflicted in fact. The explanation for this lack of pragmatism in a most pragmatic society seems to be that the slight religious overtones present throughout the *sonnō* movement were for Shōin central, and that some inner vision lay beneath his ascetic life, his emphasis on the purity of the will, and his calm self-control.

His followers, in contrast, were bound by their own interests within their particular society; they accepted Shōin's emphasis on loyalty to the Emperor, yet at the same time they remained loyal to the han. Chōshū's history from 1861 to 1868 consists to a large extent of their attempts to reconcile these two loyalties within the changing political scene. But the disciples were intensely pragmatic. In most respects they were closer to Sufu than to Shōin, and as the period advanced they approached even closer to the position of the Sufu clique, gradually becoming absorbed into the han officialdom. Perhaps the most outstanding feature of Sufu *sonnō* feeling was that, while favoring the Court over the Bakufu, it felt that the only way to advance its cause was from the base of a strong and united han, the interests of which it forwarded with equal zeal and resolution.

Adding some factors and subtracting others it would seem that the influence of Shōin has been unduly stressed by historians in the past. He was in no way representative of the main stream of the Restoration movement. But, there is no simple mathematics of ideas and influences. Even though his disciples were unwilling to emulate his action, and thought of ends and means in a very different fashion, they did recognize his inspiration. Takasugi wrote after Shōin's execution: "We, his disciples, must join hands and avenge him." [42] Kusaka voiced similar sentiments: "Do not mourn, but rather act to strike down the enemy." [43] History does many things with martyrs: some are canonized and others ignored and forgotten. Shōin, though renounced by his disciples and executed by public authority, nevertheless numbers among the former. He never became an element in a myth, yet he stood as a symbol of Japan's particular spirituality until the end of the Second World War. He was a prophet not without honor in his own country.

[42] Kumura, *Yoshida Shōin* (Tokyo, 1944), pp. 361–362.
[43] *Ibid.*, p. 362.

PART TWO

The Restoration Movement

CHAPTER VI

The Rise of Chōshū in National Politics, 1861-1863

The Nagai Uta Mediation

From the time of Perry's arrival in Japan until the death of Ii in 1860, national politics and Chōshū politics moved in separate spheres. However great the influence on Chōshū of events in the nation at large, the reaction in the han did not give rise to actual participation in national politics. The han government plodded along the course of internal reforms. And within the han, almost as great a degree of separateness existed between the han government and the nascent *sonnō* clique. Each reacted to events outside of the han, rather than to the other. This almost cellular separateness broke down in 1861.

It is dangerous to project the history of Chōshū, finding in it the history of Japan, yet I feel that 1861 was a turning point for that country as a whole. The two years during which Ii Naosuke ruled as *tairō* form a watershed between two periods with different characteristics. Before Ii, national politics were in the hands of the Bakufu; outside parties could participate only by aligning themselves with factions within the Bakufu. After Ii's death the internal politics of the Bakufu became more and more of a side issue; the Court had emerged, and national politics took place in the field of forces between the Bakufu and the Court. Before Ii, Edo had been the political center of Japan; after Ii, in most respects the center of gravity shifted to Kyoto. Again, before Ii, the *sonnō jōi* (honor the Emperor, expel the barbarian) doctrine had been a scholarly prop for the Bakufu; after his assassination it was to become the slogan of the forces supporting the Court.

The effectiveness of the Bakufu under Ii and of his Ansei Purge has been all too often underemphasized. In a large measure he succeeded in re-establishing the traditional Bakufu absolutism of the earlier Tokugawa period. Not only did he quiet or suppress the political squabbles within the Bakufu, but he also stopped the activity of even the most ambitious of the outside han. Moreover, by exerting pressure on the governments of the han he had crushed the *sonnō jōi* parties in most of the *tozama* han. Even Chōshū, where the *sonnō jōi* party was late to emerge, recalled its extremist samurai to the han and put Shōin in prison in order to avoid reprisals by the Bakufu. Only the fanatic fringe continued their plans, but this was to prove sufficient; on 3/3/60 *rōnin* from Mito and Satsuma assassinated Ii.

It is not inconceivable that, had Ii's successors continued his policies, even the fanatic fringe might have been eventually crushed. Instead they chose to placate and appease the opposition. Andō Nobuyuki and Kuze Hirochika not only released those daimyo placed in domiciliary confinement by Ii, but they also arranged for the marriage of the shogun to an Imperial princess as an outward symbol of the unity of Bakufu and Court. Their compromises, however, led to contrary results. The pent-up feelings against Ii, the Bakufu, and the Tokugawa now turned against the government headed by Andō and Kuze, contributing to the rising *sonnō jōi* movement. This movement, which was soon to reach turbulent proportions, was inaugurated pacifically by Chōshū, which emerged from the hush cast by Ii to mediate between the Court and the Bakufu.

Soon after the appointment of the Sufu government in Chōshū in 1858, it was decided that the han should enter national politics at the first possible opportunity. Participation in politics was to be based on the fundamental han policy of "loyalty to the Court, trust to the Bakufu, and filial duty to the ancestors [i.e., to the ancestors of the house of Mōri]."[1] During the period of Ii's supremacy, nothing could be done, and, even during the first year following his death, Chōshū, as well as the other han, stepped gingerly, fearing the wrath of the new government of Kuze and Andō, whose dispositions were still unclear. Chōshū was first to act; its program was based on a memorial submitted by Nagai Uta, a Chōshū samurai.

Nagai was born of a famous family, one which had been attached to the Mōri house since the earliest days of its rise in the period of Warring States. At the time he submitted his memorial, he held the office of Direct Inspector. The function of this office was to check, as a censorate of sorts, on the activities of other offices in the han; as the inspector was required to report directly to the daimyo, the post was limited to samurai of fairly high rank, usually to those in the *ōgumi* group. A stern, austere, and able man, Nagai was recognized by all in the han. Even the exacting Shōin had once ranked him among the four or five most capable men in the han.[2]

Nagai's memorial, which was partly written in collaboration with Sufu, began by presenting three reasons why Chōshū must act in the current situation. First, Chōshū was a great han, one with a special relation to the Court; therefore, it would ill become it to remain inactive when Japan was faced with a crisis. Secondly, if it wished to abide by its basic policy, Chōshū must act to resolve the differences between the Court and the Bakufu. Thirdly, if a war should break out between Japan and the foreign powers, then Chōshū itself would be endangered, and filial duty dictated that Chōshū be preserved.

Nagai then went on to say that the Bakufu, debilitated by three hundred years of peace and motivated by the pedestrian desire (*zokuron*) to avoid war, had signed the humiliating treaties with the foreign countries. Because of this, the patriots were resentful, the Court was incensed, the hearts of the people unsettled, and precious days were wasted on the question of whether to open or close the country. In spite of this, Nagai felt that it would not be wise to

[1] Suematsu Kenchō, *Bōchō kaiten shi* (Tokyo, 1921), II, 233. This was a part of the key han policy decision taken in 1858.

[2] *Ibid.*, III, 101–102.

repudiate the treaties. He argued that "for the last three hundred years the government of the Imperial country has been entrusted to Kantō [the Bakufu]. All relations with foreign countries have been handled by Kantō; therefore, it is natural for the barbarians to think that Kantō is the government of the Imperial nation."[3] Since this was the case, if Japan repudiated the treaties, the trust of the foreigners would be lost. Nagai next drew an argument from history: seclusion was a Bakufu policy; in the age when the power of the Imperial Court was at its pinnacle there were even residences for foreigners in Kyoto, hence, seclusion was not "an old custom of the Imperial nation."[4] He ended his defense of the treaties by considering the consequences of a war: "If you suddenly begin an unplanned war, using samurai who have for several hundred years become accustomed to peace, even a three-year-old child can tell you what will happen."[5]

What in fact was to be done? He proposed a "policy of expansion across the seas."[6] Japan must study the arts necessary for navigating the seas, it must build up its military, and, while trading and enriching the country, it must spread abroad the military prestige of the Imperial nation. If the Court orders the Bakufu to do this and if the Bakufu carries it out, the misfortune of the treaties will become a blessing, the shogun will be fulfilling the "barbarian-conquering" duties of his office, the Imperial glory will shine abroad, and the Court will receive tribute from the five continents.[7]

This was submitted to the daimyo on 28/3/61. It obtained the approval of the daimyo, who ordered Nagai to proceed to Kyoto to begin his mediation between the Bakufu and the Court. Nagai arrived in Kyoto on 2/5/61, and three days later he met with the Court Councillor (*gisō*), Ṣanjō Saneai. Saneai, pleased with the

[3] *Ibid.*, pp. 107–108. My description of Nagai's policy is based on three documents: the proposal made to the daimyo, the proposal made to the Court, and that made to the Bakufu. All are essentially the same in program, although each stresses those aspects of the program favorable to the party concerned.

[4] *Ibid.*, p. 110.

[5] *Ibid.*, p. 109.

[6] *Ibid.*, pp. 40–41.

[7] *Ibid.*, pp. 41–47, 60–65, 104–114.

proposals, ordered Nagai to draft them into a document for presentation to the Emperor. In due course, the Emperor approved, and on 2/6/61 he issued an order instructing Nagai to continue with his mediation. The same day Nagai set out for Edo, where he met with a similar success.

Kuze and Andō and other leaders of the Bakufu whom he saw were pleased with his proposals and ordered him to continue his mediation. This pattern of wholehearted approval, unsupported by effective action, was to persist through the subsequent months. Finally, on 8/12/61, the Chōshū daimyo, having arrived in Edo for his year of required residence, formally presented to the Bakufu a memorial containing the substance of Nagai's earlier program. The shogun approved and officially ordered that the mediation between the Bakufu and the Court be entrusted to Chōshū. On 24/2/62 Nagai was summoned to the Council of *rōjū* and questioned further about his policy; he was subsequently confirmed in his role as mediator. The Bakufu then sent a representative of the shogun to Kyoto to hasten a reconciliation with the Court. Nagai also repaired to Kyoto; he arrived on 18/3/62, and, meeting with Sanjō Saneai, reported on the progress of his mediation in Edo. Again, Nagai received Imperial approval of his actions. His program of mediation appeared to be on the verge of success.

The reasons for his early success are clear. Despite Ii's assassination, the hush cast by his rule still lay heavy over the land. The power of the Bakufu, real and imagined, was paramount, and Nagai's proposals reflected this situation. However much they were embellished with visionary or fanciful conquests of the five continents, no matter how much they chided the Bakufu for signing the treaties without the approval of the Court or expressed deference to the Court, they were, nevertheless, almost completely in support of the Bakufu policies. Nagai asked nothing of the Bakufu that it would not have done in any case, and requested the Court to accept the *kaikoku* (open the country) policy of the Bakufu. The Bakufu was offered the possibility of a "Union of Court and Bakufu" (*kōbugattai*) that would remove all resistance to its policies while leaving undiminished its power. The Court was to remain in seclusion in

Kyoto; its only voice in politics was its right to give formal approval to the efforts of the Bakufu.

Yet, at the same time, it is not difficult to see why the Court also welcomed the policy of Nagai. During the period of Ii's rule the Court had been virtually in eclipse. Now, however, Nagai not only exhorted the Bakufu to render all things to the greater glory of the Court, but also instructed it to act, at least nominally, according to the orders of the Court. This established, in principle at least, a pattern that might in the future work to the Court's advantage. Further, even if the Court—if one may speak of it for the moment as if it were a unit with a single view—had preferred to maintain its advocacy of seclusion, it was not given a choice. At this time only Chōshū had come forth to mediate; the Chōshū program was the only one available; and it offered to the Court something better than a continuation of the *status quo*.

Satsuma's Mediation and the Downfall of Nagai

In spite of its initial successes, Nagai's mediation was to end in utter failure by the summer of 1862. The failure was due to three causes: the opposition to his plan by the *sonnō jōi* adherents of Chōshū and other han, whose activities now converged on Kyoto, the alienation of Sufu Masanosuke, and, most important, the emergence of Satsuma with a competing plan for mediation between the Court and the Bakufu.

The *sonnō jōi* loyalists had begun to revive their activities just at the time that Nagai first made his proposals to the han government. The group in Chōshū, which centered on the former pupils of Yoshida Shōin, was opposed from its very inception to Nagai's program. Kusaka Genzui and Takasugi Shinsaku, the leaders of the Chōshū loyalists, were both in Edo at the time. Kusaka had been sent to Edo for the study of the English language, but he seems to have spent most of his time contacting the young loyalists of other han. When he first heard of Nagai's proposals, he wrote to Nakamura Kyūrō, an official of the Sufu clique, expressing his opposition. Kido Kōin was also in Edo at the time as an official at the Yūbikan, the han school at Chōshū's Edo residence. Like Kusaka, Kido was also hard at work contacting the *sonnō jōi* factions from various han. In particular, he tried to bring about a coalition between the officials of Chōshū and Mito; the end result of all his exertion was a blood pact signed by six relatively unimportant persons. From such obscure beginnings had arisen the *sonnō jōi* movement that was soon to become one of the major forces in the country. Many of the younger samurai of the han joined hands with the *sonnō jōi* leaders in criticizing Nagai's policies for being too favorable to the Bakufu, so that, when Nagai first arrived in Edo on 14/6/61 with the unofficial approval of the Emperor, he had to spend a good portion of his time trying to convince the young loyalists of his own han that his plan was not directed against the Court.

During this time the activities of the *sonnō jōi* samurai were also gaining momentum in the Kyoto region. Indeed, some had plotted to kill Nagai when he returned to Kyoto in 3/62 for supporting the Bakufu against the Court. Saigō Takamori of Satsuma argued that "although the idea of assassinating Nagai may seem too extreme, he is a very evil man who has come to Kyoto at the behest of the Bakufu." [8] Nagai himself was not unaware of the hostility with which he was regarded by the loyalists. At one point he had confided to a fellow inspector in Chōshū: "The Satsuma men and other *rōnin* call me an evil man, and there are many rumors of plots to kill me. Because of this, the nobles are very worried. . . . Even the Emperor has been apprised of this problem. . . . If I die now, I will have no regrets. Now is our best chance to obtain glory for the house of Mōri; we must redouble our efforts."[9] Nagai then added a note of complaint that a large part of the slander against him originated from the Chōshū loyalist samurai, who would even invite the loyalists of other han to join them in thwarting his policies.

The opposition to Nagai within Chōshū was further heightened by Nagai's estrangement from Sufu. We have seen how Sufu supported Nagai's proposals when they had been first presented in the han early in 1861; Sufu himself had composed the final memorial

[8] *Ishin shi* (Tokyo, 1939), III, 26.
[9] Suematsu, *Bōchō kaiten shi,* III, 217.

of Nagai's plan for presentation to the daimyo. But in 7/61 he went to Edo, where he became more aware of the ineptitude of the Bakufu, and, exposed to the impassioned arguments of Kusaka and Kido, Sufu came to feel that Nagai's mediation policy was not appropriate to the times. Whether or not he tried to change Nagai's views at the time is not known. Through the offices of Kido, Sufu and Nagai did meet and talk for many hours with Minobe Matagorō, a loyalist from Mito. After this meeting, Nagai returned to the han to accompany the daimyo to Edo. Sufu remained in Edo, meeting again with Minobe and Kido for further talks on the political situation.

It was from this time that Sufu began to criticize Nagai's mediation policy. Judging from later events, it is unlikely that Sufu's position had changed or that he had become imbued with the fanatical spirit of the loyalists; rather, he had become convinced of the certain failure of Nagai's mediation policy, and he wished to change to one that would succeed. On 7/9/61 Sufu left Edo in order to meet the daimyo's cortege en route to Edo and to voice his opinions. At this time he conferred with both Nagai and Masuda, the Accompanying Elder. His advice was not only rejected, but he was also rebuked for having left Edo without orders, and he was told to return to the han. On 29/1 of the following year, he was punished with twenty days of domiciliary confinement. These moves against Sufu were no doubt instigated by Nagai who felt that the opposition of one of the most important Edo officials would interfere with Nagai's own mediation program and foil his efforts to win the younger samurai to his cause.

The most decisive factor contributing to the failure of Nagai's mediation was the program launched by Satsuma in rivalry with that of Chōshū.[10] In a sense, Satsuma's new mediation was the continuation of the national role it had played in the 1853–1858 period of rule by Shimazu Nariakira. When Nariakira died in 7/58, he was succeeded by Shimazu Tadayoshi, the son of Nariakira's half-brother, Hisamitsu. In spite of his son's succession, Hisamitsu did

[10] This account of the Satsuma mediation has been taken for the most part from *Ishin shi,* III, 41–206.

not immediately gain power in the han councils of Satsuma, since Shimazu Narioki, who had been daimyo before Nariakira, still lived. Narioki died in 9/59 and the han soon came under the control of Hisamitsu, who, like his half-brother Nariakira, was an extremely competent man.

After the death of Ii, Hisamitsu began to plan for Satsuma's re-emergence on the national scene. As Chōshū had already established itself as the mediator between the Court and the Bakufu, it was a choice of either joining forces with Chōshū, which would mean taking second place, or devising a policy on which it could mediate successfully in competition with Chōshū. Satsuma chose the latter course; a policy was decided on, and Hisamitsu left Kagoshima accompanied by a thousand troops. They arrived in Osaka on 10/4/62, and, six days later, Hisamitsu and his leading retainers met with high officials of the Court. He presented to them a nine-point plan on which he proposed to mediate between the Bakufu and the Court. The Satsuma plan began with a review of the situation existing between the Court and the Bakufu: the shogun, in spite of having received an Imperial princess, was still remiss in his attitude toward the Court; the Court was not receiving the honor and homage that were its due. Satsuma was offering its services as mediator to amend this. It called for the release of Court nobles ordered into confinement at the time of the Ansei Purge; the appointment of Konoe Tadahiro to the office of *kampaku;* the release from confinement of Tokugawa Keiki, Tokugawa Keishō, and Matsudaira Keiei; the dismissal of Andō Nobuyuki, the head of the Council of *rōjū;* the replacement of Tokugawa Yoshiyori by Tokugawa Keiki as guardian of the shogun; the appointment of Keiei as *tairō;* and finally, the deputation of the *tairō* Keiei and the *rōjū* Kuze to Kyoto to attend to the wishes of the Court.

In addition, the Satsuma proposal suggested that in the future, the Imperial will should not be disclosed to *rōnin* and that the opinions of *rōnin* should be rejected by the Court. This last measure was intended to preclude any possible identification of the new Satsuma mediation with the *rōnin* movement stirring in Kyoto.

The Satsuma plan, if effected, was designed to bring about the situation that the pro-Keiki national front forces had envisioned in 1856 and 1857 before their defeat by Ii Naosuke. And more specifically, since the Satsuma program was more favorable to the Court than that submitted by Nagai, the Satsuma leaders hoped that the Court would approve of their plan of mediation and jettison that of Chōshū. Satsuma's hopes were soon realized.

Satsuma's actions in Kyoto following the presentation of the above plan can be broadly divided into two phases: the first was its suppression of extremist activities, the second was the formation of the Ōhara mission to the Bakufu. Proposals were submitted to the two Court councillors, Nakayama Tadayasu and Sanjō Saneai, who relayed them to the Emperor. The Emperor is said to have been deeply pleased with the Satsuma program. It was also reported that he had been concerned about the rising loyalist activity in Kyoto, and he immediately issued an edict to Hisamitsu authorizing its suppression.

The *sonnō* loyalists had mistakenly assumed that Hisamitsu had come to Kyoto to give support to the Court's desire to expel the foreigners from Japan. They had planned an uprising for 18/4/62, the day after Hisamitsu's receipt of the Imperial edict for their suppression. Hori Jirō, an official close to Hisamitsu, learned of the loyalists' plan and sent a message promising them that all their wishes would be fulfilled by the mediation of Hisamitsu. Hisamitsu had previously ordered the Satsuma samurai to refrain from contacting the loyalists of other han, explaining that, while he agreed with the aims of the *sonnō* loyalists, their deeds were a danger to both han and country. The loyalists postponed their uprising several times but remained steadfast in their determination to carry it out, for they felt that, even if it appeared contrary to the orders of Hisamitsu, it would help restore the prestige and power of the Court. On 23/4/62 the Satsuma loyalists headed by Arima Shinshichi left the han residence in Osaka for Fushimi to join loyalists from Kurume han. They planned to proceed to Kyoto, where they would assassinate the pro-Bakufu *kampaku* and the Kyoto *shoshidai,* the Bakufu officer at the Court in Kyoto.

Hearing of this, Hisamitsu promptly sent a party to their Fushimi lodgings with instructions to the Satsuma men that they were to abandon their plan. The loyalists refused, and fighting ensued in which many were killed. Hisamitsu sent the Satsuma loyalists back to Satsuma, and handed over the other participants to representatives of their respective han. Chōshū loyalists under Kusaka Genzui were also implicated in the plan. They had decided to join in the uprising, but chose to await a more opportune moment when they learned that Arima, the leader of the Satsuma loyalists, had been killed. Hisamitsu sent Hori Jirō to the Chōshū Kyoto residence with information about the plot and requested the arrest of the Chōshū faction. The Chōshū representative repeatedly denied that any Chōshū men had been involved in the plot and, in so doing, gave rise to some bad feeling between the two powerful han.

The second phase of Satsuma's activity dealt with the beginning of its mediation. On 6/5/62 the Court decided to accept Hisamitsu's proposals but not to send a messenger to the Bakufu until the *rōjū* Kuze had come to Kyoto. Since this decision meant that Hisamitsu's plan for a mediation mission to Edo to replace that of Chōshū would be indefinitely postponed, he immediately set to work to persuade the Court to change its decision. After considerable negotiations and several setbacks, he was successful; the Court decided to send an Imperial messenger to Edo without waiting for a Bakufu representative to arrive in Kyoto. The three-point message to be given to the Bakufu was composed by Iwakura Tomomi. The first point, which had been suggested to Iwakura by Kido Kōin, was that the shogun should lead the various daimyo to Kyoto to confer with the Court concerning the expulsion of the foreigners. The second point was a recommendation that five *tairō,* recalling the system established by Hideyoshi, be appointed from among the leading daimyo to participate in national affairs. This idea was Iwakura's own; he felt that Satsuma and Chōshū alone would be unable to control the Bakufu. The third point was the demand, taken from Satsuma's original nine-point plan, that Keiki be made the guardian of the shogun, and Keiei, the *tairō*. Ōhara Shigetomi, who had been appointed as the Imperial messenger, was carefully instructed to emphasize the third point, should difficulties be encountered on the first and second. On

22/5/62 Ōhara set out for Edo accompanied by Hisamitsu and one thousand Satsuma troops.

Now for the first time the impotence of Ii's successors was laid bare. Even before the arrival of Hisamitsu and the Imperial messenger, the Bakufu initiated a series of reforms in anticipation of the Imperial demands. Perhaps the Bakufu hoped to save face by reforming from within rather than being coerced to make concessions to the Imperial envoy; by 4/62, several former *rōjū,* including Andō Nobuyuki, had been replaced; on 25/4/62 most of the important Court and Bakufu figures purged by Ii were released, and on 3/5/62 Keiei and Matsudaira Katamori were appointed to participate in Bakufu affairs. In these changes and in the imminent arrival of Hisamitsu, Kuze foresaw the collapse of Nagai's mediation on which he had staked his career; therefore, on 2/6 he resigned from office. Ōhara arrived in Edo on 7/6, and by 28/6, in the face of opposition from the *fudai* daimyo, Keiki was made the guardian of the shogun and Keiei appointed as *seiji sōsai* (a position equivalent to that of *tairō*). All that Shimazu Nariakira had hoped and labored for years earlier had been accomplished; the mediation of Satsuma had replaced that of Chōshū, and the Bakufu was following the lead of the Court.

The Fall of Nagai and the Rise of Chōshū Sonnō Jōi Policy

In the meantime, even before the news of Satsuma's competitive mediation had reached Chōshū, opposition to Nagai among the loyalists continued to mount. For example, Kusaka, who had been sent back to the han together with Sufu the previous year, continued to agitate in favor of a more extreme position. Ever anxious to further the loyalist cause, he met with loyalists from other han, Sakamoto and Yoshimura Toratarō of Tosa, Moriyama Tōen of Satsuma, as well as others of the Chōshū party. When the first news of Satsuma's impending mediation reached the han, Kusaka misinterpreted this as Satsuma's espousal of the *sonnō jōi* cause, and, although disconcerted over Chōshū's loss of leadership, he nevertheless formed a band of Chōshū samurai who pledged themselves to leave the han for Kyoto to join other loyalists. Sufu, who was temporarily excluded from the councils of the han, seems to have encouraged these actions; he further ordered Kusaka to draft a memorial to the daimyo, lamenting that the house of Mōri, renowned for centuries for its loyalty to the Imperial House, was today following behind other han.

Concerned over the incipient mediation of Satsuma and fearing that it might lead to an uprising in Kyoto, the officials in the han tried to strengthen the power of Chōshū in the Kyoto region in the guise of sending reinforcements to the Hyōgo coastal area. Ironically enough, the reinforcements included Kusaka and others of the Chōshū loyalist party. Their inclusion may have been due to the influence of Sufu. By 19/4/62 Nagai's position had been so undercut, so to speak, by the arrival of the Satsuma troops in Kyoto that Kusaka and five other Chōshū loyalists could with impunity send a memorial to the daimyo denouncing Nagai for having insulted the Court, duped the nobles, betrayed the daimyo, and injured the military spirit of the retainers. The memorial further requested that Nagai be swiftly removed from office and punished with death.[11]

Nagai himself had left Kyoto for Edo on 14/4/62, two days before the arrival of Hisamitsu. He carried an order from the Court summoning the Chōshū daimyo to Kyoto for the final phase of his mediation; the order had been given to Nagai by Sanjō Saneai, who at the time expressed to Nagai the Emperor's warmest approval of his mediation. News of the Satsuma mediation, however, preceded Nagai to Edo; by the time of his arrival on 21/4, the understanding that his mediation would miscarry had awakened strong currents of opposition within Chōshū. Sufu was recalled to Edo and again participated in the highest councils of the han government. And on the same day that Nagai arrived in Edo, Kido Kōin was appointed to a high office under the Accompanying Elder. This was the first time that any of Shōin's disciples had attained an office on the policy-making level. It is significant that, although the honor had been obtained as soon as the failure of the Nagai mediation became certain, it was still achieved through the ordinary channels of bureaucratic advancement by one with a hereditary rank commensurate with the position.

[11] Suematsu, *Bōchō kaiten shi*, III, 257–259.

The effect of Satsuma's appearance in Kyoto and the installment of the anti-Nagai officials was immediately reflected in Chōshū's negotiations with the Bakufu. In order to fortify its own position against the rival mediation of Satsuma, Chōshū began to reinterpret its own program: more emphasis was placed on the respect due to the Court by the Bakufu and a more critical attitude was adopted toward the disposition of the Bakufu. Sufu and others of his clique in Edo drew up a list of questions to be put to the Bakufu and declared that, if satisfactory answers were not given, Chōshū would withdraw its mediation. Was the Bakufu willing to apologize for its past disrespect to the Court; or, observing that Japan had suffered a serious loss of national prestige because of the Bakufu's desire to avoid war, would the Bakufu be willing to fight should it become necessary? The intent behind these questions was obvious: they were calculated to strengthen the position of the Chōshū daimyo against the day of his return to Kyoto, where he had been ordered to go in the last Imperial communication carried by Nagai to Edo.

This move by Chōshū to reinterpret its own position and at the same time to steal a march on Satsuma was advanced one step further when Sufu proposed that the shogun go to Kyoto to confer with the Court officials and thereby establish a truly national policy. This proposal was communicated to Kuze, the Bakufu *rōjū,* on 2/5/62, with a note adding that Chōshū would be forced to end its mediation if this were not accepted. If this stratagem had succeeded, the Bakufu would then have been in direct contact with the Court, Chōshū would have scored a triumph, and Satsuma's advance would have been checked. Chōshū's stratagem, however, came to nothing. On 16/5/62 Kuze met with Nagai, and told him that because of financial difficulties the shogun would not be able to go to Kyoto at that time. Chōshū's last hope for the success of its mediation was now dead. A conference was hastily called at which it was decided that, abiding by the Emperor's wish, the mediation would be continued even though the inability of the shogun to go to Kyoto would make virtually impossible mutual trust between the Court and the Bakufu. On the same day Nagai informed Kuze of the results of the conference.

Eleven days earlier, the Court had sent a message to the Chōshū representative in Kyoto accusing Nagai of slandering the Court in his original proposals. The message also stated that Nagai's "open the country and cross the seas" policy would involve a change in the national polity. The han government was thrown into consternation. They immediately asked for a clarification: when and in what manner had Nagai slandered the Court? A definite answer was never given, but Sanjō Saneai later labeled as invidious Nagai's comparison of the halcyon days when the Court invited foreigners to Japan with the troubled present when the foreigners had forced open the country to the disgrace of the Court. Actually, as everyone was quite aware, the details of the charge were unimportant; the Court, well over a year after the mediation had begun, was merely using this as a pretext to disembarrass itself of Nagai's mediation, which had in fact been already replaced by that of Satsuma. Similarly, Chōshū's request for clarification of the charges was primarily intended to secure an advantageous position for future maneuvering. And, in order not to hinder the prospect of the han, Nagai Uta presented a letter to the daimyo on 16/5/62 in which he took complete responsibility for the document which was said to have slandered the Court; he asked that he be punished for having brought down the wrath of the Court on Chōshū.[12] Two weeks later Nagai was ordered to return to the han where, relieved of his position, he was ordered into domiciliary confinement. On 15/11, barely seven months after the time he had been made chief mediator between the Court and the Bakufu, Nagai was ordered to commit suicide in atonement for "his egotism and sins against the Court and Chōshū."[13] To such an extent had the tides of opinion changed within the han.

Chōshū's Adoption of the Loyalists' Policies

Having been superseded by Satsuma and having failed to obtain concessions from the Bakufu, the daimyo of Chōshū and his cortege

[12] *Ibid.*, p. 262.
[13] *Ishin shi*, III, 39.

of officials and samurai grimly set out for Kyoto. Nominally, they were merely following the orders that had been given to Nagai, but, in fact, they hoped somehow to reinstate themselves in the affections of the Court. Daily conferences were held en route to decide what policy the han should take; the conferences may have been influenced by reports of a new surge in the activities of *rōnin* loyalists that had broken out since the departure of Hisamitsu and the Imperial messenger for Edo. The daimyo of Chōshū and his party arrived in Kyoto on 2/7/62. But, as the han officials had not been able to agree on a new policy, the daimyo, pleading illness, put off his appearance at the Court.

Heated conferences continued; almost a week elapsed without a decision. The officials seemed to fall into two groups. Kaneshige Jōzō and Yamada Uemon, members of the Sufu clique, upheld a moderate position, whereas Nakamura Kyūrō, a member of the Sufu clique, Kido Kōin, and others took a more extreme position. The conferences were also attended by four Elders: Mōri Chikuzen, Mōri Ise, Masuda Danjō, and Ura Yukie, and by Sufu himself. The ultimate decision probably rested with Sufu, subject of course to the approval of the Elders and the daimyo.

Those favoring the *sonnō jōi* policy argued that from the very beginning the Court had desired to expel the barbarians (*jōi*) but that this had been thwarted by pressure from the Bakufu. Therefore, Nagai's program of mediation was not in accord with the wishes of the Court, and Chōshū had been entrusted with the mediation only because it had come forth in advance of other han. It was plain that the Emperor favored a *jōi* policy; therefore, they should strive for its acceptance by the Bakufu as the basis for national unity. To this, the moderates rejoined that Chōshū was indeed a great han, and it was proper for it to attend to the affairs of the nation. It was fitting to obey the Imperial will; but, in order to obey it, it must first be ascertained. A serious policy decision must not be made on the spur of the moment. Moreover, it would be most difficult to break the treaties and expel the barbarians, because of the military weakness of the Bakufu and the various han. Indeed, were it attempted, a situation intolerable to the Emperor might arise.

But, the extremists answered, the national polity had been insulted, and, if a *jōi* policy were not adopted, the Emperor would be humiliated before his ancestors. Therefore, casting aside all questions of profit or loss, success or failure, Chōshū must offer itself to the call of duty. The moderates replied that, if hostilities were begun before adequate defenses were established, then, in case of defeat, the national polity would be irreparably hurt. The extremists countered that hostilities were necessary to create a sense of crisis, which would lead to improved defenses. If fighting did break out, the old habits formed during long years of peace would be swept away, a new military order would rise, and, even if a temporary defeat were suffered, Japan would recover.

The moderates then cautioned that, since this decision was a matter of gravest importance to the han, it would not do for the few officials accompanying the daimyo to make it themselves. The other officials and retainers in the han should be consulted too. The extremists were impatient: the daimyo was in Kyoto, the Court was waiting to see what position Chōshū would take; if they dallied any longer, they would be laughed at by Satsuma and the other han; they should therefore unite now with the several hundred loyalists in Kyoto and carry out the will of the Emperor.[14]

The final decision, which was taken on 6/7/62, favored the extremists. The 1858 statement of Chōshū's basic policy, as one of "loyalty to the Court, trust to the Bakufu, and filial duty to the ancestors [of the house of Mōri]," was maintained, but the balance between the three obligations was now lost. In the words of the han Elder, Ura Yukie: "Our basic policy remains one of 'loyalty to the Court, trust to the Bakufu, and filial duty to the ancestors [of the house of Mōri].' But, if a situation should arise in which our loyalty [to the Court] were threatened, then it might [be necessary] to neglect our trust [to the Bakufu]."[15] In the light of previous events, it seems virtually certain that this interpretation of Chōshū's "basic policy," or better still, reinterpretation, was due to the exigencies of

[14] Suematsu, *Bōchō kaiten shi,* III, 302–328.

[15] Seki Junya, *Hansei kaikaku to Meiji ishin: han taisei no kiki to nōmin bunka* (Tokyo, 1956), p. 128.

the situation rather than to the strength of those in the han councils who favored the *sonnō jōi* cause. The *sonnō jōi* program had been adopted by a traditional bureaucratic clique, not because they favored it, not because of the strength of the *sonnō jōi* faction in Chōshū, but solely because it was the only program remaining with which they could compete successfully against the mediation of Satsuma. Ironically, the position favored by the Sufu moderates differed little from that of the Satsuma leaders. Yet, had the han adopted this as its new policy, it could have hoped for nothing more in national politics than a poor second place, supporting Satsuma yet excluded from its councils. By adopting the position of the loyalists, Chōshū could reasonably hope to replace Satsuma in the affections of the Court.

We have suggested that this abrupt change or even reversal of han policy can be viewed as the adoption of the program of the loyalist clique by the reformist clique of Sufu. That Sufu, who in 1858 had attacked Yoshida Shōin as a noisy reckless student, would only four years later adopt his views, is a telling commentary on the rapid development of the pro-Emperor movement within Japan. But it must be stressed that, in terms of han politics, the adoption of the extremist policy did not mean that the Sufu group had merged with the *sonnō jōi* group. It was rather a coalition, each group complementing the other. The Sufu group controlled the han bureaucracy, whereas the extremists, with the exception of Kido and one or two sympathizers among the Sufu group, enjoyed little power. By utilizing the power of the han, the loyalists hoped to further their cause. The Sufu government, on the other hand, needed the services of the Chōshū loyalists in order to contact the *rōnin* loyalists of other han and the extremist party at the Court.

This sudden reversal of policy also provoked a reaction in the han. It will be remembered that the 6/7/62 policy decision was made by a handful of high officials who were thoroughly acquainted with the newest developments in Edo and Kyoto. The same changes were only dimly perceived by the officials in the han and vaguely whispered among the body of retainers, many of whom were increasingly alarmed by the progressive radicalization of han policy. This was particularly true of the conservatives, who formed the backing of the Tsuboi-Mukunashi clique. This group felt, perhaps justly, that their opposition, the Sufu group, which had failed in its first attempt at national mediation, was now rashly committing the destinies of the han to the dangerous policies of the extremists solely to save its own fortunes. As a result, the Tsuboi-Mukunashi clique, which had been silent since 1858, began to group with other conservatives in the han in vocal opposition to the new policy.

They maintained that, as Chōshū's basic policy was one of loyalty to the Court and trust to the Bakufu, to support the *jōi* policy of the loyalists would harm Chōshū's attitude of trust to the Bakufu. They also charged that the Sufu faction, in its eagerness to obtain "the glory of achievement," had changed han policy without regard for "the tradition of the han." They further contended—what to an observer one hundred years after the event seems an obvious truth—that Nagai was being sacrificed due to changes in the general climate of opinion for which he was not responsible, and asked that his punishment be postponed. The Resident Elder in Chōshū was unable to refute these charges and sent to Kyoto for advice. The Chōshū party in Kyoto, replying that first and last the Imperial edicts should be obeyed, ordered in the name of the daimyo that the conservatives be silenced. This was done: members of the Mukunashi clique were ordered into domiciliary confinement, and five minor officials, who had supported the *zokuron* faction, were dismissed from their posts.[16]

Chōshū's Sonnō Jōi Mediation and the Displacement of Satsuma

With fresh confidence, Chōshū moved to implement its new policies. Its first move was to dissociate itself from the earlier policy proposed by Nagai: on 12/7/62 Mōri Chikuzen, an Elder, was sent to the Court to inquire further into the slander incident. The following day Mōri Ise was sent to the Court to reject the mediation of Nagai but, at the same time, to confirm Chōshū as mediator between the Court and the Bakufu. Three days later the Chōshū daimyo was

[16] Suematsu, *Bōchō kaiten shi*, III, 380–393.

summoned before the two Court councillors, Sanjō and Nakayama, and other Court personages. He was informed that the slander incident had been disposed of to the full satisfaction of the Court and was given anew an Imperial edict instructing Chōshū to mediate in cooperation with the Ōhara mission of Satsuma.

Restored to the good graces of the Court, Chōshū began pressing the Court to adopt an even more extreme position, one that would exclude Satsuma and leave Chōshū unchallenged in the field. On 20/7/62 the daimyo of Chōshū expressed to the Court his doubts concerning the Ōhara mission. He pointed out that of the three commissions Ōhara had been entrusted with, namely, the establishment of a five *tairō* system, the attendance of the shogun in Kyoto, and the appointment of Keiki and Keiei, he had accomplished only the last. The daimyo went on to suggest that Chōshū join in this mediation in order to rectify this failing. The Court approved; on 27/7 the daimyo was ordered to return to the Court, where he was given an Imperial order: the heir was to proceed to Edo and the daimyo to remain in Kyoto. Significantly enough, at the time of the Nagai mediation the daimyo had gone to Edo and the heir had remained in Kyoto, but the political importance of the two areas had now been reversed.

The Chōshū heir, Mōri Motonori, was entrusted with an order dated 2/8/62 instructing the Bakufu that all who had died for the Imperial cause from the time of the Ansei Purge to the present were to be reburied with honors, and that those among the living who had suffered for the Imperial cause were to be reinstated. The heir left for Edo the following day accompanied by Sufu, Kaneshige, Yamada, and Kido. Before leaving, Yamada proposed to the daimyo that the Chōshū heir should meet with Hisamitsu soon after his arrival in Edo, since the success of the Chōshū mission depended on the cooperation of Satsuma. The daimyo agreed to this and bade his son to stop outside of Edo the night before his arrival so that the next morning, immediately on entering the city, he could call on Hisamitsu and, explaining the purpose of Chōshū's mediation, ask for his cooperation. It is highly probable that the moderate Yamada, in proposing a *rapprochement* with Satsuma, hoped to temper the extremism of Chōshū's new policy. For, if a coalition with Satsuma were realized, the restraining influence of Hisamitsu coupled with moderate opinion within the Sufu government might have reversed the decision already taken in favor of an extremist policy.[17]

But there were already too many obstacles to such a coalition. Both han wished to dominate the national politics of the day. Chōshū's Nagai mediation was opposed by Satsuma's mediation based on Bakufu reform. When Chōshū proposed that the shogun go to Kyoto, Satsuma had replied that it was too early for the shogun to visit the Court and that Keiei should be delegated instead. When Satsuma argued that the issue of whether to open or close the country could be settled after the development of adequate military preparations, Chōshū insisted that the abrogation of the treaties and the expulsion of the foreigners were the first necessary steps toward formulating a national policy. Point for point, the two han countered and competed with one another.

Moreover, the order entrusted to the Chōshū heir included among "those who had given their lives for the Imperial cause," "those who have fallen recently at Fushimi."[18] These were the Satsuma loyalists who had been cut down for directly disobeying the orders of Hisamitsu, and Hisamitsu, not unnaturally, resented this. It appears that the inclusion of this reference was either an inadvertent slip or had been requested by Chōshū, since the two Court councillors who obtained the order were closer to the position of Satsuma than to that of Chōshū. It is possible, however, that it was included deliberately by loyalist nobles at the Court in order to prevent the formation of a coalition between the two han; the position of such nobles would obviously have been more enhanced by the mediation of Chōshū alone than by a joint mediation of the two han.

As a final blow to the prospects of a coalition, Kido, a known extremist and the least propitious of agents, was sent ahead to contact Hisamitsu. The feeble attempts to form a coalition were without issue: when the Chōshū heir called at the Satsuma Edo compound, he was put off with an excuse, and further attempts were abandoned.

[17] *Ibid.*, pp. 394–430.
[18] *Ishin shi*, III, 130.

The edict from the Court was presented to the Bakufu, and the Chōshū heir remained in Edo to further the cause of Chōshū. Meanwhile the most important developments were taking place in Kyoto.

It would be very difficult to characterize the Court at this time. Perhaps the most common attribute of the Court nobles was their dissociation from any real responsibility in the affairs of the nation—although, of course, a few were actively intriguing in national politics. Some advocated a safe if unexciting support of the Bakufu, and others felt the Bakufu should be replaced by Imperial government, but the majority held views somewhere between the two extremes. Not only did the range of political opinion vary, but, divorced as they were from the realities of responsibility, opinions within the Court were subject to kaleidoscopic change. Moreover, since the Court had little power when isolated, it was dependent on the strength it could derive from those han which chose, largely for their own ends, to enlist in its cause. Consequently, the official position of the Court was usually close to that of the han giving it the strongest support. Having adopted the loyalist position, Chōshū now tried to bring the Court in line; it contacted those at the Court who favored the loyalist position, and by giving these men its support Chōshū strengthened their voices in the councils of the Court and so hoped to change the official Court policy.

Chōshū's first overt move took place on 2/8/62. Nakamura Kyūrō, an official of the Sufu clique sympathetic to the loyalists, presented a memorial to the Court councillor Nakayama questioning the "true position of the Court" on six points; the most important point concerned the Court's views on the treaties. Was it the opinion of the Court that these should be speedily abrogated, even if such an act should lead to war? On 7/8 Nakamura again visited Nakayama and was told that the Emperor wished that both treaties be revoked. This opinion unleashed a flood of activity by the Chōshū loyalists.[19]

Kusaka Genzui, the leader of the Chōshū loyalists, sent a memorial to the daimyo demanding that "the guilt [of the Bakufu] for having signed the second treaty against the Emperor's wishes be made

[19] Suematsu, *Bōchō kaiten shi,* III, 331–334.

clear."[20] Moreover, "the lack of respect [on the part of the Bakufu] for over two hundred years must be corrected, the veneration of the Emperor be established, and the duties of rulers and subjects rectified."[21] Similar demands were also voiced by Takasugi Shinsaku, in which the entire system of Tokugawa rule was called into question. The loyalists of other han who had been temporarily suppressed by Hisamitsu also rallied under the new and positive leadership of Chōshū, which gradually came to dominate the Kyoto scene. On 27/i8/62 an Imperial edict was sent to Chōshū asking it to mediate between the Court and the Bakufu in order to obtain the Bakufu's acceptance of the *jōi* policy of the Court. The Court was now officially committed to the extremist position of Chōshū. The following day Chōshū's daimyo was visited by the daimyo of Tosa, who, having been shown the edict, agreed to join with Chōshū in its support of the Court; this further strengthened the Chōshū position in Kyoto.

In the meantime Hisamitsu, having successfully completed his mission, returned to Kyoto on 7/i8/62. His return was to have been a triumphant one; he had achieved all that had been hoped for when he set out; the Bakufu was now in the hands of Keiki and Keiei, both of whom were sympathetic to the Court. Arriving in Kyoto, however, Hisamitsu found that his good spirits were unwarranted. The loyalists, whom he had suppressed before leaving for Edo, had not only resumed their activities, but, led by Chōshū, they had gained control of the Court. Hisamitsu also learned to his dismay that during his absence the Court had sent edicts to ten other han asking their help on behalf of the Court.

Two days after his return to Kyoto, Hisamitsu presented a memorial to the Court containing his views: he argued that the Court should not be influenced by the views of lowly men, that there was no point in having two or three differing edicts, and that the Court would do well quietly to watch, for a while at least, the actions of the Bakufu. On 21/i8 Hisamitsu visited the Konoe residence where he further expressed his earlier suggestion that the Court forbid direct

[20] *Ishin shi,* III, 265.
[21] *Ibid.,* p. 266.

contacts between its nobles and persons of lowly rank. Now that the tendency of the Bakufu to compromise had been demonstrated, he continued, various daimyo come running to gain fame as "mediators," whereas, earlier in the year, not one had come forth. It is not in the Court's interest to accept the services of these men without discerning whether they were good or bad, their opinion true or false. Any attempt to expel the barbarians without real power will only invite the fate that had overtaken the Ch'ing dynasty in China. Hisamitsu forwarded his document to the Emperor, but nothing came of it. Chōshū, at the head of the loyalists, steadily tightened its grip on the Court, and, on 28/i8, a gloomy and disgruntled Hisamitsu left Kyoto to return to Satsuma far to the south. The field was now clear for the further development of Chōshū's *jōi* campaign.[22]

On 16/9/62 the loyalists of Chōshū and Tosa met together with several loyalist Satsuma *rōnin* and decided that they should request the Court to send to the Bakufu an Imperial edict ordering the enforcement of the Court's *jōi* policy. Two days later their memorial was sent to the Court, and on 21/9 the Court, agreeing with the loyalists' memorial, appointed two new Imperial messengers: Sanjō Sanetomi and Anenokōji Kintomo. These two messengers can be taken, perhaps, to epitomize the situation in Kyoto at the time. They had for some time been in contact with the loyalist party of samurai; they represented the extremist fringe among the Court nobles; and both were young, twenty-six and twenty-four. The more moderate nobles, such an Sanjō Saneai and Nakayama, who had hitherto held key positions within the Court, vigorously opposed their appointment; on 22/9 Sanjō wrote to Nakayama, asking rhetorically, "How is it that we have such extreme messengers. This is truly a worrisome thing."[23]

Heartened by the Court's approval of a *jōi* mission to the Bakufu, the loyalists next petitioned the Court on 5/10 for the creation of Imperial troops to protect the Court from the barbarians. Step by step the *sonnō jōi* adherents were moving to strengthen the Court while detracting from the power and prestige of the Bakufu. The

[22] *Ibid.*, pp. 272–275.
[23] *Ibid.*, p. 278.

extremist nobles, now ascendant within the Court, responded by appending the new proposal to the *jōi* order to be carried to Edo. On 27/10/62 the two Imperial messengers, accompanied by the daimyo of Tosa, arrived in Edo.

The Bakufu was in a quandary: it had to uphold the treaties in order to avert war with the foreign powers; it also had to expel the foreigners to prevent internal strife. Its response betrayed the weakness that had come upon it since the death of Ii: it promised to carry out the *jōi* policy desired by the Court, while well aware that this was impossible. Before the arrival of the Imperial messengers, the Bakufu had been divided by the opposing views of Keiei and Keiki. Keiki favored a *kaikoku* policy whereas Keiei advocated expelling the foreigners. Keiei had resigned once in support of his views, and, at the time the Imperial messengers arrived, he was on the verge of agreeing with the *kaikoku* policy of Keiki. The Imperial edict led him to reverse his views; he now argued that the Bakufu must carry out the orders of the Court, and, if unable to do so, it must be ready to return the political power of the country to the Court. On 2/11 the Bakufu declared itself willing to carry out the orders of the Court. Thereupon Keiki resigned, saying that it was the duty of the hegemon to state clearly that he could not do the impossible; to accept the Imperial order, knowing that it could not be carried out, would be to deceive the Court. He was, however, persuaded to remain and receive the Imperial messengers in the faint hope, as some Bakufu officials put it, that the Bakufu envoy could afterwards be sent to the Court to effect a change of its *jōi* policy.

On 27/11 the Imperial edict was officially delivered to the Bakufu. Subsequently, conferences were held in which the differences in the Court's and the Bakufu's positions as well as the discords within the Bakufu noted earlier came to light. Sanjō Sanetomi demanded that the Bakufu announce its decision to carry out the *jōi* policy; Keiki rejoined that it should not be announced until after the shogun's visit to Kyoto in the spring of the following year. The Imperial messenger next queried whether the Bakufu, having agreed to implement the Imperial order, had any definite plans for expelling the foreigners. Keiki replied that a policy could only be called a policy when it was kept secret and refused to divulge the Bakufu's plans. The Court's request for Imperial troops was also refused.[24]

Meanwhile Konoe and other moderate nobles close to Satsuma wrote to Hisamitsu, requesting him to return to Kyoto to aid in the regulation of affairs. Hisamitsu answered that the situation in Kyoto was not propitious to his idea of mediation. Some word of this plan of the moderates seems to have reached the Chōshū party, for on 7/12 the Imperial messengers suddenly left Edo for Kyoto. The daimyo of Tosa and the Chōshū heir also returned to Kyoto at this time. Shortly before leaving for Kyoto, the Chōshū heir called on Keiei and gave him a message expressing dissatisfaction at the Bakufu's rejection of the demand for Imperial troops. The message further asserted that this rejection strongly suggested that the Bakufu was still not free from its old habits of temporizing.[25] The extremist mission to Edo was not an unqualified success, but, in achieving its main objective, it did goad the Bakufu one step deeper into the dilemma from which it would never really escape.

Kyoto in 1863: the Chōshū Sonnō Jōi Movement[26]

During the first months of 1863, two parties sedulously plotted, maneuvered, and wrestled for power in Japan. The dominant *sonnō jōi* group was composed of extremist nobles, *rōnin,* Chōshū, and for a brief interval, Tosa. The opposing moderate party, which was also called the *kōbugattai* party, since it advocated a balance in the power between the Court and the Bakufu, was made up of high-ranking nobles, Tokugawa Keiki, Matsudaira Keiei, Shimazu Hisamitsu, and several other daimyo. It is not our purpose here to describe all the machinations of the various factions in the two parties at this time; only the major events and those most directly concerned with Chōshū will be outlined.

The struggle between the two parties was not confined to the

[24] *Ibid.,* pp. 302–304.

[25] *Ibid.,* pp. 305–306.

[26] This section is based on the following: Suematsu, *Bōchō kaiten shi,* IV, 35–203; *Ishin shi,* III, 353–394; Tōyama Shigeki, *Meiji ishin* (Tokyo, 1951), pp. 81–115.

central arena in Kyoto. In fact, the configuration of forces in Kyoto itself was often determined by the outcome of many smaller struggles. The case of Tosa can be given as an example of an outside struggle, vitally affecting the strength of the Chōshū party. The new situation brought about by Perry, the treaties, and the evolution of national politics in Japan had given rise to three cliques among the politically active samurai in Tosa. The dominant clique in the Tosa government was the reform clique headed by Yoshida Tōyō, who advocated reform within the han, support of the Bakufu, and the opening of Japan. This was opposed on the one hand by a clique of high-ranking samurai, who opposed internal reform, supported the Bakufu, and favored *jōi*. The third clique was the *sonnō jōi* loyalist group led by Takechi Zuizan. Many of Takechi's group wanted to leave Tosa to join with other like-minded samurai in Kyoto. Takechi, wisely realizing how easily any uprising of unattached samurai could be suppressed, restrained them and concentrated instead on gaining control of the han, which could then be used as a base for a loyalist campaign. He first presented his ideas to Yoshida. Yoshida replied that Tosa differed from Chōshū or Satsuma in that it had a special relation to the Bakufu and, therefore, could not plot against it. Takechi next tried to form an alliance with the clique of high-ranking samurai with the intention of shaking the power of Yoshida. This also failed. Finally, on 8/4/62 Yoshida was assassinated by a member of the *sonnō jōi* clique, and soon after, as Takechi had foreseen, the high-ranking conservatives gained control of the han government.

The indomitable Takechi now bent his efforts to involve the han in national politics. Specifically, he wished the young, seventeen-year-old daimyo to go to Kyoto where he could be dominated and controlled by the loyalists. In Tosa, Takechi urged the han government to enter national politics; in Kyoto, Tosa loyalists urged the Court to issue an order for mediation to the daimyo. The Imperial order went out to Tosa, and, after some controversy, the daimyo left the han, arriving in Kyoto on 25/8/62. Three days later Tosa joined with Chōshū to help advance the Imperial cause.

Takechi's success, however, was short-lived. In the spring of

1862, Yamanouchi Yōdō, the former daimyo, was released from the domiciliary confinement into which he had been ordered by Ii. Angered by the assassination of the moderate Yoshida, whose ideas he had favored, he set about planning the downfall of Takechi. An order was sent from the Court, presumably by a moderate official, ordering Yōdō to replace the youthful daimyo as Tosa's representative in Kyoto; he arrived on 23/1/63 and immediately began suppressing the Tosa loyalists. Many were sent back to the han under arrest, and the Tosa-Chōshū coalition was undone. Takechi was too powerful to arrest, because of his connections at the Court, but a few months later, when Chōshū lost its footing in Kyoto he was sent back to Tosa, where he was ordered to disembowel himself.[27]

Chōshū now stood alone against the *kōbugattai* daimyo. One expedient by which it attempted to tighten its hold on Kyoto was the creation of new offices at the Court. Even before the defection of Tosa, on 9/12/62 the Court had created the position of "Court Advisors on National Affairs" (*kokuji goyōgakari*). Twenty-nine nobles, high-ranking ones as well as some lower-ranking extremist nobles, were given this appointment; they formed an advisory council to the Court on matters of policy in national affairs. In a sense this new office was merely the institutional reflection of the Court's new voice in national affairs. On the other hand, it might not be an exaggeration to call this the first embryonic development in the direction of what was eventually to become the Meiji government.

In 13/2/63, two months after the change within Tosa, the loyalists reinforced their control of the Court by the further creation of two new offices: Officials for Participation in National Affairs (*kokuji sansei*) and Advisors on National Affairs (*kokuji yoriudo*). Fourteen officials, mostly extremists, were appointed to these positions, which immediately superseded the earlier office. A few nobles sympathetic to the *kōbugattai* position did hold some offices, but they had no power, for the extremist nobles, who were in control of the inner circle of the Court and the person of the Emperor, only called on those sympathetic to themselves.

[27] Suematsu, *Bōchō kaiten shi*, III, 341–342; *Ishin shi*, III, 212–241.

The loyalists' control of the Court was pushed forward one more step on 20/2/63 when a Court order was issued permitting commoners to petition the Court and enabling samurai to participate in some Court councils. Possibly this order was promulgated to bolster the loyalists against the day the shogun would arrive in Kyoto. It effected a chain of authority extending from Emperor to extremist nobles, down to extremist samurai that completely by-passed, and therefore made obsolescent, the traditional apparatus of Court government.

By sundry methods the *kōbugattai* forces tried to counter the burgeoning strength of the Chōshū-extremist noble alliance. Earlier in 1862, Hisamitsu had sent a message to the Bakufu suggesting that its difficulties would soon be solved if he were permitted to join its councils. The Bakufu told him to go to Kyoto by 20/1/63 to consult with Keiei, Yamanouchi Yōdō, and the moderate nobles on a unified *kōbugattai* program. Hisamitsu answered that the time had not come for effective action in Kyoto. Matsudaira Katamori of Aizu arrived in Kyoto on 24/12/62, Keiki arrived on 5/1/63, and Keiei on 4/2/63, but they were not able to accomplish anything.

On 19/2/63, Keiei, Keiki, Katamori, and Yōdō conferred and decided to memorialize the Court, asking the Court either to confirm the political authority of its hegemon, the shogun, or to take over itself the task of government. On 4/3/63 the shogun arrived, for the first time in two centuries, in Kyoto. Three days later the Court officially confirmed the stewardship of the shogun. This gesture merely symbolized the new power of the Court; other than this nothing was altered, and the loyalists continued to agitate for an increasingly radical national policy.

On 14/3/63 Hisamitsu finally arrived in Kyoto, where he told the other leaders of the moderate faction what should be done: in short, that the ill-advised *jōi* decision should be reversed, the Court should disentangle itself from the violent extremist *rōnin,* the extremist nobles should be removed and the extremist samurai punished, the majority of the daimyo should be sent back to their han, and the new Court officials should be abolished. The moderate nobles rather halfheartedly agreed to his views. Hisamitsu felt, and rightly, that the moderate nobles feared the extremists in spite of the shogun's presence in Kyoto. Unable to foresee any effective action against the extremists, Hisamitsu left for Satsuma only four days after his arrival. Keiei, Yōdō, and other *kōbugattai* leaders also left soon afterward, leaving the extremist forces with an almost undisputed field. On seeking for reasons why the *kōbugattai* party could do nothing against the extremists, it appears that they were not yet willing to resort to force, partly because of the strength of Chōshū, and partly because the extremist strength of Chōshū was backed by Imperial prestige.

One other aspect of life in Kyoto in the early part of 1863 that must be mentioned in order to explain the unwillingness of the moderate nobles to back Hisamitsu was the campaign of threats, terrorism, and assassination carried on by the extremist samurai. This began shortly after Hisamitsu left Kyoto with the Ōhara mission and rose to a crescendo during the period of Chōshū's ascendancy. The Kyoto *shoshidai* and the Kyoto *bugyō,* the usual police powers of the city, were so cowed that the loyalists, as well as certain other lawless elements, could do virtually as they pleased. Toward the end of 1862 the situation had so deteriorated that even members of the Chōshū loyalists were constrained to post notices on Sanjō Bridge in the heart of the city: "Seizing money or trying to avoid the repayment of debts are not the actions of virtuous [i.e., loyalist] samurai, and anyone caught in such actions will be arrested and punished." [28]

As one can imagine, such notices had little effect, especially where the loyalists were concerned; their campaign became increasingly violent through the early months of 1863. On 14/1 Hayashi Suke, an official of Kyoto, was killed in his house. Eight days later, Ikeuchi Daigaku, a Confucian scholar, was killed on his return from the residence of Yōdō of Tosa. Ikeuchi's ears were then severed from his head and tossed into the compound of Sanjō Saneai. Attached to them was a note to Sanjō, to the effect that Ikeuchi had once been virtuous but had since joined the ranks of the evil pro-Bakufu officials. The implied threat, of course, was that the same thing might happen to the moderate Sanjō if he did not support the *sonnō jōi*

[28] *Ishin shi,* III, 252.

extremists. On another occasion the hands of another assassinated moderate were sent—one each it appears—to the residences of two other moderate nobles, one of whom was Iwakura Tomomi. On 7/2 the head of still another moderate was placed outside the house of Yōdō of Tosa; to it was attached a list of demands headed by a statement that this was "Heaven's revenge." [29] Apparently, the head had originally been intended for Matsudaira Keiei, but, upon finding that his residence was too carefully guarded, the loyalist deputies of Heaven had picked Yōdō as second choice. On 22/2 six loyalist samurai put on display in the city the heads of three wooden statues of early Ashikaga shoguns which they had appropriated from a temple; to the heads was affixed a placard: "The traitors Ashikaga Takauji, Yoshiakira, and Yoshimitsu. . . . Today there exist traitors more villainous than these evil rebels. . . . If the evil practices existing since the time of Kamakura are not abolished . . . the patriots of the country will rise up and avenge them."[30] The prominent *kōbugattai* military figures were carefully protected and could express their moderate views with impunity, but like-minded Court nobels were more vulnerable to intimidation.

Would this justify, then, the often expressed view that in early 1863 the power in Kyoto was in the hands of the loyalist samurai? They could initiate actions within the Court and could practically speak in the name of the Emperor. That they held a real measure of power is certain. Many of the daimyo in Kyoto at this time were mere figureheads: the Chōshū daimyo, weak and easily swayed by those about him, was no exception. Yet one must also bear in mind that even during the period when extremist nobles and samurai dominated the Court, the loyalist samurai were extremely vulnerable to actions by their own han government. The swift suppression of the Tosa loyalists by Yōdō in Kyoto is a ready example. And to no less an extent, the Chōshū loyalists were dependent on the Chōshū government for their strength and sustenance. Had the Chōshū government withdrawn its support and left Kyoto, a political vacuum would have been left in which the isolated *rōnin* and a helpless Court

[29] *Ibid.*, pp. 358–359.
[30] *Ibid.*, p. 363.

could not have long maintained their extreme *sonnō jōi* position—a vacuum which the Bakufu or *kōbugattai* forces would soon have filled. It was obviously not in the interests of the Chōshū government to withdraw; it had everything to gain by staying. In speaking of the Chōshū government too, it must be remembered that it was still almost entirely controlled by the Sufu bureaucratic clique. Thus, even when the loyalist movement had all the appearances of an autonomous force, the power of Chōshū and the authority of the Imperial name stood behind it. Had any one of these three elements been missing, the movement would have foundered.

Concomitant with the rise in the power of the Court, changes were taking place in the doctrine of the loyalists. Where the early *sonnō jōi* thinkers such as Nariaki had advocated veneration of the Emperor in order to unite feeling within Japan against the foreigners, the situation was now reversed. Now the loyalists stressed anti-foreignism in order to advance the position of the Emperor and, one should add, to undermine the position of the Bakufu, for this was clearly the intention of the loyalist movement at the time. It is hard to say what ultimate end, if any, the loyalists may have envisaged. The trend of their demands in 1863, beginning with the establishment of Imperial troops and leading up to the proposal that the Emperor himself lead the troops of the han against the foreigners, would suggest that they ultimately hoped for the re-establishment of the power of the Court assisted by a loyal Chōshū, but this is at best hazy, since they totally neglected to relate their abstract political goals to the social and political realities of the Japan of their day. Tōyama Shigeki has described their goal as a "dreamed infantile restoration of power to the Court." [31] He goes on to say that it was totally devoid of any constructive elements. In spite of this extremely negative character, its immediate successes were immense. Itō Hirobumi spoke of it afterward: "The *jōi* argument broke out like bees swarming from a broken beehive. . . . If one speaks logically of the things [that happened then], they are impossible to understand . . . but emotionally, it had to be that way."[32] The very success of the

[31] Tōyama, *Meiji ishin*, p. 95.
[32] Suematsu, *Bōchō kaiten shi*, III, 429.

Chōshū *sonnō jōi* movement, however, brought it face to face with the dilemma which earlier had reduced the Bakufu to a state of paralysis. It now had to initiate action against the foreigners; the actions which it took soon led to the downfall of the Chōshū movement.

The Deadline for Jōi and Chōshū's First Shimonoseki Attack

For some time the extremist faction in Kyoto had been demanding that the Bakufu set a definite date on which the expulsion of the foreigners would begin. On 8/2/63, for example, a group of twelve extremist nobles sent joint petitions to the Court and to representatives of the Bakufu asking for the rapid commencement of the *jōi* policy. The Bakufu party in Kyoto, consisting of Keiki, Keiei, Katamori of Aizu, and Yamanouchi Yōdō, which had originally hoped to wean the Court from its emphasis on *jōi,* was again put on the defensive; on 14/2/63 they sent a letter to the Court stating that the expulsion of the foreigners would begin during the middle of the fourth month. Two months later, on 16/4, the daimyo of Chōshū sent a memorial to the Court asking that a date be fixed so that Chōshū could prepare for action. Four days later, the Bakufu informed the Court that the expulsion of the barbarians would begin on 10/5/63, and on the following day the representatives of the various han were informed of this decision.

That they should set such a date, while fully aware that the foreigners could not in fact be expelled, is indicative of the total paralysis of will with which Keiki and Keiei were afflicted. Originally given power by Hisamitsu because of their pro-Court sympathies, they were now benumbed by the force of the demands made by the new extremist Court, and hence, they agreed to impossibilities merely to obtain a few days' reprieve. (It is also not inconceivable that they were waiting for Chōshū to precipitate a crisis which would deflect the anger of the foreigners against that obstreperous han.)

The Bakufu officials in Edo informed the foreign representatives that the Court had issued an order for their expulsion, without, of course, divulging that the Bakufu itself had set the date. The foreign representatives said that this was regrettable for Japan and that it would not be tolerated. The Bakufu then sat back to witness the collision of what seemed to it an irresistible force and an immovable object.

Chōshū, in the meantime, hastened its preparations for the execution of the *jōi* order. As there were no foreigners in Chōshū, Chōshū directed its attention toward the foreign ships that frequently skirted its shores at the Shimonoseki Straits. One thousand men were put to work building forts on the littoral of the straits; a military specialist on Western learning was invited from Nagasaki to advise the han on these matters. The seat of han government was moved from Hagi, which was vulnerable to foreign attack, to Yamaguchi, nearer to the center of the han. The daimyo returned to the han on 2/4/63. On 21/4 the Chōshū heir, accompanied by loyalist leaders such as Takasugi, Kusaka, Nomura, and others, also returned to the han in order to participate in the first action against the foreigners.

On the appointed day Chōshū alone of all the han in Japan carried out the *jōi* order. On 10/5/63 it shelled a United States merchant ship en route to Shanghai from Yokohama; on 23/5 it attacked a French ship passing through the straits. The foreign response was swift. On 1/6 a United States warship entered the straits; it was bombarded by the Chōshū forts and, in answer, it shelled the Chōshū forts and sank two gunboats which Chōshū had only recently bought at Nagasaki. Four days later two French warships appeared in the straits. The Chōshū forts, not completely recovered from the earlier attacks, held their fire. In spite of this, they were again shelled and French troops subsequently landed and destroyed the forts, burned the ammunition, confiscated most of the weapons, and set fire to the dwellings in the vicinity. A lively if somewhat confused account of the attack is given in the following letter:

On the first day of the sixth month, a single United States ship destroyed two gunboats at Hagi; on the fifth day it destroyed all the forts both at Hagi and Chōfu [a section of the Shimonoseki coast] and half of the Hagi castle. Seven or eight hundred persons were killed, thus throwing Shimonoseki and Hagi into utter disorder. . . . Everyone was making preparations to flee. I have heard that even peasants and merchants gnashed their teeth murmuring, "Are the samurai such good-for-nothings?" [33]

These attacks did not convince the han of the infeasibility of their *jōi* program; this was to come as a consequence of the more massive action of the four-nation fleet in 1864. They did, however, convince the han of the absolute necessity for a thorough military reform on Western lines.

The han called on Takasugi Shinsaku to carry out this reform. At the time Takasugi was in domiciliary confinement for having gone to Kyoto without the permission of the authorities. His appointment was of singular significance: it was the first time that a member of the *sonnō jōi* clique was given official position for his specialized knowledge of some aspect of Western learning. This combination of extremist political views and Western learning, which gradually gained prominence, was to characterize the most important officials of the early Meiji government.

Two days after his appointment, just as the disquieting news of the second Shimonoseki defeat reached Yamaguchi, Takasugi proposed a mixed militia of samurai and peasants which would be formed separately from the regular army units. Since these units were subsequently to play a crucial role in Chōshū history, we will digress here to consider in detail the rise of opinion favoring their formation.

Before the arrival of Perry, Murata Seifū had written that the possession of arms by the peasantry must be prohibited since it would "result in the loss of distinction between upper and lower classes." [34] By 1854, however, his view had changed: "Even peasants and merchants . . . should be ordered to engage in rifle practice, and it would also do no harm to priests and bonzes." [35] In 1860 Sufu, Murata's successor as the head of the Chōshū reformist clique, wrote: "Though it is said that a military system has been achieved, it is not enough that only the high-ranking samurai with incomes of five hundred

[33] Tanaka Akira, "Bakumatsuki Chōshū han ni okeru hansei kaikaku to kaikaku-ha dōmei," unpublished thesis submitted in 1956 to the Bungakubu Kenkyūka of Tokyo Kyōiku Daigaku, pp. 416–417.

[34] Tanaka Akira, "Chōshū han kaikaku-ha no kiban: shotai no bunseki o tōshite mitaru," *Shichō,* 51:10 (Mar, 1954).

[35] *Ibid.*

koku are trained. From the lower samurai down to the *ashigaru* and peasants, rifle troops must be organized in every part of the han."[36] His opinions were embodied in official han policy with the promulgation in 12/60 of the following order: "Selecting from among the villagers, rifle units are to be trained and peasant troops to be established . . . for use, should an emergency arise."[37] But, until the foreign attacks in 1863, nothing had been done. In spite of the earlier orders, training peasants in the military arts was generally regarded as a preparation for attacks which in fact would never occur. And when in 1863 the unexpected did take place, Takasugi was called on to form the new units.

Some have suggested that the demands for peasant troops were a consequence of the small size of the samurai component within the total population. It must be remembered, however, that this was considerably higher than the national average. Of the total of 47,000, about 12,000 were men capable of rendering military service.[38] This is not too great a number, but it is still a greater number of samurai than was actually used by Chōshū at any time during the Bakumatsu period. Why then, one must ask, did the han decide to use peasant troops rather than to utilize fully the talents of its hereditary military class? The answer is that of the 12,000 samurai only a very small number were available for the sort of military units which the Sufu government desired to establish. The Chōshū *shi* were extremely conscious of their ranks, their prerogatives of command, and of their own archaic military skills. The Chōshū *baishin* (rear vassals), comprising about one-half of the 12,000, lived for the most part on the country fiefs of their samurai lords, under whose command they would fight in time of war. The *sotsu,* who, in a sense, fell in between the two groups, were also jealous of their minutely graded ranks and distinctions, and were loath to adapt themselves to new formations and disciplines. The rifle, for example, had been symbolic of the *ashigaru,* one of the lowest stratum of samurai. To participate in one of the new rifle units formed by Sufu would be to identify

[36] *Ibid.*
[37] *Ibid.*
[38] Umetani Noboru, "Meiji ishin shi ni okeru kiheitai no mondai," *Jimbun gakuhō,* 3:44–45 (Mar., 1953).

oneself with a low status group. Moreover, most samurai, influenced perhaps by a tradition of feudal romances, still looked on war as an opportunity for individual heroism and personal glory. The idea of drill and centrally regulated movements of troops they found repugnant. The peasants, on the other hand, were aware more of the prestige of arms than of gradations among different types of arms; they had no prejudices against military innovations. Thus, feudal commitments in thought and organization, rather than numerical limitations in the size of the class, had created the need for additional troop units. The leaders of Chōshū were unable (and, as we shall see later, unwilling) to change the social structure of the han. The limited use of peasants enabled them to form a new type of military without disturbing the traditional society.

The formation of peasant soldiers in local areas of the han enabled the han to utilize as a training cadre the 1299 samurai who were living outside of the castle town by Bakumatsu times due to straitened personal finances.[39] In fact, their utilization was one substantial reason for the formation in so short a time of such a strong irregular force.

Actually, two new types of military units were formed: one, a peasant militia composed entirely of peasants who were trained during the hours in which they were not engaged in agriculture; the other, mixed peasant-samurai forces that engaged in full-time military training and served as auxiliaries to the regular han army.[40] It was the latter type of unit that soon became the center of extremist agitation in the han. With its formation, the loyalist leaders, who hitherto had been completely dependent on the dominant Sufu clique, gained an organized autonomous power, a force within which their ideas could be effectively disseminated and upon which they could rely to maintain their extremist views, even if the tide of opinion in the han should turn against them.

The reverberations of Chōshū's *jōi* action were felt immediately

[39] Kimura Motoi, "Hagi han zaichi kashin dan ni tsuite: kakyū bushi shiron no ichi mondai," *Shigaku zasshi,* 62:37 (Aug., 1953).

[40] This very significant distinction between the two types of militia was first made by Seki Junya in "Bakumatsu ni okeru hansei kaikaku (Chōshū han): Meiji ishin seiritsuki no kiso kōzō" (II), *Yamaguchi keizaigaku zasshi,* 6:74–75 (May, 1955).

in both Edo and Kyoto. The Court praised Chōshū for its fortitude and enjoined other han to follow its example. The Bakufu was promptly beset by the foreign ministers demanding reparations for damages and guarantees for future safe passage through the Shimonoseki Straits. The Bakufu received renewed demands from the Court for the expulsion of the foreigners.

Since the foreigners were a more immediate threat than the Court, the Bakufu sent a note to Chōshū forbidding it to attack foreign ships. The message was rejected. Thereupon, the Bakufu sent an official messenger to Chōshū on one of the ships it had purchased from abroad. When the vessel appeared off the shore of Chōshū, it was fired on as a foreign vessel, although the Chōshū samurai were well aware that it was a Bakufu ship. It was then boarded by the Kiheitai, the earliest of the new auxiliary military units formed by Takasugi, who seized the ship as a replacement for those sunk by the foreigners. The indignant representative of the Bakufu was sent on to talk with officials of the Chōshū government. His message was delivered, but the han would not agree to comply with it. Soon after, an assassination of the messenger was attempted by certain young samurai. Alarmed, the messenger asked the han government to help him escape from Chōshū. He was given a small boat, but the loyalists pursued the vessel and killed the ill-fated messenger. This occurred toward the end of the eighth month in 1863; the incident further aggravated relations between Chōshū and the Bakufu, but it was already overshadowed by events in Kyoto.

The Fall of Chōshū and the End of the Sonnō Jōi *Movement*

The momentum of Chōshū's venture in Kyoto seemed to drive it into an ever more extreme position. During the middle months of 1863, Imperial troops were formed, increasingly acerbic notes were dispatched to the Bakufu, and finally the loyalists began to demand that, since the Bakufu was reluctant, the Emperor himself should lead the troops of the various han to drive the foreigners from the land. On 11/7/63 Masuda Danjō, one of the most influential of the

Chōshū Elders, officially requested that the Emperor lead the military of the country against the barbarians. The Court replied that it would take some time to decide on such an important matter. Masuda answered that Chōshū was already following the *jōi* orders of the Court, and that, if the Emperor did not take the lead, the spirits of the people would languish. The following day the daimyo of Inaba, Bizen, Awa, and Yonezawa were asked their opinions on the matter. They replied that the prestige of the Court would suffer if it acted according to the ideas of a few loyalists; it should first sample a wider range of opinion and then decide.[41]

The loyalists answered moderation with a new wave of terror. The daimyo themselves were surrounded by their vassals and secure from attack, but others who echoed their opinions were waylaid: these assaults were proclaimed the "vengeance of Heaven." When Keiei returned to Kyoto, bills were posted denouncing him as "the enemy of the Court," and on 26/7/63 his lodgings were set on fire.[42] On 13/8 an Imperial edict was issued sanctioning a campaign by the Emperor against the barbarians. This was a critical juncture in the development of the Chōshū *sonnō jōi* movement, for, had such a campaign been realized, it would have taken from the Bakufu its last fragment of authority and at the same time would have assured Chōshū of a dominant position in the inner councils of a resurgent Court.

A vital question that cannot be answered at the present time is that of the relation between the Sufu clique and the extremist clique in the summer of 1863. Was Sufu so beguiled by the dazzling successes of Chōshū in Kyoto that he had become blinded to the potential strength of the *kōbugattai* opposition? Was he himself for a more moderate position but unable to restrain the loyalists of his own han? Did the daimyo and the Elders support the loyalists rather than Sufu because of their recent successes? Or did Sufu's dependence on the loyalists for their contacts among the extremist nobles enable Kusaka and Takasugi rather than Sufu to guide

[41] *Ishin shi,* III, 543–547.
[42] *Ibid.,* pp. 547–548.

Chōshū's Kyoto policy? It has been maintained that the Sufu clique and the extremist group had fused into a single force by the summer of 1863. But, in view of subsequent actions within the han, the two groups were clearly distinct in both action and doctrine.

Whatever the internal structure of the movement, its radicalism soon provoked a reaction from the *kōbugattai* forces led by Satsuma and Aizu. Satsuma in particular was not only concerned with the general danger threatening the country, but also with its own declining prestige. Its earlier enmity toward Chōshū was heightened when on 29/5/63 its samurai were forbidden to enter the Court; Satsuma had also been deprived of its station at one of the nine gates of the Imperial Palace. Aizu was a Tokugawa branch han and a staunch supporter of the Bakufu throughout the Bakumatsu period. When the edict for an Imperial campaign was issued, these two han joined to carry out a *coup d'état* at the Court. Approached by *kōbugattai* nobles in 16/8/63, the Emperor, while hesitating, gave the project his approval. Two days later at dawn, the coup was executed. Satsuma and Aizu troops secured the nine gates at the Palace, admitting no one without an official summons. The *kōbugattai* nobles were called to an Imperial conference, and Imperial edicts were subsequently sent to other daimyo calling for troops to strengthen the defenses of the Palace.

The coup proved a complete success. When news of it reached the Chōshū residence in Kyoto, the Chōshū forces were taken completely by surprise. Two men were hastily ordered to look into the matter, and they found that even the gate entrusted to Chōshū was now held by the Satsuma troops. Four hundred Chōshū troops assembled opposite the Sakaimachi Gate, and these were soon joined by one thousand "Imperial troops," formed of loyalist *rōnin* only a few months earlier. The Satsuma troops drew up in battle formation around the gate. Satsuma held an Imperial edict that had been issued as soon as the success of its coup became apparent. As the two forces faced each other impatiently, an edict was sent to the Chōshū forces warning them not to resort to violence.

At this juncture one of the ranking retainers of the daimyo of Yonezawa was sent by the Court to persuade Masuda, the Chōshū

Elder, to withdraw his troops. In a letter to his son, the retainer described the scene:

The two lines were facing each other, poised with rifles and cannon. . . . Squads were formed, the lines were drawn, both sides wore armor, and holding themselves in readiness, each glared at the other. . . . Chōshū did not falter in the least . . . and did not draw back although faced with a desperate situation; it was truly heroic. Among the Chōshū forces were youths of fifteen or sixteen wearing headbands of white silk and carrying Western rifles; their eyes betrayed no fear of the huge army confronting them. You, my son, should take them as your model. . . . Chōshū, which alone advocated righteousness and loyalty, had set up several tens of cannon, but all pointed downwards so as not to point towards the abode of the Emperor.[43]

All Kyoto was thrown into confusion; fighting was expected at any moment. A contemporary account compared the scene to a scroll from the *Heike monogatari:* young and old, rich and poor, the wives of nobles, and even princesses seized their belongings and ran pell-mell crying through the streets of the city.[44]

Eventually, outmaneuvered, outnumbered, and caught in so unfavorable a situation, the Chōshū forces had to withdraw. An agreement was reached concerning the mode of troop withdrawal whereby the Satsuma troops withdrew within the Palace gate as the Chōshū troops moved off to the east of the city. This was intended to spare Chōshū from too great a loss of face. Chōshū's position, however, was impossible, and its forces had no alternative but to retreat to the han. Iwakura Tomomi spoke of the situation as follows: "Satsuma and Chōshū are like the tiger and the dragon." The Satsuma-Aizu coup "arose solely out of a power struggle between the two han."[45] The change the coup brought to national politics was immense. The seemingly irresistible momentum of the Chōshū *sonnō jōi* movement had been halted. The Bakufu, no longer pressed to expel the foreigners, had been granted a last minute reprieve, and Satsuma and Aizu now dominated the Court through the faction of *kōbugattai* nobles.

[43] *Ibid.*, pp. 572–573.
[44] *Ibid.*, pp. 574–575.
[45] Tōyama, *Meiji ishin*, p. 131.

CHAPTER VII

The Decline of the Chōshū *Sonnō Jōi* Movement, 1863-1864

The Reaction within Chōshū to the Satsuma-Aizu Coup

The Satsuma-Aizu coup signified a new phase in Bakumatsu politics. First, the Chōshū *sonnō jōi* partisans had been driven from Kyoto and in their place the *kōbugattai* movement led by Satsuma was now supreme. Second, the earlier diplomatic instruments of power had been replaced by military force. It is perhaps more than coincidental that just at this time the various political cliques within Chōshū also began to align themselves with different groups in the han military.

News of the Kyoto debacle arrived in Chōshū on 23/8/63, and immediately precipitated a political crisis. The Tsuboi-Mukunashi clique, dormant since 1862, once again began to attack the Sufu government. As they had predicted, Chōshū's fortunes had been reversed and the han brought low. Chōshū's moment of glory had been purchased at too high a price: it was now totally excluded from national politics, out of favor with the Court, and in danger of being punished by the Bakufu for its past rashness. Could the Sufu government answer these charges, they demanded. One unsympathetic account of the conservative reaction to the news of Chōshū's fall began: "In general, since last year when the seat of government was moved to Yamaguchi, Hagi has become a nest of unemployed conservatives. In spite of the fact that these had been put down for a time, they are now becoming more and more critical . . . venting their spleen on their old enemies."[1]

The Tsuboi-Mukunashi clique was joined at this point by the Sempōtai (literally, spearhead troops), an elite unit of the regular han army recruited, according to the *Bōchō kaiten shi,* from the strong young samurai of the upper-ranking houses of Hagi. Actually, a detailed analysis of the class composition of the Sempōtai has yet to be made, but it is clear that they did support the Tsuboi-Mukunashi clique and that they did resent the formation in 1863 of the auxiliary militia by Takasugi Shinsaku. And it was this latter force which now became the military base of the Sufu-loyalist alliance in Chōshū.

The opposition between the Sempōtai and the *shotai* (auxiliary militia) dated back to the time of the first *jōi* attacks on foreign vessels in the spring of 1863. The Sempōtai had at the time been ordered to man certain positions along the Shimonoseki coast. Loyalist samurai just back from Kyoto, in the hope of participating in the first *jōi* attack against the Western powers, asked the han government to station them in attack positions in the same general area. The han government was aware that the loyalists were quite likely to provoke the Sempōtai with taunts of being "spineless samurai,"[2] therefore, hoping to prevent a clash, it sent them instead to a separate area "to reconnoitre the dispositions of the enemy."[3] Three months later on 16/8/63 a few members of the Kiheitai were attacked by the Sempōtai samurai. In retaliation, the entire Kiheitai raided the Sempōtai camp at Kyōhō Temple, and the Sempōtai, in turn, struck back at the Kiheitai camp at Amida Temple. The incidents themselves were inconsequential, but they do indicate the tension existing between the two groups on the eve of the Satsuma coup in Kyoto.

Five days after the news of the Satsuma-Aizu coup reached

[1] Suematsu Kenchō, *Bōchō kaiten shi* (Tokyo, 1921), V, 58.

[2] Tanaka Akira, "Chōshū han kaikakuha no kiban: shotai no bunseki o tōshite mitaru," *Shichō,* 51:21 (Mar., 1954).

[3] Suematsu, *Bōchō kaiten shi,* IV, 238–239.

Chōshū, Mukunashi and other members of the Tsuboi-Mukunashi clique proceeded from Hagi to Yamaguchi where, accusing the Sufu clique of having fallen short of its duty to the han, they demanded a change of officials. The following day this demand was reiterated by several hundred *zokuron* or conservative troops (presumably made up in part, at least, by members of the Sempōtai), who gathered about the residence of the daimyo to voice their opinions. In particular they demanded the removal from the han government of Sufu Masanosuke, Maeda Magoemon (a member of the Sufu clique), and Mōri Norito. Ura Yukie, the han Resident Elder, wrote in his diary at the time: "Word had reached me that they had resolved to kill the three if their requests were not heeded . . . therefore, the three were replaced."[4] In their places, Tamaki Bunnoshin, Sugi Norisuke, and Kamiyama Hōden were appointed to office. It must be noted that these three were moderates, not conservatives, but men who were not identified with Sufu's past Kyoto policy. That Kaneshige Jōzō, a moderate member of the Sufu clique was promoted at the same time, indicates that the coercive power of the conservative troops was not unlimited. And yet the action by the conservative troops had established in the han the same sort of precedent which had been established on the national scene by Satsuma and Aizu: bureaucratic position now depended, to a certain degree at least, on military power.

The conservatives also demanded that the Kiheitai be disbanded in punishment for its attack on the Sempōtai. Sufu opposed this on the grounds that this would only incite fresh violence, since the Kiheitai would be aware that the order was caused by the pressure of the conservatives. The han government accepted his view and merely ordered the Kiheitai to proceed to Aihonomura in Ogōri, a district of Chōshū. Instead, on hearing of the changes in the han government, the Kiheitai marched into Yamaguchi and restored the officials dismissed a few days ago. By 9/9/63 the government was once again in the hands of the Sufu clique. And, in recognition of the Kiheitai action, Kusaka and Takasugi were also given official appointments. Orders were issued prohibiting the samurai of Hagi to enter Yamaguchi, and several conservatives were punished. Tsuboi

[4] *Ibid.*, V, 59–60.

was ordered to commit suicide, Naitō Hyōe and Tsuboi's son were banished, and others were ordered into domiciliary confinement.[5]

One consequence of the rapid shift back and forth within the han government was to accelerate the polarization of political forces in Chōshū. The conservative Sempōtai, which was not representative of the han regular army as a whole, became linked more firmly than ever with the Mukunashi clique. More important than this, however, were the changes taking place in the relation between the Sufu and loyalist cliques. As we mentioned earlier, before being driven from Kyoto, the Sufu clique had depended on the loyalists for their contacts both at the Court and with samurai from other han, while at the same time the loyalists needed the Sufu clique, which represented the power of Chōshū. Driven from Kyoto, the Sufu clique had less reason to favor the loyalists, since the latter's contacts were no longer so important. But the vigorous rise of the Tsuboi-Mukunashi opposition had forced the Sufu group to depend on the *shotai* which were led by loyalists, and, as a result, their need for loyalist support was, if anything, greater than before. The two groups, which were not completely distinct even in the Kyoto period, now began to amalgamate: extremist representation in the han bureaucracy, for example, increased when Takasugi and Kusaka entered the government, joining Kido. Yet the fusion was never complete. Loyalists remained in the leadership of the *shotai,* and the Sufu clique continued to dominate the han government. Not only did the Sufu group hold the chief policy-making positions, but they also controlled the Elders and daimyo in the same sense in which one or another of the factions of nobles controlled the Emperor. One must not overemphasize the degree of political polarization, however, for at the time the han was clearly not divided into two camps: the majority of samurai were still neutral or at best faintly inclined toward one or the other group.

The Relations between Chōshū and the Court after the Coup

After recovering its hold on the han, the Sufu loyalist government immediately turned to the problem of regaining the good offices of the Court. The more responsible opinion in the han, from Sufu

[5] *Ibid.*, pp. 66–70.

down to Kusaka and Takasugi, felt that Chōshū should, if possible, negotiate a return to grace. The first mission sent to Kyoto was headed by an Elder, Negoro Kazusa, who carried a letter from the daimyo asking the Court why Chōshū, which in accord with the Emperor's wishes had led other han in expelling the barbarian, had been dismissed from duty at the gate of the Imperial Palace. Negoro's party arrived in Osaka on 13/9/63 and asked the Kyoto *shoshidai,* the Bakufu official in charge of Kyoto, who had regained some power since the *kōbugattai* coup, for permission to enter the city. This he flatly refused, bidding the Negoro mission to wait in Osaka for further instructions. The mission then received a message from the Court ordering it to relay its communication to the Court, but denying it entry into Kyoto. This was done and on 23/10 Negoro returned to the han reporting that his mission had failed.

As the Court had expressed a "distrust of Chōshū"[6] due to the dispositions of its troops at the time of the *kōbugattai* coup, it was next decided to delegate to Kyoto a second Elder, Ibara Shukei, with a full report on the actions of the Chōshū troops in the coup. This report, in effect, was an extended apologia for all its actions since the arrival of Perry. Copies of the message were also sent to six daimyo who had been sympathetic to Chōshū in the past. Again protracted negotiations took place as to whether Ibara should be permitted to enter Kyoto. The Court was in favor of granting him permission, but since the *kōbugattai* daimyo and the Bakufu officials opposed it, permission was finally refused. Some of the samurai accompanying Ibara suggested that they force their way into the city but prudent counsel prevailed. On 11/12/63 Ibara handed his documents to representatives of the Court, and on 21/1/64 he withdrew from Fushimi to Osaka. Throughout the negotiations, Chōshū had acted on the assumption that the Court would at least be willing to absolve it from blame. This would not restore Chōshū to its former position, but it would open the way for a renewal of contacts with the Court, and put it in a favorable position for action should an opportunity present itself. The *kōbugattai* forces were determined,

[6] *Ishin shi* (Tokyo, 1939), III, 631–632.

however, to prevent Chōshū from regaining a foothold in Kyoto, and all its attempts were rebuffed.

In the meantime in Chōshū, the leaders of the *shotai* were sounding the clarion for war. Satsuma and Aizu, they argued, had seized power by a military coup and were now issuing false edicts in the name of the Emperor. Chōshū, which alone had been true to the Court, was being treated as a culprit and not even allowed to send its representatives to the Court. Therefore, the han must send its troops to Kyoto to stage a countercoup against those who would distort the will of the Emperor.

To understand the role of the *shotai* in Chōshū history after 1863, it is important to realize that they were as much schools of *sonnō jōi* thought as they were military units. What Yoshida Shōin's school had done in 1858, they now did on a larger scale. From the beginning they attracted the most radical, sanguine, and adventuresome spirits in the han. These were knit into tight, strictly disciplined military units, the soldiers of which were instructed in the scriptures of the Chōshū *sonnō jōi* movement, in the writings of Yoshida Shōin, for example, as well as in the use of arms. Their leaders in most cases were former disciples of Shōin or men closely associated with them. Both *shotai* leaders and *shotai* soldiers were isolated from the affairs of the han, from official routine and governmental responsibility. Both were eager to prove their mettle as a fighting unit. As a consequence, their messianic creed turned inward, and they became even more extreme in their views.

In contrast to this *shotai* extremism, Takasugi, Kido, and Kusaka, coming under the influence of the Sufu moderates and tempered by the responsibility of their new bureaucratic posts, had come to agree with Sufu's strategy: that the han must negotiate peacefully until an opportunity for effective military action presented itself. To implement this policy and restrain the *shotai,* the han government sent to their leaders a series of assuasive letters explaining the actions of the han. One such letter, sent on 21/11/63, reasoned:

In this time of trouble for the Imperial nation, we are planning for the expulsion of the foreigners and the revival of Imperial fortunes. In order to render loyal service to the Court, which is our long-cherished wish, the government of the han must be firmly established. There are many "virtuous groups" in the han at this time holding strategic positions. [Among these] discipline must be strictly enforced so that the han will not be weakened. Therefore, the *shotai* should take as their chief aim the desire to excel in loyal service by faithfully adhering to their regulations.[7]

A second device used to restrain the *shotai* was their dispersal throughout the han. The Yūgekitai of five hundred men was sent to Mitajiri, the Kiheitai of three hundred men was sent to Akamanoseki, the Hachimantai of one hundred men was stationed in Yamaguchi, the Shūgitai of fifty men was sent to Ogōri, the Giyūtai of fifty men was sent to Kaminoseki, and other *shotai,* as well as peasant militia units were organized in almost every area of the han. The ostensible purpose behind this dispersion was to strengthen the defenses of the han; actually, however, the authorities were just as concerned to prevent the emergence of a "critical mass" of *shotai* opinion in Yamaguchi.

Yet while wishing to restrain the *shotai* the han government was also concerned to strengthen the han military and to arouse the military spirit of the samurai class as a whole. The entire han was charged with an awareness of its danger: it spoke of itself as being engaged in a "*jōi* war."[8] The usual ceremonies were put aside, periodic inspections of the troops were called, the daimyo, Mōri Takachika, attended the conferences of han officials in armor, and on 15/1/64 the daimyo read to his assembled retainers not only the customary "Statutes of the Manji Period" (*Manji seihō*), but also the military regulations laid down by the founder of the han.

Within such an atmosphere of tension it is not remarkable that during 1/64 Kijima Matabei, the leader of the largest of the *shotai,* the Yūgekitai, went to Yamaguchi to petition the han government for permission to lead his troops to Kyoto. Efforts were made to convince him that his proposal was premature, but he remained adamant. Fearing that he might leave without permission, Takasugi

[7] Suematsu, *Bōchō kaiten shi,* V, 153.
[8] *Ibid.,* pp. 162–163.

Shinsaku was sent to reason with him on 24/1/64. Kijima, however, was unmoved and announced that if the han government would not cooperate, the Yūgekitai would do it alone. Forty-nine-year-old Kijima further argued with the twenty-four-year-old Takasugi: "Today is the feast day of Sugawara no Michizane. Therefore, let us draw lots in order to divine whether or not it is auspicious to set out [for Kyoto at this time]." [9] In addition he upbraided Takasugi with these words:

The trouble with all of you is that you read too many books. Because you read books, the word "afterwards" is always uppermost in your minds. How can one have an army with such lukewarm ideas! If you read books, you should learn first that, when the lord is humiliated, the vassals must [determine to] die. And what in fact has happened at the present? Have not our daimyo and his heir, who are peerless in loyalty and sincerity, been barred by an Imperial edict from the Court for an offense which they did not commit? "If the lord is disgraced, the vassals must die;" does this not hold true of the present? We, the vassals, cannot dally for even one day. Is it perhaps because you have a new fief of one hundred and sixty *koku* stuck between your teeth that, in spite of being a sympathizer, you are now mouthing this sort of dilatory argument? [10]

Takasugi countered with a proposal that the two travel to the Osaka-Kyoto area to ask the opinion of Kusaka, Kido, and other Chōshū officials there. If they felt that now was the time to send troops to Kyoto, then he, Takasugi, would support Kijima, but if they felt that it was still too early, then Kijima would restrain his troops. Kijima agreed to this; they arrived in Osaka on 2/2/64 where Kido, Kusaka, and the others emphatically declared that it was too early to attempt any action. Thereupon, Kijima returned to the han to pacify his troops.

Takasugi stayed on in the area in order to plot, unsuccessfully, the assassination of Hisamitsu of Satsuma. After a time he received word from the Chōshū heir urging him to return to the han; he left with Kusaka, arriving back in the han on 29/3/64, whereupon he was promptly put in the han jail "for having recklessly left the han

[9] *Ishin shi,* IV, 10–11.
[10] *Ibid.,* p. 11.

without waiting for an order from the daimyo." [11] This was typical of the petty and arbitrary exercise of authority that characterized the upper reaches of the han government. The Confucian rationale in the daimyo's mind was, no doubt, that if the *shotai* were under strict orders not to leave the han without permission of the daimyo, then all the more, Takasugi as a member of the government ought to set an example.

In spite of the strict orders issued by the han government, and in the face of constant injunctions from their own extremist leaders, some individual members of the *shotai* continued to slip off to Kyoto. As a result, roads leaving the han were watched, and all ships entering or leaving the han were carefully inspected. The feeling among the members of the han government that it was still too early to send military forces to Kyoto was strengthened by the return to Chōshū of Kusaka, who argued that time was on the side of Chōshū. He pointed out that little by little the Court was coming to realize the drawbacks of the *kōbugattai* position, and with this the prestige of Chōshū would surely rise.

Kōbugattai Politics in Kyoto

The events which took place in Chōshū after the *kōbugattai* coup constituted only one very small drama on the periphery of the Tokugawa political scene. Even in eclipse, Chōshū was still the symbol or focus of a *sonnō jōi* movement with adherents in every han, but the center of the political stage was now held by the *kōbugattai* forces. They had made their entry by military means; they were now faced with the task of implementing their moderate program of joint-rule by "Court and Camp."

Their first steps were designed to sweep away the innovations introduced by Chōshū. The Imperial troops, the *shimpei,* were disbanded; extremist samurai were either driven from Kyoto or placed under strict surveillance; and the new Court offices, the *kokuji goyōgakari* and the *kokuji sansei* were abolished. And, of course, the

[11] *Ibid.,* p. 13.

kōbugattai nobles were restored to power in the Court. After the coup, many daimyo in Kyoto, especially those who had been sympathetic toward Chōshū, began to raise doubts concerning the legitimacy of the new powers. The Imperial edicts contradicted one another; after all, how was one to know what the Emperor truly thought? To quiet such doubts a conference of all the daimyo and high nobles in Kyoto was called on 26/8/63 at which the Emperor spoke as follows: "Until now there have been many arguments as to whether [certain Imperial edicts] were true or false. All those from the eighteenth [18/8/63] on are truly in accord with my will; therefore, concerning this issue the various han may now unite without any misunderstanding." [12] At this conference also, Katamori of Aizu was rewarded for his services to the Court, another sign of Imperial approval of the *kōbugattai* coup. These moves, however, were merely preliminaries; the ultimate success of the *kōbugattai* movement depended on the formation of a new political body which would be constituted of both Court and military personnel.

On 5/12/63 Hisamitsu addressed the leading *kōbugattai* figures gathered in Kyoto:

> The vacillation and indecision of the nobles is truly indescribable. Frequently they even fail to carry out matters on which they have already decided. Faced with such a situation, no matter how resolute the decision of the military houses may be, it is of no avail, and grave matters are almost impossible to execute. Therefore, wise daimyo must be summoned to the Court and added to the number of Imperial Councillors [*gisō*].[13]

Those assembled agreed with Hisamitsu's proposal, and Satsuma set about to convince the Court of its necessity. There had been no precedent for such a proposal, and this led some nobles to criticize it; others argued with more pertinency that their own power was contingent on their mediate position between the Emperor and the military powers, and that, if the already powerful daimyo were made councillors, then, they, the nobles, would suffer. These arguments,

[12] *Ibid.*, III, 642.
[13] *Ibid.*, p. 660.

however, were soon overcome, and on the last day of 1863, Tokugawa Keiki, Matsudaira Keiei, Matsudaira Katamori of Aizu, Date Munenari, and Yamanouchi Yōdō of Tosa were ordered to participate in the councils of the Court as participating daimyo (*sanyo* daimyo). Court rank and title were prerequisites for such appointments; Hisamitsu himself had never been daimyo, and, lacking these requirements, he was only given the appointment two weeks later, after having been given necessary rank and title. There were some who were hostile toward Hisamitsu and who hinted uncharitably and perhaps naively, at the time, that he had devised the entire scheme to obtain a long-coveted title and rank.

Officially recognized by the Court, the group of *sanyo* daimyo became, at least for the duration of its existence, the most powerful political organ in the country. Its formal organization included Keiei and Keiki, the most powerful if not the most representative figures in the Bakufu, some of the daimyo of the most powerful han, and the highest Court nobles. This indeed was the union of "Court and Camp" which many had long advocated. And yet in practice, the real locus of power came to rest at a point somewhere between the *shimpan* daimyo, Keiki and Keiei, and the *tozama* powers, Tosa and Satsuma. Questions of policy usually were decided in private conferences, and not infrequently, official meetings at the palace merely rubber-stamped what had been determined beforehand.

Not the least of the problems confronting the new council was that of Chōshū. Although banished from Kyoto, Chōshū's continued existence and activity as the nerve center of the *sonnō jōi* movement was a thorn in the side of the *kōbugattai* leaders. Especially those closest to the Bakufu looked on Chōshū as a dangerous foe because of its earlier activities. They could not criticize Chōshū for its *sonnō jōi* policy, since it had at one time been accepted by the Court, and since many in Japan tended to view this doctrine as erring only on the side of virtue. Nor could they criticize Chōshū for its behavior before the coup, as this would focus attention on the changes they had instituted after gaining control. Therefore, they leveled their attack on Chōshū's actions at the time of the coup: why had Chōshū, in spite of an Imperial order relieving it of guard duty at the Sakaimachi gate of the palace, drawn up its troops as if to contest by force the decree of the Emperor?

Sundry plans and proposals were made concerning Chōshū, but no decisive measure was taken during the early months, for the various parties were still jockeying for position and none wished to risk the rigidity of a decisive stand. On 21/9/63 the Bakufu decided to summon the daimyo of Chōshū and his heir to Edo, but this decision was later put aside. During the period when the two Chōshū missions were waiting in Osaka, the Court was constantly holding conferences and consulting with different groups, but, here too, the last months of 1863 passed without decision on the question of Chōshū.

The actual development of any form of "Chōshū policy" at the Court had its beginning in a very minor incident. On 24/12/63 the Kiheitai sank the *Nagasaki-maru,* a Bakufu steamship which had been lent to Satsuma. Chōshū claimed, with tongue-in-cheek, that it had mistaken the ship for a foreign one and attacked it according to the *jōi* wishes of the Emperor. The Satsuma reaction to this was immediate and violent: some even advocated burning Chōshū's Kyoto residence. However, at a meeting on 2/1/64 Keiki argued:

> There will be a time in the future to rectify the misdeeds of Chōshū; it is not appropriate to begin today with such a trivial incident. . . . Let Chōshū be ordered to surrender the seven nobles [extremist nobles who fled from Kyoto with the Chōshū troops] and Masuda [the Chōshū Elder] who precipitated the outbreak [i.e., who directed the Chōshū troops on the morning of the coup]. If they refuse to comply with this order, then the Bakufu can send a punishing army.[14]

To this Matsudaira Katamori of Aizu added that "the shogun ought personally to assume command of the expedition without borrowing forces from the *tozama* han."[15] Keiki replied that this would be very difficult to carry out with the Bakufu in its present condition and proposed as an alternative that the daimyo of Kii lead the

[14] *Ibid.,* p. 667.
[15] *Ibid.,* p. 667.

expedition as the representative of the shogun. He reminded his fellow daimyo, however, that they should first exhaust all peaceful means of settlement. Relatively concrete plans for an expedition were drawn up. Date Munenari wrote in his diary, under the heading "Decisions of the secret talks concerning Chōshū" dated 11/1/64, "The Lord of Kii was named the commander, Katamori the lieutenant commander, and nine other han were listed to form the body of the army." [16]

Not all of the nine han were interested in chastising Chōshū. On 10/1/64 the daimyo of Inaba sent a memorial to the Court asking it to proceed with the expulsion of the foreigners and requesting lenience for Chōshū. The daimyo of Bizen entertained similar views. These daimyo, however, were only minor figures. On 8/2/64 Keiki, Keiei, Date Munenari, Shimazu Hisamitsu, and Yamanouchi Yōdō, all *sanyo* daimyo, met at the Nijō castle, where in concert with the shogun and his *rōjū,* who had recently arrived in Kyoto, they resolved to order Chōshū to surrender the seven nobles and to send representatives to Kyoto to discuss its past misdeeds. If Chōshū refused, an army carrying an Imperial edict was to be sent against it. On 11/2/64 orders were dispatched to the various han which were to join in this action. An order to Chōshū was also drawn up. The daimyo Kuroda, however, wrote to the throne contesting this policy, and subsequently, the Court withdrew its backing. Again confusion descended, a series of conferences was held, and the various daimyo were asked their opinions. Finally, on 5/3 it was decided that the daimyo of the Chōshū branch houses and the Chōshū Elders should be summoned to Osaka.

When Chōshū received the order, it had already been apprised of the secret orders issued by the Bakufu for its subjugation, and most of the officials in the han had concluded that it was best to comply. However, Kido, Kusaka, and other Chōshū samurai spying out the situation in the Kinki region counseled delay. They argued that Satsuma's prestige was slowly declining while that of Chōshū was on the rise; it was to Chōshū's advantage to wait. To substantiate

[16] *Ibid.*, p. 668.

this opinion, Kusaka returned to the han on 19/3 reporting in detail on the situation in Kyoto. His counsel was accepted, and, in order to stall for time, it was decided once again to request permission for Chōshū representatives to enter Kyoto.

Kusaka's advice was well-founded. As time wore on, tensions immanent in the configuration of *kōbugattai* forces from their inception became more pronounced. The *sanyo* daimyo were powerful because of the powers they represented and because of their position at the Court. But they did not command an administrative apparatus by which they could implement their decisions; for this they had to look to the Bakufu. The Bakufu, a stronghold of entrenched *fudai* officials who were jealous of their prerogatives and suspicious of the newly risen outside forces, tended to thwart or hamper the program of the *sanyo* daimyo. They were hostile toward Shimazu Hisamitsu and even distrustful of Keiki and Keiei, who more often than not found themselves in the middle of disagreements between the Satsuma leader and the *rōjū*.

Not the least of these disagreements were those dealing with the question of Bakufu reform. The *tozama* members of the *sanyo* daimyo felt that such reforms were basic; without them the *kōbugattai* movement could not succeed. Shimazu Hisamitsu had often attacked Chōshū on the count that its extreme *sonnō jōi* policies would stir up trouble in the country, making Japan an easy prey to foreign powers. To this Chōshū had habitually rejoined that the Bakufu was ineffectual and only by bringing the strong han to the support of the Court could the foreign problem be solved. Therefore, after the coup Hisamitsu and other *kōbugattai* figures had to make good their claims that the Bakufu could be reformed. Consequently, when they were rebuffed by the Bakufu, they proposed on 13/2/64 that they be permitted to participate in the councils of the Bakufu. Already admitted to the inner sanctum of the Court, why should they not join in Bakufu deliberations as well? The Bakufu resisted and protested, but in vain, and finally they were admitted.

The only result of this move was to bring into the open and magnify the fundamental opposition between the two groups. Many in the Bakufu viewed the Satsuma-Aizu coup as a restoration of their own power, and, even when the need for reform was glaringly apparent, they felt that to follow the lead of Hisamitsu would somehow reduce their own stature. Moreover, hoping to regain prestige as well as power, the Bakufu was turning to an antiforeign position in order to attract to its own support the *jōi* feeling widespread throughout the country. It hoped, for example, to negotiate the closing of the opened port of Yokohama. The *tozama kōbugattai* daimyo, and in particular Satsuma (owing to the foreign bombardment of Kagoshima in 1863), recognized that the expulsion of foreigners was impossible, and they found themselves opposed to the Bakufu on this point as well. The Bakufu also intended to increase the number of its controls and the degree of supervision of foreign trade in order to secure more completely the revenues from this new source of income. Satsuma and the other han opposed this extension of Bakufu controls, since they too wanted a share of these revenues.

The unrealistic appeal to the popularity of the *jōi* slogans and the Bakufu's unwillingness to reform soon disillusioned the *sanyo* daimyo, and, unable to accomplish their ends, they resigned. Yōdō of Tosa resigned on 28/2/64 and the others followed on 9/3, returning to their respective han. Soon after, most of the other daimyo who had hurried to Kyoto on the orders of the Court "to assist in national affairs" [17] returned disenchanted to their han, leaving Kyoto and the Court in the eager hands of the shogun and his entourage. Left without support, the submission of the Court was inevitable: on 20/4/64 the Court, for the second time, issued an edict entrusting the Bakufu with the affairs of the land. Having been reconfirmed in his authority, the shogun saw no point in remaining in Kyoto, and so, with the Court's permission, he left Kyoto with his officials and repaired to Edo on 7/5. The authority and power of the Bakufu were again rising. Though the Court nominally had the right to issue orders to the Bakufu on matters of policy, Matsudaira Keiei, speaking for the Bakufu, suggested that it would be better for the Court not to exercise this right for a time, since it had so recently entrusted the functions of civil government to the Bakufu. No longer

[17] *Ibid.*, p. 706.

supported by the military houses and under the control of the Kyoto Bakufu officials, the Court had no choice but to agree, while ruing the sudden decline in its fortunes.

The situation foretold by Kusaka in 3/64 had now been realized: Hisamitsu and the *sanyo* daimyo had left in disgust, the Court after a brief and heady taste of power had been again reduced to its former impotent state, and the Bakufu was again acting as if it were the sole source of political authority. When Kido visited Sanjō Saneai in Kyoto on 17/4, he deplored the fallen estate of the Court; to this Saneai replied: "Indeed, the Court has lost face vis-à-vis Chōshū. The power of the high-ranking nobles is small, the prestige of the Bakufu great, and the military preparations of the various han inadequate, and so there is no other course but to rely on the Bakufu. This is truly regrettable."[18] Chōshū hoped that this re-evaluation on the part of the Court would lead to its reinstatement. These hopes were dashed when on 10/5 a message arrived from the Court canceling its earlier order to Chōshū to send representatives to Osaka. This message was, doubtless, initiated by the Bakufu; it stated that the Court could no longer handle the affair of Chōshū, since it had entrusted to the Bakufu all the affairs of the nation. With this communication Chōshū realized the impossibility of regaining its position in Kyoto by peaceful means. Those who had argued in favor of a countercoup now became even more vociferous in their demands.

The Countercoup

The division in the han between those who opposed a military expedition to Kyoto and those who favored it was curious in that it cut across previously existing clique lines. Sufu opposed it, but so also did Takasugi; Nakamura Kyūrō, an important member of the Sufu clique, favored it, as did Kusaka. The seven extremist nobles were among the strongest partisans of the expedition. They agitated tirelessly, demanding that Chōshū return to Kyoto and expel the traitors surrounding the Emperor. Most of the *shotai* leaders favored

[18] *Ibid.*, IV, 25.

an expedition to Kyoto, but, even here, there were a few opposed to it.

Kijima Matabei, of whom we spoke earlier, was perhaps typical of these leaders. A swordsman of considerable fame in the han, he was influential as the leader of the largest unit of auxiliary militia (the Yūgekitai) even though he held no office within the han. A little earlier Kijima had been persuaded that it was too early to strike only after Takasugi had led him to Osaka to confer with the Chōshū representatives in the field. By 5/64, this time supported by Kusaka, Kijima again had come to feel that the Chōshū armies should take to the field, and, incensed at the passivity of the Chōshū government, he appeared one morning in Yamaguchi determined to strike down Sufu if Sufu would not yield. He made his way to the government building and, calling for Sufu, began to climb the stairs leading to Sufu's second floor office. Entrance into the offices of the government without official permission was strictly forbidden; Hirosawa Saneomi, a large and muscular samurai who was on duty with Sufu at the time, threatened to kick Kijima down the stairs if he persisted. Kijima replied that he intended to mount the stairs even if he had to use his sword. The two began to grapple when Sufu appeared and invited Kijima to enter.

The conversation which is reported as having taken place between the two was only recorded some years later and may, accordingly, suffer from the embellishments of time, but there is no reason to think that it is not substantially true. Kijima, with tears in his eyes, pleaded with Sufu:

We have always been friends. Believing in your wisdom and fortitude, I have never opposed you, but rather have held you in esteem second only to the daimyo. But at this time, we cannot wait a day longer. As vassals we must take upon ourselves the troubles of the daimyo—the matter of Chōshū's duty at the Sakaimachi Gate and that of the Bakufu's violence. I, together with the *shotai,* have applied to the [han] government for an order from the daimyo to go to Kyoto to restore the former state of affairs [i.e., that existing before the Satsuma-Aizu coup] and to appeal the innocence of our daimyo to the Court. But to this day, the permission has not been granted. I have heard that this is so because you believe that if, for the ostensible purpose of vindicating the name of our daimyo, the hot-headed and enthusiastic *shotai* set out for Kyoto, they would in fact clash with the soldiers of Matsudaira, Ii, and the like, who guard the palace gates according to the orders of the Bakufu. And that, since this might possibly create even greater difficulties for our daimyo, you feel it would be wiser to wait for an opportunity [some other time] and because of this, it [the expedition] has been delayed. Since this is the case, I have come here to try to obtain our long-cherished object [the expedition] even at the risk of crossing swords with you.[19]

To this Sufu is said to have replied: "I appreciate your loyalty to the daimyo and to the Court. However, first and last, you are a soldier, and, no matter how intense your desire to serve the Emperor, you are not the sort of person with the talent or knowledge to deal with men of other han, to answer Bakufu officials about previous entanglements, or to obtain a pardon for the daimyo by an appeal to the Court."[20] Sufu went on to suggest that Kijima was the sort who would inevitably lose his temper in the middle of negotiations and begin name calling.

In such a case how would you excuse yourself to the daimyo? Slicing your stomach into two or three parts according to the inner principles of *bushidō* would not alleviate one bit the troubles weighing on the house of Mōri. Are you not a middle-ranking samurai of the house of Mōri? You must think mainly of its welfare! I do not differ from you in the least in loyalty to the Emperor, yet as an official in the government of the house of Mōri, I cannot endorse an action that might leave a smear on the record of our daimyo. . . . As my friend you must reconsider things and act your age. I am not one who wants to dampen the ardor of the *shotai* to serve the Emperor.[21]

Sufu had won his point. Kijima is said to have shed tears anew and, conceding that Sufu was right, confided that "even at home my wife complains that I am too attached to military matters."[22]

Yet the arguments and views of Kijima were shared by an ever growing segment of han opinion, particularly within the *shotai*. In

[19] Suematsu, *Bōchō kaiten shi,* V, 324.
[20] *Ibid.,* p. 324.
[21] *Ibid.,* pp. 324–325.
[22] *Ibid.,* p. 325.

many ways, the arguments of those in favor of a military expedition to Kyoto were similar in character to those voiced earlier by Yoshida Shōin, and differed from the more rational views of those who later went on to establish the Meiji government. Kido and Yamagata, as well as Takasugi, opposed the idea of a countercoup at this time.[23] Kijima, like Shōin, demanded action because there ought to be action; he gave little thought to its consequences. Chōshū, the champion of the Court and the epitome of loyalty, had been unjustly banned from Kyoto after a treacherous Bakufu coup. Was Chōshū to do nothing? Variations on this argument were to be repeated over and over in the *shotai* and in the han at large. It was unworthy of a samurai to be over-solicitous about the consequences of his action so long as the action itself was virtuous. More and more members of the han government were moved by these arguments, and, as always, the daimyo and his Elders were unable or unwilling to resist the dominant opinion.

On 25/5/64 the letter canceling the talks that were to be held in Osaka had arrived. This, together with the warlike spirit pervading the han and the news from Kyoto of the Court's dissatisfaction with the existing state of affairs, tipped the balance in favor of an expedition. On 27/5 Kunishi Shinano, an Elder, was ordered to lead troops to Kyoto for "an appeal" to the Court. A few days later another Elder, Fukuhara Echigo, was ordered to lead troops to Edo for "an appeal" to the Bakufu.[24] The intent of the second order was no different from that of the first: Fukuhara, ostensibly on his way to Edo, would be in the vicinity of Kyoto at the same time

[23] Yamagata's argument at this time was as follows: "With the present domestic and foreign problems, an audacious attack on Kyoto would have some merit, no doubt, but it cannot be called a sound policy. I advocate an alliance with the clans of Aki and Bizen, and the winning of the inhabitants of those ten provinces, which were in the past under the Mōri clan, to our point of view. Then an impressive advance, displaying great power, could be made by land and sea. Any who oppose and obstruct this advance should be met decisively. What we must consider is that if we undertake a long sea expedition we will be advancing like a wild boar losing our maneuverability and leaving our rear vulnerable. So, rather than occupy Hikeyama [in Bungo], we should advance along the northern coast, occupying the important points and establishing advanced bases as we go." Roger Hackett, "Yamagata Aritomo 1838–1922: A Political Biography" (diss. Harvard University, 1955), p. 49.

[24] *Ishin shi,* IV, 36.

as the forces under Kunishi. He was being sent to reinforce Chōshū's military potential in case any fighting broke out. On 4/6 it was decided that the Chōshū heir would also lead a contingent of troops to Kyoto, and, in preparation for this, the heir led his troops in a series of maneuvers within the han.

On 14/6/64 news came of Bakufu police action against loyalist samurai in Kyoto, in which several Chōshū samurai had been slain, among them prominent members of the Chōshū loyalist group. This news spurred on preparations within the han, and a third Elder, Masuda Danjō, was also given orders to proceed to Kyoto with a contingent of Chōshū troops.

On the same day Sufu Masanosuke, the only remaining obstacle in the way of the expedition, was ordered into domiciliary confinement for fifty days. The charge against him was that he had visited Takasugi Shinsaku, who was still in prison at this time for having gone with Kijima to the Kinki area, and that, while shouting words of encouragement to Takasugi, he had drunkenly brandished his sword at the prison guards. It was against the law to speak with anyone in prison, and it was unbecoming of an official to act as he did; hence, even for Sufu such a punishment was, supposedly, warranted. The explanation is unconvincing; it certainly did not convince Sufu's colleagues. Kaneshige Jōzō, one of the original members of the Sufu clique, defended him: "Even after a long drinking bout, Sufu would get up and continue his work when there were matters to be attended to, showing no signs of unusual feelings;" this, he continued, was so well known as "to have been recorded on a stele." [25] Since this was the case, Kaneshige suggested that the real reason for Sufu's erratic behavior was that he had become depressed with the course of events in the han and that this resulted in his outburst at the jail. This may be true, but it does not explain why Sufu, who tried to enter the prison to speak with Takasugi on 6/5, should not have been ordered into confinement until 16/6. Yoshitomi Tōbei, a rich peasant who was very active in the *sonnō jōi* movement in the han, suggested that the han government did not want to punish Sufu but the conservative elements in

[25] Suematsu, *Bōchō kaiten shi*, V, 335.

Hagi had made so much of the incident that the government was obliged to take action.[26] Although these two reasons may have contributed to his confinement, it is more likely that the compelling reason for it was to remove the last obstacle to the Kyoto expedition, which Sufu had continued to oppose.

On 1/6/64, the day after Sufu's temporary removal from the government, the first contingent of Chōshū troops left for Kyoto. By 6/7 all the troops under the three Elders were on their way. The first group arrived in Osaka on 22/6 and immediately left for Fushimi. One observer wrote that "all were arrayed in battle-dress; anyone watching along the road could not help being astonished."[27] When one guard detachment of Kii samurai asked the Chōshū troops why they were wearing their full battle-dress, they were told with a straight face that the troops of Chōshū always wore full battle-dress in readiness for barbarian attack. The samurai of Kii were few and those of Chōshū many, therefore, they were permitted to pass into Fushimi. The following day letters were sent to both the Court and the Bakufu stating that the group of samurai under Fukuhara, the Elder, were en route to Edo, but, having heard of the Ikedaya incident (in which the Chōshū samurai were killed), they had become concerned over the situation in Kyoto and had decided to remain in Fushimi for two or three days. Letters were sent to the Court asking again for the expulsion of the foreigners and the recognition of Chōshū's innocence, and to the representatives of the various han asking them to mediate on Chōshū's behalf.

By 14/7 all the troops under the command of the three Elders had arrived in the outskirts of Kyoto, where they took up strategic positions. And while negotiations continued, the Chōshū troops prepared their weapons and watched for an opportunity to retake the Court from which they had been ousted less than a year before. To quote the words of one historian, "the Imperial will was now to be contested by force of arms."[28] And in the meantime, while

[26] *Ibid.*, p. 336.
[27] *Ishin shi*, IV, 42.
[28] Tōyama Shigeki, *Meiji ishin* (Tokyo, 1951), p. 119.

the camp fires of Chōshū flickered around the city, the Bakufu officials within hastened their defenses. Samurai of the various han in Kyoto were mobilized; some were posted at the entrances of the city and others doubled and redoubled the defenses of the palace, the inevitable target of any attack. Also, within the city, the representatives of the Court, Bakufu, and various han debated Chōshū's demands for recognition.

Many nobles, feeling that the foreigners ought to be driven from Japan, were secretly sympathetic to Chōshū. They asserted that warfare would break out if Chōshū's demands were refused, and that the Court should recognize Chōshū. Other nobles, backing Matsudaira Katamori of Aizu, replied that Chōshū was seeking redemption with threats and force. If these demands were granted, then Bakufu prestige would suffer and the gains of the earlier coup be lost. Resist Chōshū now or it would be uncontrollable later. Keiki took a more temperate attitude: Chōshū desired the expulsion of the foreigners, but so did many other han; if we attack Chōshū now, many han will feel that our action is arbitrary, and this will cause disharmony within the country. Keiki therefore proposed that the Bakufu win with persuasion, while showing the various han the moderate rationality of its judgment, and that Chōshū be attacked only if it persisted in its demands.

Keiki's views were accepted; he was summoned to the Court on 29/6/64 and given complete authority to deal with the Chōshū troops. To facilitate his labors, he was also given an Imperial edict saying that the Satsuma-Aizu coup of the previous year was in complete harmony with the Imperial will and that the Chōshū troops should withdraw and await an Imperial order. They were not to enter the city itself. The edict was sent to the Chōshū commanders, who rejected it as the handiwork of the Bakufu traitors who surrounded the throne.[29]

The range of opinion among the han was as wide as that within the Court. Inaba han, the most sympathetic toward Chōshū, called a meeting of the han in order to bring about the acceptance of Chōshū's requests. Other han such as Bizen, Awa, Yonezawa,

[29] *Ishin shi,* IV, 58–59.

Chikuzen, and Matsuyama sent memorials to the Court and the Bakufu, asking that Chōshū be treated with clemency. Others, however, asked that it be subjugated as punishment for its numerous offenses. Satsuma in particular was vindictive; in spite of its dissatisfaction with the Bakufu (that would eventually lead it to join with Chōshū), Satsuma not only refused to mediate on Chōshū's behalf, but it also refused to attend the meeting called by Inaba han. Saigō Takamori and the Satsuma Elder, Komatsu Tatewaki, writing to Ōkubo Toshimichi, who was in Satsuma at the time, on developments in Kyoto, urged that troops be sent quickly from the han in the event of a general mobilization for the subjugation of Chōshū.

On 6/7/64 the Court told Keiki to secure the aid of other han for the rapid withdrawal of the Chōshū troops. As the letters from many han were favorable toward the Chōshū cause, the Chōshū Elder Fukuhara considered the possibility of withdrawing his troops to Osaka to await the arrival of the Chōshū heir. Kusaka agreed to this plan, but it was vetoed by an overwhelming majority of the *shotai* leaders, and, as these leaders were in actual control of the troops, Fukuhara and Kusaka could do nothing. As a concession to the Court, one very small forward outpost was withdrawn, but the majority of the troops and leaders were determined to attempt a countercoup.

The deadline set for the withdrawal of the Chōshū troops was 17/7/64. That day and the next passed without action: the *kōbugattai*-Bakufu forces in Kyoto did not wish to begin hostilities, and the Chōshū forces saw no opening by which they could reach the palace. In desperation an attack was begun on the morning of 19/7, but it was totally repulsed. Only one of the three contingents from Chōshū, that led by Kijima under the orders of Kunishi, even reached the palace. This gave the skirmish its name, the "Disturbance at the Forbidden Gates" (*Kimmon no hen*). Kijima's troops momentarily overcame the troops of Aizu which they encountered, but they were soon routed when reinforcements of Satsuma and Aizu samurai arrived. Fukuhara's troops were stopped at the outskirts of Kyoto and forced to retreat, and those under

Masuda did not even join in the fighting but left meekly for Chōshū when word came that the others had been defeated. By the afternoon of the same day the fighting had ended, the disgraced Chōshū troops were on their way to the han, and Kusaka, Kijima, Maki, Terashima, and other *shotai* commanders who, had fate been kinder, might have become Meiji leaders, were dead.

The Four-Nation Attack[30]

The defeat in Kyoto was not all. Nemesis seemed to pursue Chōshū relentlessly; while still reeling from this blow, it was struck again at Shimonoseki by the combined fleets of Great Britain, France, Holland, and the United States. In 1863, when Chōshū launched its attacks on foreign shipping in the Shimonoseki Straits, the United States and France had retaliated by destroying Chōshū's Shimonoseki forts; this, they thought, would put an end to Chōshū's attacks. The attacks persisted, and Great Britain, France, Holland, and the United States gave to the Bakufu a joint communiqué stating that they themselves would undertake direct punitive action against Chōshū if the Bakufu could not bring it under control. The Bakufu, of course, could do nothing to prevent Chōshū's attacks on merchant shipping but issue a flow of wholly ineffectual orders. The foreign powers, however, were not united on the actual measures to be taken, in spite of their joint communiqué. The United States and Holland held that military action should be taken to preserve a strict observance of the provisions of the treaties. Great Britain and France, in accordance with instructions from their home governments, opposed any military action and contented themselves with the promotion of trade.

On 24/1/64 Sir Rutherford Alcock, the British minister, returned to Yokohama from England with the authority to take at his own discretion direct military action. Messages were sent to the other foreign powers; the United States and Holland were willing to

[30] The account of the Four-Nation Expedition is based almost entirely on the following: Suematsu, *Bōchō kaiten shi,* VI, 96–173; *Ishin shi,* IV, 302–362.

join in a joint action, but France demurred. This seems to have been due partly to a resentment of English leadership in diplomatic affairs and partly to the hope of obtaining special commercial privileges by supporting the Bakufu more strongly than the other Western powers. On 22/3/64 Roches, the new French ambassador, arrived at Yokohama; he finally agreed on 25/4 to form a common policy front with the other powers, although without any commitment to military action. Reports to his government at the time indicate that he opposed foreign military intervention on the grounds that it would lower the prestige of the Bakufu. Soon afterward, however, he discovered that the Bakufu was not at all averse to a foreign attack on Chōshū; therefore, he declared his willingness to join with the other nations. A plan for the attack on Shimonoseki was drawn up on 19/6/64, and the Bakufu was notified that the four-nation fleet would attack Chōshū without warning if the former belligerence of that han were not disavowed within twenty days.

In the meantime Itō Hirobumi and Inoue Kaoru, two young Chōshū samurai studying in England, had heard of the projected attack on their han. They speedily returned to Japan, and, meeting with Alcock, they asked that the attack be delayed to give them time to persuade the Chōshū government of the folly of its *jōi* policy. The English, who saw this as an opportunity to enter into direct negotiations with Chōshū, provided a warship to carry the two samurai back to their han. On 23/6 Itō and Inoue returned to Chōshū, and, at an audience with the han government, they recounted what they had seen and heard during their travels. They argued that the han's *jōi* policy was futile and that it must be abandoned. The han government was unmoved; the foreign warship returned to Yokohama and preparations were made for the attack.

On 2/8/64 the foreign fleet appeared off the coast of Chōshū. The news of the defeat of the Kyoto expedition had reached Chōshū on 23/7. Disturbed by the gravity of its plight and fully remembering the foreign attacks in 1863, Chōshū took steps to avoid combat. Orders were sent to Akane Taketo, the head of the Kiheitai, and to Maebara Issei, stationed at Shimonoseki, to refrain from hostilities. Itō and Matsushima Jōzō were sent to Himeshima, where the foreign fleet was stationed, to negotiate for its withdrawal, but they arrived too late, the foreign ships had already sailed for Shimonoseki. Maeda and Inoue were then sent to negotiate with the foreign fleet off Shimonoseki. They were instructed to explain to the foreigners that Chōshū itself was willing to allow foreign ships peaceful passage on the Shimonoseki Straits, and that Chōshū had attacked foreign vessels up to this time only because it had been ordered to do so by the Bakufu and the Court. On their way Inoue and Maeda were forced to stop and waste several hours persuading the Kiheitai not to fire on the foreign ships. They sent ahead a messenger saying that a peace mission was coming and asking the foreign fleet to delay its attack. They were given two hours in which to appear, but their attempts to convince the Kiheitai took too long: just as they were setting out in a small boat to contact the foreign fleet, the foreign ships began to bombard the Chōshū forts. The Chōshū forces answered in kind.

This took place on 5/8/64, and by the following day Chōshū's forts had been completely demolished and its troops defeated by foreign landing parties. It sent the first peace mission on the eighth. This mission was headed by Takasugi Shinsaku. It had taken the threat of a foreign attack to get him out of jail; he was freed on 23/7 and appointed Director of Military Affairs or *gummugakari* on 3/8, the day after the appearance of the foreign fleet. Takasugi was seconded by Sugi Norisuke and Watanabe Naizō, while Itō Hirobumi and Inoue Kaoru were sent along as interpreters. In response to the foreign accusations, they answered again and again that Chōshū had merely been following the orders of the Court and the Bakufu. On 14/8/64 a treaty of peace was signed between Chōshū and the foreign powers. Chōshū agreed to ensure benevolent treatment of foreign ships and to supply provisions when needed, to build no new fortifications on the Shimonoseki coast, and to pay an indemnity to cover both the military expenses of the expedition and the foreign powers' indulgence for not having razed the city of Shimonoseki when it lay at their mercy.

Historians have often stressed the sudden change within Chōshū brought about by the four-nation attack, and, in particular, they have emphasized the adaptability of the Chōshū samurai and the quick reversal of their attitudes toward foreigners. Seen in historical perspective, this is true in a very general way, but it needs serious qualification. First, most of the han leaders had been well aware of the superiority of Western arms for several years before the attacks. Second, it would be more accurate to say that those who were already inclined toward the West had now begun to come to the fore, rather than to say that all became pro-Western. Some did begin to turn to the West at this time, but we must stress that the majority of politically active samurai remained *jōi* in opinion.

For example, during the negotiations with the foreign powers, the Chōshū negotiators were in constant peril of assassination. Helpless against the strength of the foreigners, some members of the loyalist group turned their ire against the members of their own group who now seemed ready to abandon the program for which they had worked so long. Thus, when Takasugi and Itō returned from talks with the foreigners to report to the Chōshū heir, they were warned that certain samurai who had fought in the Kyoto countercoup were planning to assassinate them. It will be recalled that at the time of the countercoup Takasugi was in jail and Itō abroad. Consequently, both relinquished their part in the negotiations and hid themselves in a peasant's hut in the country. A second group of negotiators had to be appointed, and it was this group that eventually came to terms with the foreigners. Inoue and Yamagata, who were less intimidated by the threats and perhaps less controversial as individuals, called Itō and Takasugi out of hiding and, rebuking them for having placed their personal safety first at a time when they were needed by the han, persuaded the two to return to their duties. The final team sent to sign the treaty consisted of two Elders, Shishido and Yamada, two *shotai* commanders, Tarusaki Yahachirō and Watanabe, and two interpreters, Inoue and Itō. While these men signed the treaty, most of the other members of the government were busy trying to mollify the irate *shotai* samurai.

After the conclusion of the treaty, Ōmura Masujirō proposed that a party of Chōshū samurai be sent to Yokohama along with the foreign fleet in order to improve relations between Chōshū and the foreign powers. The han government accepted the proposal and appointed Ibara Shukei, an Elder, to head a delegation. The delegation departed with the foreign fleet soon afterward; in Yokohama they conferred with the English and American ministers, who finally "agreed" that the Chōshū indemnity should be paid by the Bakufu. On 23/9/64 the delegation returned to Yamaguchi to report to the han government on the negotiations and the situation in Edo.

The conclusion of the special treaty with the Western powers brought the Chōshū movement to the nadir of its fortunes. It had been weakened first by the exclusion of the *sonnō jōi* partisans from Kyoto after the Satsuma-Aizu coup, second by the failure of the countercoup, and third by the defeat at Shimonoseki. Branded after the countercoup as the "enemy of the Court," Chōshū found it difficult to maintain that it alone was the Court's true champion. Yet even in their fallen and outcast state, the Chōshū samurai staunchly maintained that they alone represented the true will of the Emperor—on the grounds that, in spite of all, the Emperor still believed in expelling the foreigners from Japan. Once the treaty with the foreigners was signed, however, Chōshū had negated the second *jōi* tenet of its creed, and could no longer legitimate its resistance on this ground. It was for this reason that the treaty negotiations with the foreign powers had been so beset with difficulties in spite of an obviously hopeless situation. It was this treaty that brought the *sonnō jōi* movement to a close.

In addition, the necessity to find new grounds of legitimacy led to the rise of the *tōbaku* (overthrow the Bakufu) movement. The shift from the *sonnō jōi* to the *tōbaku* movement was due, basically, to the change in Chōshū's political position rather than to any change in political sentiment. In the new situation it was easier to maintain, negatively, that the Bakufu officials in Kyoto were traitors distorting the will of the Emperor, than to argue, positively, that Chōshū represented the true Imperial will. Apart from this there was very little difference between the two movements except, as already noted, for the greater *tōbaku* emphasis on military means and its greater dependence on extremist power. Even the gradual change from *jōi* to *kaikoku* signified, within the confines of the Bakumatsu period, only a re-evaluation of means rather than a change in the ends desired. Since 1862, at least, the obverse of pro-Court sentiment had been anti-Bakufu sentiment. The same basic han feeling concerning the necessity for action on the national scene continued from one movement to the other. The two movements must be seen as two phases in the same historical process.

Perhaps the greatest change in the transition from the *sonnō jōi* movement to the *tōbaku* movement was that the men who were positively inclined toward the West now began to come to the fore in han politics. Men such as Kido, Takasugi, Itō, Inoue, or Ōmura, who combined strong, almost reactionary, feudal values with a willingness to adopt Western techniques, now spoke with greater authority within the han. Takasugi was called in as a specialist in Western military science, Ōmura gained official status for the first time, and Itō and Inoue, who had hitherto been totally insignificant, now entered the fringes of the loosely knit body of men who controlled the affairs of the han.

The First Chōshū Expedition

The failure of the Chōshū countercoup in 1864 appeared to the Bakufu as an unparalleled opportunity to regain the military leadership of Japan. No sooner had the Chōshū troops fled Kyoto than conferences were begun to determine its punishment. Fuel was added to these discussions when a Satsuma samurai found in the wake of the fleeing army a military order bearing the seal of the Chōshū daimyo. That it was an order concerning the discipline of the Chōshū troops and without relation to the action in Kyoto was brushed aside as irrelevant by the vindictive Bakufu. The menace of Chōshū could now be removed once and for all. Tokugawa Keiki had the document forwarded to the Court on 27/7/64 as concrete evidence implicating the Chōshū daimyo in the attempted coup. The following day the subjugation of Chōshū as the "enemy of the Court" was ordered. Keiki transmitted this order to twenty-one han, instructing them to ready their troops for military action.

We now have the spectacle of the Bakufu, armed with an Imperial edict, leading the han in military action against Chōshū. Less than a year ago, when Chōshū was master in Kyoto and the Bakufu humiliated at every turn, such a state of affairs would have been difficult to foresee. Bolstered by the exercise of its traditional authority, the prestige of the Bakufu now rose to the highest mark reached since the death of Ii. Even in the months after the Satsuma-Aizu coup the Bakufu had lacked a rallying point about which its hegemony could be exercised. Now it had such a point, and, had it been willing to compromise, it might have been able to recreate a national front such as had existed from 1853 to 1856. Unfortunately, the Bakufu "learned nothing and forgot nothing": in attempting to recreate the absolute authority of bygone days, it soon dissipated its recent gains.

The first blunder was to issue an order reviving the *sankin kōtai* system of compulsory biennial attendance in Edo. When the various daimyo on one pretext or another refused to comply, the Bakufu was unable to enforce it, thus demonstrating that its authority was titular and not actual. Its newly found prestige was further tarnished by the bickering, irresolution, and delays that attended the organization of the expedition. At the heart of these was the antagonism and mutual distrust existing between the *fudai* Bakufu bureaucrats in Edo and the *shimpan* Bakufu party in Kyoto. On 2/8/64 the latter group announced that the shogun would personally lead the troops of the various han in the attack against Chōshū. Five days later, this order was countermanded; instead, Tokugawa Keishō, the former daimyo of Owari, was assigned to the duty. Keishō, however, declined on the grounds of ill-health. Orders had already been issued to the various han and their troops had been mobilized accordingly; all was in readiness except the leader. The Bakufu party in Kyoto deplored this situation and, backed by Satsuma and Kumamoto, urged the shogun to lead the army. The Edo officials continued to favor Keishō who continued to decline the appointment. On 4/9/64 Keishō finally left for Kyoto, although en route he issued a statement saying that the shogun should lead the army in the fashion of Ieyasu. In Kyoto he agreed to lead the army on the condition that he would be given complete authority over the expedition. His terms were met, and on 3/10/64 a conference began to discuss the military aspects of the expedition.

The quarrel over leadership had also become entangled during the eighth month with the question of whether it would be proper to launch an attack on Chōshū when Chōshū was under attack by the foreign powers. Daimyo sympathetic toward Chōshū, such as Asano Shigenaga of Aki han, sent memorials to the shogun criticizing the expedition. Ikeda Yoshinori of Inaba han wrote to the shogun: "One speaks of Chōshū and yet it is Chōshū of the Imperial country. One speaks of the people of Chōshū and yet they are also the subjects of the Imperial country. Therefore, [for Chōshū] to be invaded by foreign nations is shameful to the Imperial country. . . . As this bears on the prestige of the nation, let us punish the misdeeds of the house of Mōri after the withdrawal of the foreign ships."[31] In contrast to this, on 17/8/64 samurai from Satsuma, Kumamoto, Aizu, Fukui, and other han bent on punishing Chōshū resolved that the expedition should continue as planned but that the Bakufu ought to negotiate with the foreigners for the cessation of hostilities on the grounds that the Bakufu itself would chastise Chōshū.

Arguments concerning the expedition were further complicated by differences of opinion on what should be done with Chōshū after its subjugation. Some, such as the leaders of Bizen, argued that the daimyo of Chōshū should be treated with great lenience, since he had expressed his remorse concerning the action in Kyoto. Others, such as Matsudaira Keiei, contended that Chōshū's attempted coup surpassed in villainy the actions of the Ashikaga or the Hōjō, and that Chōshū must be severely punished. Many han did not care what happened to Chōshū but nevertheless favored a policy of lenience in the hope that, thereby, the duration of the

[31] *Ishin shi,* IV, 144–145.

expedition would be shorter, their troops returned sooner, and the cost of the whole affair made lighter. Tokugawa Keishō and the majority of his advisors felt that it would be best to obtain Chōshū's surrender without recourse to arms. One of them, in a masterpiece of understatement, ventured that if fighting began and the Bakufu troops were beaten, it would deal a grievous blow to Bakufu prestige.[32]

One of Keishō's most important advisors delegated to his staff from Satsuma was Saigō Takamori, who was destined to play a singular role in the negotiations with Chōshū. His early views were harsh and uncompromising: on 7/9/64 he wrote to Ōkubo proposing that not only should Chōshū's domains be diminished but that the house of Mōri should be transferred to a smaller han in eastern Japan.[33] Shortly after this, however, having been apprised no doubt of the situation within Chōshū, he wrote to Ōkubo: "As the conditions in the han [Chōshū] appear to be very complex, I wonder if there is not a way of leaving the punishment of those responsible for the [Kyoto] violence in the hands of Chōshū men?"[34] And by 24/10/64 this view had progressed to the point that he could write: "The enemy is divided into two camps, that of violence and that of righteousness; this is truly a blessing."[35] Saigō's plan of campaign was to utilize the camp of righteousness, that of the Mukunashi conservatives, against the camp of violence, the Sufu-loyalist alliance responsible for the Kyoto countercoup. It would indeed be naive to interpret this change wholly in terms of Saigō's increasing awareness of the internal situation in Chōshū. In part Satsuma's initial antagonism had been modified by the opinion that it would be too great a waste of samurai power to engage in an internal war when Japan was still faced with the foreign threat. In part it was undoubtedly due to a reconsideration of what Satsuma's position would be vis-à-vis the Bakufu if Chōshū were totally removed from the national balance of power.

[32] *Ibid.*, p. 146.
[33] Tōyama, *Meiji ishin*, p. 150.
[34] *Ishin shi*, IV, 147.
[35] *Ibid.*, pp. 147–148.

By 11/64 all was in readiness: 150,000 samurai from the chief han of Japan were poised on the borders of Chōshū awaiting the signal to attack.

The Reaction within Chōshū

Driven from Kyoto, defeated at Shimonoseki, and now facing an overwhelming army on their perimeter, the Sufu clique and their extremist allies found themselves in an impossible predicament. The question of what steps should be taken in such a situation split the alliance between the Sufu clique and the loyalist group. The latter advocated a policy of "armed submission."[36] Since such a policy would be totally unacceptable to the Bakufu commander, it was tantamount to a declaration of resistance. The Sufu clique, on the other hand, only too aware that their policies had brought disaster to the han, now cast about for some middle of the road policy that would preserve the han without injury to the house of Mōri. The most vocal clique at this juncture proved to be the conservative Mukunashi clique. They argued that their bureaucratic opponents, the Sufu clique, must be removed from office and punished for their past misconduct. They advocated a new policy of submission and trust, which they hoped would induce the Bakufu to treat Chōshū clemently. The Mukunashi clique was supported by groups of conservatives, chiefly those of Hagi, who had been silent since 1863. Yet it would be an exaggeration to depict the conservatives as a popular force in Chōshū. Majority opinion continued to champion the Sufu activist position, which had been the han policy since 1862. The strength of the conservatives derived from the situation and from the approaching Bakufu army led by such men as Saigō, who planned to use the Chōshū conservatives to crush the Chōshū *sonnō jōi* forces. But even under these circumstances the rise of the conservatives was slow.

The first shift toward a conservative government occurred on 26/7/64 when a conference of the leading Chōshū officials decided

[36] Suematsu, *Bōchō kaiten shi,* VI, 174–226. A discussion of the policy of armed submission and the rise of the *zokurontō* is given.

that those responsible for the Kyoto attack should be removed from office in order to placate the Bakufu. Consequently, the three Elders who had led the Chōshū troops to Kyoto and a few officials of the Sufu extremist coalition, such as Nakamura Kyūrō, who had actively encouraged the coup and led troops in Kyoto, were removed from office. These Elders were sent for "safekeeping" to Tokuyama, one of Chōshū's branch han, and the rest were placed in the custody of relatives.

The second step taken by the officials of the han was to release Sufu from confinement so that he might again participate in han affairs. As we noted earlier, Sufu had conveniently been sentenced to fifty days of domiciliary confinement on the day before the Chōshū forces set out for Kyoto. Sufu's position in the han at this time was unique: he had virtually dominated Chōshū politics since 1858, and yet, having opposed the Kyoto coup, he bore no responsibility for its failure. All this would seem to be a perfect background for leadership at this critical point, but Sufu found himself lacking the necessary authority. To quote his own words: "What is unfortunate is that my rank is low. Since it is low, no matter what ideas I may have, I can do nothing at a time of trouble such as the present.... The many Elders put their heads together but with such incompetents as these [nothing can be hoped for]."[37] When we consider that Sufu had been the leading figure in Chōshū for many years, what does such a statement mean? What Sufu probably meant was that his rank was low relative to that of the Elders who, surrounding the daimyo, were the ultimate wielders of authority in the han. Previously, when his policies had been successful and he had had the support of the three "extremist Elders," he had not been disturbed by his "low rank." (Actually, as we noted earlier, by any definition Sufu was a high- or, at least, middle-ranking samurai.) But now, in a time of crisis, the remaining Elders were unwilling to commit themselves to any action in support of the administration that had brought such disaster to the han. Even though Sufu had opposed the Kyoto action, his clique was identified with it and he found himself unable to obtain the Elders' support.

[37] *Ibid.*, pp. 57–58.

It was at this point, that Kaneshige Jōzō, a member of the original Sufu reform clique, suggested to Sufu that the only man with sufficient ability and authority to cope with the situation was Kikkawa Tsunemoto, daimyo of the branch han of Iwakuni. To this Sufu is said to have replied, "What you say is true. If Lord Kikkawa is employed, this matter can probably be settled. Lord Kikkawa is a good man, and although he is weak in initiative, he is able to conserve. Since today is a time for conservation, let us first straighten out affairs in the han. When the time arrives for initiative, a man of initiative will then arise." [38] On 3/8/64 Sufu went to Iwakuni with Shimizu Seitarō, the only Elder who remained friendly towards the Sufu faction. Three days later they met with Kikkawa. Sufu asked Kikkawa his opinion concerning the Bakufu army: would its commander be satisfied with the execution of the three Elders who led the Chōshū forces to Kyoto? If these three were punished, would the daimyo be spared? Sufu added that he himself, having been an official in the past, was responsible for the present position of Chōshū and would be willing to forfeit his own life if it would help the daimyo. Kikkawa replied that the punishment of the three Elders should suffice, and he promised to negotiate with the leaders of the Bakufu army on behalf of Chōshū.

In the meantime the city of Yamaguchi, to which the han government had moved from Hagi in 1862, had become the focal point in the power struggle between the Sufu-extremist alliance and the Mukunashi clique. The former, having lost its top-level backing, came to rely increasingly on the strength of the *shotai,* whose extreme loyalist spirit seems to have continued undiminished. The *shotai* leaders sent to the daimyo a stream of memorials in which they insisted that Chōshū's past actions had been righteous. Realizing the strength of the Bakufu forces, the *shotai* commanders were willing to make a token submission, but they were reluctant to lay down their arms. They asked their opponents: if the evil Bakufu leaders demand the death of our daimyo, are we to comply? It may seem paradoxical to assert that the Sufu clique, while rejecting the loyalist program, had come to rely more heavily than ever on the

[38] *Ibid.,* p. 58.

loyalist militia. Yet, in the face of rising conservative strength, the Sufu clique hoped by this means to maintain their control of the government while adopting a more moderate program. Sufu's appeal to the conservative Kikkawa was symbolic of the new trend in his thought. The *shotai* had no choice but to support Sufu since the only alternative was a Mukunashi government which would have promptly disbanded them. The loyalist clique had a few representatives in the han government, such as Kido and Takasugi, but, in the climate of opinion created by the approaching Bakufu armies, their influence was slight. The only course left to them was to support their erstwhile allies in the hope that the future would provide other opportunities for action.

The Mukunashi clique, on the other hand, was supported by high-ranking samurai houses in Hagi and, as in 1863, by the Sempōtai. The latter elite guard unit had become, in the light of the two great defeats of the *shotai* in 1864, even more belligerent and indignant that the demands they had made the year before had not been carried out. Soon after the defeat of the *shotai* in the attempted Kyoto countercoup, Sempōtai members began to filter into Yamaguchi, where they congregated at the Enryōji, a temple on its outskirts. From this camp they dispatched memorials to the daimyo urging him to carry out a complete change of the high officials of the han. Only a new government and an attitude of complete submission toward the Bakufu armies, they argued, would serve to save the han.

During the weeks that followed neither side gained a decisive victory. The Sufu-loyalist government was unable to curb the growing conservative strength, but the conservatives were equally unable to unseat their opposition. On 20/8/64 Mōri Izumo, a partisan of the conservatives, was appointed to the Council of Elders. The Sufu extremist officials countered this by instituting a reform of the government in which civil and military offices were combined. The consequence of the reform was to increase Takasugi's power, and, for the first time, to give official position to Inoue Kaoru and several other loyalist military figures. Orders to disband were sent to the conservative troops encamped at Enryōji, but these were not heeded.

In the face of this reform, the conservatives advanced another step. Okamoto Yoshinoshin, a member of the Mukunashi clique, was newly appointed to official position on 1/9/64, and, shortly thereafter, Mōri Ise and Mōri Noto, both conservatives, to the Council of Elders. It will be remembered that the Council of Elders, in spite of having lost its early legislative function, was still of importance in times of crisis when the rule of a particular bureaucratic clique was called into question. And because this was such a time, the appointment of the conservative Elders had more importance than it otherwise might have had. On 8/9/64 Kikkawa, the daimyo of Iwakuni, arrived in Yamaguchi for conferences with the Chōshū daimyo. He was visited in turn by loyalists, moderates, and conservatives; each party tried to convince him of the correctness of its own position, but, as Sufu had foreseen, Kikkawa favored the conservative position. When the han government did not accept his views, he asked on 16/9/64 for permission to return to Iwakuni. The han government bade him stay on in Yamaguchi; it could not agree with him nor could it do without him.

The government thus teetered between the two groups. On 25/9/64 the daimyo called a general conference of Elders and high officials to decide on han policy. At the conference, Mōri Ise and others of the conservative camp reiterated their argument that Chōshū could be saved only by adopting an attitude of complete penitence for its past misdeeds. To this Inoue Kaoru replied that the han must be prepared to fight if the terms offered by the Bakufu proved too harsh. The samurai of Chōshū must not accept the execution of their daimyo or the diminution of the Chōshū domains. The debate lasted many hours; the final decision of the daimyo favored the Sufu-loyalist government. But on the same night, two events took place which reversed the decision. Inoue Kaoru, on his way home from the conference, was set upon by assassins and severely wounded. And at about the same time, Sufu Masanosuke committed suicide. In a sense Sufu's suicide was a recognition of the fact that the han could not possibly maintain its loyalist program in the face of the powerful Bakufu armies. It was also a personal atonement reflecting Sufu's consciousness of his own responsibility for the plight of the han. For the loyalists with their religio-political philosophy, even the failure of their program could be rationalized as one additional count against the evil Bakufu. For Sufu, however, the han was the final end, and the failure of his policies could not be excused. For some weeks he had been depressed and brooding. At one point, even when aware that the daimyo had called for him, he had not gone out because of shame over the consequences of his policies. He felt, it was reported, that all those bearing the responsibility for the Kyoto countercoup should commit suicide, and he intended that his own actions should set an example for the others and for future generations.[39]

With Sufu gone and Inoue close to death, the tide turned in favor of the conservatives. On the following day Shimizu Seitarō, the only Elder who still supported the loyalists, resigned, and changes of personnel took place in many of the important offices. On 3/10/64 it was decided that the government should return to Hagi, the center of the conservatives, and the daimyo, the symbol of authority, returned to the Hagi castle, accompanied by Kikkawa, who had overseen the shift in the government. Within the first month of conservative rule, Maeda Magoemon of the Sufu clique as well as the other loyalist officials had been dismissed from office. Satisfied that the government was in conservative hands, Kikkawa returned to Iwakuni on 17/10/64. On 22/10 Mukunashi, who had been in confinement since the previous year, was pardoned, and two days later he was appointed to one of the high offices in the han government. Miyake Chūzō, another member of the Mukunashi clique, was appointed to office along with others sympathetic to the conservative position. These appointments secured the conservative grip over the government and the daimyo. An order was issued to disband the *shotai,* but it was ignored by the members, who continued to group in and about Yamaguchi, hoping to find some opportunity for renewed action. The inability of the *shotai* to act underlines the impossible situation of the former Sufu-loyalist alliance. The new government was now ready to begin negotiations with the Bakufu armies. As Saigō had hoped, "the righteous" in the han, with the

[39] *Ibid.,* pp. 209–220.

help of the Bakufu armies, had already moved to punish the "men of violence."

The Uneasy Peace

On 20/10/64 the branch han daimyo, Kikkawa Tsunemoto, met with a samurai and two monks sent to Iwakuni by Tokugawa Keishō, the leader of the Bakufu expedition. He was informed that Keishō was disposed to treat Chōshū with the greatest leniency if only Chōshū's daimyo were to give adequate proof of his submission. The messengers argued that a bloodless settlement would not only benefit the house of Mōri, but also the entire country. Kikkawa agreed with this and sent two of his vassals to Osaka to ask Keishō if the execution of the three Elders responsible for the countercoup would constitute an adequate proof. Kikkawa wrote: "As the time set for the attack draws near . . . the daimyo and his samurai, united in distress, weep tears of blood. Therefore, without waiting for orders, the heads of the guilty Fukuhara, Masuda, and Kunishi, are being readied for inspection, and their staff officers will also be executed."[40] Keishō replied that the execution of the three Elders would suffice to prove the sincerity of Chōshū's change of heart, and on 3/11/64 sent out an order postponing the attack which had been scheduled to begin on 18/11.

This order, however, was subsequently cancelled because of opposition by the field commander at the operational headquarters in Hiroshima. At this point Saigō Takamori, now a staff officer with the Bakufu army, arrived in Hiroshima. Saigō's presence in the Bakufu headquarters was important because of the role he played in deflecting the Bakufu army from its original punitive objectives. Why he could do this, why he could influence the Bakufu commander, Tokugawa Keishō, as he did, is difficult to explain, except, perhaps, by the force of his personality. Satow spoke of Saigō as the most compelling figure he had met during his years in Bakumatsu Japan.[41] In any case, soon after his arrival in Hiroshima, Saigō

[40] *Ishin shi,* IV, 193–174.

[41] Sir Ernest Satow, *A Diplomat in Japan* (London, 1921), p. 267.

proceeded boldly to Iwakuni in Chōshū itself where he met with Kikkawa and suggested, as Keishō had earlier, that Chōshū, as proof of its submission, send the heads of the three Elders to Hiroshima and execute the four staff officers responsible for the 1864 countercoup. At this time Saigō also returned to Chōshū ten samurai who had been the prisoners of Satsuma since the time of the attempted countercoup. (This could be cited as another example of han autonomy: the prisoners that Satsuma took, Satsuma kept, and Satsuma returned, as it pleased.) Some have spoken of this gesture as the first step in the gradual amelioration of Satsuma-Chōshū relations.

Kikkawa sent a messenger to the han government informing it of Saigō's "suggestions." He also counseled the han to swiftly carry out these measures as proof of its submission. Hearing of this, about seven hundred and fifty *shotai* samurai gathered in Yamaguchi on 4/11 to draw up a memorial to the daimyo. They argued that it would be wrong to execute the three Elders, that the present slothful government would make it appear as if the daimyo himself were guilty, that the daimyo should return to Yamaguchi, that officials of worth should be appointed, that the defenses of the han should be strengthened, that the Emperor should be honored, and that all foreigners should be driven from Japan. Although they were unable to act, even in the face of the advancing Bakufu army, the ideological unity of the *shotai* was preserved.

The Chōshū government had little time in which to act. Chōshū was already surrounded by the Bakufu armies, which were to attack on 18/11. To effectively delay the attack, orders would have to leave Hiroshima by 14/11. Kikkawa, therefore, sent another messenger to Hagi asking that the three Elders be executed by 12/11. The conservative government agreed. On 11/11 the three "extremist" Elders, Masuda at the age of thirty-four, Kunishi at twenty-four, and Fukuhara at fifty-four, were ordered to commit suicide, and on the same day, the four staff commanders of the ill-starred Kyoto expedition were executed at the Hagi prison.

Messengers were then sent from Chōshū to Hiroshima; the heads of the three Elders were presented; it was reported that the four staff officers had been executed, and the Bakufu representatives were informed that Kusaka, Terashima, and Kijima had died during the fighting in Kyoto. Thereupon, the Hiroshima headquarters issued orders to delay the attack on Chōshū until further notice. On 16/11 Kikkawa met with Bakufu representatives in Hiroshima and requested that the daimyo of Chōshū be treated with clemency since he had not been involved in the Kyoto incident. If he had not been involved, the Bakufu representatives queried, why had an order bearing his seal been found? Kikkawa explained that the daimyo, "deceived by violent retainers, had handed it over even while doubting" the wisdom of his act.[42] He was next asked why the Chōshū troops accompanying the Chōshū heir to Osaka had worn battle-dress. Kikkawa replied that a rumor of a possible foreign attack had been circulated and battle-dress had been ordered so that the Chōshū troops might assist in repulsing the attack. Why then, the Bakufu envoys pursued, did the Chōshū heir turn about and go back to the han when he heard of the defeat of the Chōshū troops in Kyoto? To this Kikkawa replied that the Chōshū heir, astonished at the news of the attack, had returned to the han for further orders from the daimyo. The Bakufu representatives were persistent: where were Kido, Takasugi, and Sasaki? Kikkawa answered that their whereabouts were unknown but that an investigation would be made.

On the same day Tokugawa Keishō arrived at Hiroshima; he inspected the heads of the Chōshū Elders and sent a report on Chōshū's submission to the Bakufu and the Court. On 19/11 Kikkawa was summoned before the Bakufu commander who ordered him to obtain a confession of guilt from the daimyo and the Chōshū heir, to have the Yamaguchi castle destroyed, and to hand over the five extremist Court nobles who had fled to Chōshū in 1863. If these demands were fulfilled, then the Bakufu army would disband. Kikkawa transmitted the relatively light demands to the government of Chōshū.

The conservative government was gratified to learn that the daimyo was to go unpunished, and it immediately set about to fulfill the conditions laid down by Keishō. A statement of guilt was forwarded and

[42] *Ishin shi*, IV, 162.

the headquarters of the former Sufu-loyalist government in Yamaguchi was destroyed. The government, however, was unable to deliver up the five nobles since they were in the hands of the *shotai,* which would not recognize the authority of the Mukunashi government. In fact, upon hearing on 15/11 that the conservative government had ordered their subjugation, the *shotai* had left Yamaguchi, which was poorly situated for defense, and proceeded, seven hundred and fifty strong, to Chōfu, one of the branch han in Chōshū, whose daimyo they hoped to win to their cause. They were accompanied by the five nobles. Aware that this could become a stumbling block in the course of the negotiations, Saigō Takamori hastened to Shimonoseki to confer with the *shotai* leaders and the nobles. On 15/12 Saigō persuaded the latter to leave Chōshū; they agreed to go to Chikuzen han within two weeks time, and although they did not actually leave Chōshū until 14/1/65, Saigō, speaking for the Bakufu army, declared himself satisfied on this point.

The way was now clear for the conclusion of the expedition. The sole obstacle remaining was the problem of the *shotai.* As they had formed the main body of the attacking forces in the attempted countercoup, their continued organized existence in the han was inimical to the purposes of the Bakufu expedition. When questioned by Bakufu inspectors concerning the *shotai,* the conservative Elder Mōri Ise answered that this was a very difficult matter with which to deal. The Bakufu inspector then warned that "the troops of Owari would strike them down if Chōshū had not the strength to do so." [43] Feeling that its position would be irreparably damaged in the eyes of the Chōshū samurai if it called in outside troops against the *shotai* who, though dissident, were nevertheless samurai of Chōshū, the conservative government decided to use the han army to force their dissolution. This decision satisfied the Bakufu inspectors who announced that by and large the han was in a state of submission.

A small outbreak took place at Shimonoseki when several small *shotai* under Takasugi and Itō attacked a local government office for money and supplies. But the majority of the *shotai* had abstained and, considering that even now the Chōshū army was moving to

[43] Suematsu, *Bōchō kaiten shi,* VI, 409.

suppress them, the Bakufu headquarters under Keishō saw no reason to prolong the expedition. Besides, many of the participating han were grumbling at the expense of maintaining their forces in the field. Consequently, on 27/12/64 Keishō announced that the objectives of the expedition had been accomplished, and the troops left for their separate han. The first Chōshū expedition thus came to an end.

CHAPTER VIII

The Chōshū Civil War

The Origins of the War

The first Bakufu expedition against Chōshū ended on 27/12/64 with the disbandment of the Bakufu army. Though its success was to prove illusory in the long run, over the short term at least, it had all the appearances of a decisive victory. The days when Chōshū was a threat to the Bakufu seemed past: controlled by the conservative Mukunashi government, its former leaders liquidated, its regular army even now moving against the obstreperous *shotai* which had attacked the han offices at Shimonoseki, there seemed little possibility that Chōshū, beaten and humiliated at every turn, could once again regain its position as the leader of the pro-Court forces of all Japan. Or, so at least thought the leaders of both the Bakufu army and the Mukunashi government. And yet, both were wrong; both had misjudged the temper of the han. The Bakufu army had been fooled, and the Chōshū government had deceived itself, into thinking that a government formed under the pressure of an alien army could endure when that pressure was removed. Perhaps the Mukunashi clique was deluded by its possession of the daimyo, thinking that this would guarantee their survival as well as confer on their rule a titular legitimacy.

Many have viewed the uprising of the *shotai* against the Mukunashi government as an uprising or even a rebellion of lower samurai against a legitimately constituted han government. Actually, on observing the situation within Chōshū at this time, one could argue with greater plausibility that there existed not one but two han governments, each of which in its own right possessed a certain legitimacy. One, of course, was the Mukunashi government at Hagi. The other was an activist pro-Court government whose existence was only potential, but it was an idea to which the majority of Chōshū samurai were partial and an idea which had the military support of the *shotai* forces. Or, to use another image, the pro-Court government existed only in the sense in which a compound exists that is broken into ions in a chemical solution. The ions which in Chōshū would later crystallize into a new pro-Court government were the few remaining members of the old Sufu clique, moderate samurai who had previously given their support to the Sufu government, and the *shotai*. The last group, originally a united force under the Sufu government, was now broken apart, and each militia unit acted as a free agent according to its particular internal inertia. The reason why each unit could act as a free agent lay, first, in the nature of the chain of command under which they were organized, and second, in the degree of autonomous power vested in their commanders (*shireikan*). *Shotai* members, possibly because it was felt that they were less responsible than the samurai of the regular han military units, were placed under commanders who in turn were responsible to the "generals" (*rōjū*) of the han military. Members of the conservative Sempōtai, on the other hand, were directly responsible to the "generals."[1] And, organized by Takasugi, the *shotai* were led by extremist samurai. Moreover, the *shotai* commanders were often given funds and equipment by the han government, and thus were practically autonomous as long as the funds lasted. Therefore, with the change in government, the *shotai* could continue to agitate as isolated focuses of resistance within the han.

One should also note that the *shotai* forces could lay claim to a legitimacy at least equal to that of the Mukunashi government. They represented the principles for which the entire han had fought in the recent past; they upheld the martial pride of the samurai

[1] Tanaka Akira, "Chōshū han kaikakuha no kiban: shotai no bunseki o tōshite mitaru," *Shichō,* 51:18 (Mar., 1954).

creed, which found repugnant Chōshū's abject submission to the Bakufu army; and, by involving the han in national politics, they sought to regain the glory that had belonged to the house of Mōri before the Tokugawa victory, in 1600. In the eyes of the *shotai* forces, the Mukunashi clique had sold the honor of the han to maintain its material existence—a transaction it felt called on to repudiate. In such a situation the control of the daimyo by the Hagi government was also less important than it might appear. The *shotai* held that the daimyo was being forced against his true will to cooperate and that to rebel and retake the daimyo was no more wrong than their unsuccessful attempt to retake the Emperor in the countercoup of 7/64.

On the other hand, in spite of having come to power only as a caretaker in the face of the approaching Bakufu armies, the claims of the Mukunashi government were not inconsiderable. They were a traditional bureaucratic clique of the han; they had ruled from 1855 to 1858. They undoubtedly upheld what they felt to be the best interests of the han. Since 1858, and especially since 1861, they had predicted that the activist policies of the Sufu clique would bring the han to ruin. They had remained prophets in the wilderness; yet from 1861 on their auguries had been realized: the han had witnessed a steady decline in its fortunes, culminating in its humiliation by the Bakufu army. Moreover, the Mukunashi group could truly say that only their prompt actions had saved the han from invasion and perhaps even from dissolution. By calling on Kikkawa and by committing suicide, even Sufu had recognized the need for a new government. And, as the leaders of the new government they had not only preserved the han but had also saved the two Mōri. Was this not, after all, the substance of feudal loyalty?

As we have already noted, the conflict between the two opposing factions began when Takasugi led an attack on the han offices at Shimonoseki. Takasugi had returned earlier on 25/11/64 from Chikuzen, where he had been hiding since the attempted countercoup in Kyoto. He remained for several weeks in Shimonoseki, however, feeling that it was still too early to strike. Then, on 13/12/64 he proceeded to Chōfu, one of the branch han to which the majority of the *shotai* had withdrawn. Meeting with the commanders of the various *shotai,* he tried to persuade them that the time had come for action. The other *shotai* leaders, however, protested that the time was still too early and tried to dissuade him from action until a later date. Only the Yūgekitai under the leadership of Ishikawa Kogorō was willing to join in an uprising. This was the same unit which earlier under Kijima had been so importunate concerning the countercoup in Kyoto. Takasugi then went to Shimonoseki where he secured the support of Itō Hirobumi.[2] Itō at this time led the Rikishitai, a small unit composed largely of professional wrestlers, which he had been given to command during the peace negotiations after the four-nation attack on Shimonoseki a few months earlier.

On 16/12/64, Takasugi, Ishikawa, and Itō led their units against the government offices at Shimonoseki. The purpose of the attack, in Itō's words, was to "raise troops and go to Shimonoseki, to take it, and to use it as a base for operations against the *zokuron* party [the Mukunashi government]." The attack met with no resistance and the government offices were quickly occupied. Subsequently, Takasugi crossed over to Mitajiri with a party of volunteers, where he seized an armed sloop belonging to the han. Again in Itō's words, "the *zokuron* troops were certain to come to attack us in Shimonoseki, therefore we planned to anchor this ship off shore and use it as a floating fort to strike down those who come by land."[3]

That Takasugi should have attacked the Mukunashi government at the very moment that Bakufu inspectors were questioning the government about the *shotai* is difficult to comprehend. To have waited until the Bakufu army withdrew, as did the other *shotai,* would have been a far more prudent course. On the one hand, the premature attack indicates the degree of animosity which the extremist leaders harbored toward the han government for having executed the most important members of the Sufu clique; on the other hand, it illustrates the extent to which the *shotai,* no longer restrained by the rational calculations of bureaucratic leaders, had become independent actors under their various commanders. This

[2] Suematsu Kenchō, *Bōchō kaiten shi* (Tokyo, 1921), VI, 404.

[3] *Ibid.,* p. 406.

"premature" attack by Takasugi and others might have been, although it was not, a situation in which "for the want of a nail a kingdom was lost." As we have noted before, the Bakufu inspector offered to loan the Mukunashi government several detachments of Owari soldiers to put down the *shotai.* The government refused, saying that it would do it alone. Had the inspector insisted, or had the government requested such military aid, or had the troops under Takasugi been so numerous that the Bakufu army would have felt it necessary to interfere, the various *shotai* would most certainly have been forced to dissolve. And had this taken place, then the pro-Court forces of Chōshū would have had no rallying point and Chōshū might have continued indefinitely under the conservative government of Mukunashi, even until some "other restoration" by other han.

It is also interesting to note that of the three leaders who were responsible for this uprising, two, Takasugi and Itō, were samurai who only a few months earlier had been forced into hiding in order to escape assassination at the hands of *shotai* ideologues for their role in the treaty negotiations with the representatives of the four powers. This is symbolic, first, of a certain change in the thought of the leadership of the extremist movement, and second, of the change involved in the transition from the *sonnō jōi* to the *tōbaku* movement, that is to say, of the loss of the *jōi* element. However, one must not make too much of the change at this time. The han had signed its own treaty with the powers, the Court would soon be forced to follow suit, and as time went on, more and more of the extremist leaders saw the handwriting on the wall. Yet, at the time of the *shotai* uprising, the bulk of the *shotai* samurai and even most of their leaders felt that the treaty had been obtained under duress; they remained unchangingly *jōi* in thought and feeling. Having studied abroad, Takasugi and Itō were very much the exceptions to the rule.

The leadership by Itō of a small unit of troops in the Shimonoseki attack also symbolizes a certain shift in the group which led the extremist movement. Itō was a Chōshū *sotsu* who could never have risen under ordinary circumstances. His early career pattern contains a number of significant elements. He had been a student of Yoshida Shōin; he studied abroad, he acted as translator in the negotiations at Shimonoseki, and he had been put in command of a *shotai*. Too much cannot be made of this; Itō was still extremely unimportant: his Rikishitai was one of the smallest of the *shotai* and he was never to gain an important bureaucratic position in the han. Nevertheless, his career was under way.

After the first attack on Shimonoseki, Takasugi's forces withdrew to the countryside where they camped at local temples. The Bakufu army disbanded on 27/12/64, and a few days later, on 2/1/65, Takasugi again raided Shimonoseki with a party of thirty Yūgekitai troops. He seized various official goods of the han and left behind a manifesto concerning "striking down traitors" and announcing that he could not live under the same sky as the "evil party."[4] As we noted previously, the government's reaction to Takasugi's initial attack was to mobilize certain units of the regular han army. The Mukunashi goverment forwarded to the daimyo for approval an order for the "subjugation" of the *shotai*. The daimyo changed "subjugation" to "pacification," approved the order, and appointed Mōri Senjirō, the commander of the conservatively oriented Sempōtai, to lead the han forces against the *shotai*.[5] They set out on 25/12/64; their aim was to secure the submission of the *shotai* without actual combat in much the same way that the Bakufu army had obtained Chōshū's submission without battle. Upon receiving word of Takasugi's second attack on Shimonoseki, the army advanced to the village of Edō on the road to Shimonoseki. Edō was approximately fifty miles from Shimonoseki. Only fifteen miles distant on the same road, however, was the village of Isa where seven hundred and fifty "neutralist" *shotai* forces were encamped. The han army units wanted to proceed to Shimonoseki but they were afraid to pass through Isa for this would leave a potentially hostile force in their rear. Consequently, they summoned the leaders of the Isa *shotai* and ordered them to disband and return to the han the weapons that had been given to them at the time of their formation. The *shotai*

[4] *Ibid.*, VII, 16.
[5] *Ibid.*, VI, 430–431.

leaders readily expressed their willingness to obey the directives of the han but asked for a short delay. During this delay, on the night of 6/1/65, while negotiations were still in progress, one hundred of the Isa *shotai* soldiers attacked and completely routed the four-hundred-man army sent out by the government. In a declaration said to have been given to the government forces only moments before the attack, the *shotai* argued: "Since the affair in Kyoto last fall [the unsuccessful countercoup], Mukunashi and other evil men have gained office. They have put aside honest, loyal samurai; they have killed or thrown into jail several tens of honest officials . . . and, changing the basic policy of the han, they have lost sight of duty [*meigi*] before all the land and have brought a hitherto unexperienced shame to the han. . . ." [6]

Why had these *shotai* who had earlier been cool to Takasugi's proposals now taken the offensive themselves? First, and most important, the withdrawal of the Bakufu army opened the way for action. Second, since Takasugi's first attack, another wave of executions and arrests of former government personnel had taken place in Hagi; this appeared to the *shotai* as a nefarious move by the Mukunashi clique to wipe out the opposition within the han. Third, the above executions undermined the prestige of Akane Taketo, the leader of the Kiheitai who had advocated cooperaton with the Mukunashi government. Suspected of intrigue, he was forced to flee from Chōshū, and his place was taken by Yamagata Aritomo, whose position was much closer to that of Takasugi. Finally, the *shotai* at Isa had no choice but to fight or disband, and they had already decided against the latter.

The unexpected defeat caused great consternation to the Mukunashi government, which, for the first time, realized that it was literally fighting for its life. Martial law was declared in the castle town, and a general mobilization of the han troops was ordered. On 10/1/65 a new and more powerful army marched out to subdue the *shotai*. On 11/1 and 12/1 seven hundred *shotai* troops fought with almost twice the number of government troops; the battle ended without victory for either side.

[6] *Ibid.*, VII, 23–24.

In the meantime each side strove to strengthen its forces. The Isa *shotai* were reinforced by the units of Itō and Ishikawa, which had fought under Takasugi at Shimonoseki. They were also joined by the Kōjōtai recruited in the Ogōri and Yamaguchi regions and led by Inoue. With this newly found strength the *shotai* began to advance toward Hagi, defeating the regular troops of the han in almost every battle. The Mukunashi government was now frantic in its efforts to bolster its forces. Attempts were made, though with little success, to mobilize peasants as soldiers. The families of the three Elders who had been executed for their part in the Kyoto countercoup were summoned to Hagi; their fiefs which had been confiscated were restored in a vain attempt to mobilize their vassals for the government cause. The Mukunashi clique also sent messengers to the branch han urging them to send troops to support the han government. Here they did meet with success. Kikkawa of Iwakuni, who had been so instrumental in the founding of the conservative government, personally led his vassals to the aid of the government. In spite of this, the *shotai* forces now based on Yamaguchi continued to advance until on 16/1/65 they drove the government troops from Sasanami, a village halfway between Yamaguchi and Hagi.

Meanwhile, the reverberations reaching Hagi of the growing number of *shotai* victories caused a gradual fragmentation of the political base on which the Mukunashi government rested. Not only was the Mukunashi government, which had been unpopular from the start, fighting a fratricidal war and bringing on the han devastation such as not even the Bakufu army had done, but it also was losing the war. This was unforgivable. Moreover, the military base of the Mukunashi government was weak in Hagi. The Sempōtai, the "claws and fangs"[7] of the conservative government, together with all the troops which the conservatives trusted, were away fighting the *shotai* in the south. The only active supporters of the Mukunashi government in Hagi were a group of older samurai who, bound by a common animus against the *shotai,* gathered at the Shōkōji (a temple) and encouraged the government to hold out against its enemies. This faction, however, was of negligible military

[7] *Kanjōtai shimatsu,* one document among others in the category entitled "Shotai rui" in the Chōshū Archives at the Yamaguchi-*ken* Library in Yamaguchi.

significance. The only military forces in or about Hagi were not ardent supporters of the conservatives. As one account relates, "the four approaches to Hagi were guarded by troops who either would not submit to the arguments of that party [the Mukunashi faction] or were too old or weak to be of use [in battle]." [8] These troops were not sent to the front lines because the Mukunashi group feared that they would be uncontrollable if dispersed on the various battle lines, and, in spite of their being almost under the very eyes of the Mukunashi government, spies and police were constantly sent out to report on their activities.

The Mukunashi government had good reason to distrust these troops posted at the entrances to Hagi. A certain group had been meeting secretly in Hagi almost from the beginning of the civil war in the hope of rising against the Mukunashi government. This group was also in contact with Kaneshige Jōzō, one of the few surviving members of the original Sufu clique, who was now on duty at the Kawachi entrance to Hagi. When the group asked Kaneshige for advice, he replied that he was considering three possible policies: one, to go to the castle and ask the daimyo what should be done; two, to enter the castle with samurai and hold it against the Mukunashi government; three, to take the daimyo to a temple outside the city and to await for further developments there. Upon hearing of the *shotai* victory on 6/1/65, this group wanted to rise against the Mukunashi government, but Kaneshige warned them that it was still too early. By 16/1/65, however, Kaneshige agreed that the time had come. Seventy or so samurai went to the castle, and after expressing their views to the daimyo they retired to the Kōkōji, a temple on the outskirts of Hagi. As the daimyo had instructed them to remain peacefully and to await further orders, they called themselves the Peace Assembly (*chinsei kaigi*). Others, hearing of this group, came to join, and soon their numbers increased to over two hundred men. As a samurai group, its primary character was military; it was essentially a group of fighting men who had resolved not to participate any longer in a senseless civil war. This Peace Assembly can be thought of as a

[8] *Ibid.*, p. 3 (actually, the pages were not numbered; this was the third page of the narrative).

group whose political position was approximately that of the former Sufu clique. They were sympathetic to a policy of han participation in national politics but were not so partisan as to rise in armed rebellion against a duly constituted government headed by the daimyo. Their sympathies, however, were clearly with the *shotai*. In a series of memorials they demanded: a change in officials, a clean-up of the han government, the end of the civil war, and the conservation of the han's strength. They wrote:

> Though for over a month several thousand men have been sent out to subjugate the *shotai* . . . there are no signs of victory; on the contrary, the strength of the *shotai* increases day by day. It will be extremely difficult to destroy them as they are supported by the samurai and even by the peasants in the various regions [of the han]. Moreover, burdened with orders and tired, the masses may rebel. . . . This is truly a time of crisis.[9]

The Mukunashi government, of course, ordered the Peace Assembly to disband, but its orders were ignored since it no longer had the means to enforce them.

The social composition of the Peace Assembly may be taken to illustrate the complexity of the Chōshū civil war. If one accepts the simple view of the civil war as a class war between low-ranking *shotai* soldiers and high-ranking conservative samurai, then one would expect the Peace Assembly to be an in-between group, too high to join with the *shotai* but too low to support the Mukunashi government. Actually, this was not the case. The *Bōchō kaiten shi* gives the names of thirty-one of the "charter members" of the Peace Assembly who met secretly in Hagi. Of these thirty-one samurai, I have been able to discover the stipend or fief *(kokudaka)* of twenty-five: the average stipend was 182.2 *koku,* a considerably higher figure than that of the bureaucratic cliques, both of which averaged about one hundred *koku.*[10] This would seem to indicate that the Peace Assembly was composed of relatively higher-ranking

[9] Suematsu Kenchō, *Bōchō kaiten shi,* VII, 46–48.

[10] The names of the *chinsei kaigi* samurai are given in Suematsu, *Bōchō kaiten shi,* VII, 207–210. Their *rokudaka* (stipends) were entered in the *Kyū Chō-han shoshin ichiran sōkō* in the Mōri House Library of the Yamaguchi Bunshokan.

samurai than the Mukunashi government. Can the rebellion of such a group be accounted for in terms of class interest? Only if the Mukunashi government can be shown to have represented a group of still higher-ranking samurai. The following tabulation shows the composition of the Chōshū samurai class (excluding *baishin*):[11]

Name of Rank	Number of Families
Ichimon	6
Eidai karō	2
Yorigumi	62
Ōgumi	1378
Funategumi	29
Enkintsuki	216
Jishagumi	87
Takajō, Ujō	7
Mukyūdōri	512
Sentō	50
Zempu	25
Edo kachi	75
Chi kachi	55
Sanjūnindōri	55
Jinsō	39
Osaka sentō	1
Shiko	118
Ashigaru and below	2958
Total	5675

Of the 661 families with stipends or fiefs of over one hundred *koku,* all but 22 were in the top four strata. The Mukunashi government cannot be said to have represented the top two strata, the eight Elder families, since at least half of them sympathized with the *shotai.* It may have represented the third stratum, the *yorigumi,* whose fiefs ranged from two hundred and fifty to five thousand *koku,* but no evidence has yet been uncovered to show that this was the case. (If the group of older samurai who met at the Shōkōji to lash the government on to greater exertions could be shown to be members of this stratum, this would indicate a clearer class differ-

[11] Kimura Motoi, "Hagi han zaichi kashin dan ni tsuite," *Shigaku zasshi,* 62:34 (Aug., 1953).

ential than seems, at the present, to have existed.) And, the members of the Peace Assembly (or at least those whose stipends I have uncovered) were still of higher rank (or of higher income) than the majority of the government troops who fought against the *shotai*. As we shall see, class attitudes were certainly involved in the civil war, but they alone do not suffice to explain it.

The Fall of the Mukunashi Government.

Caught between the military force of the advancing *shotai* and the maledictions of the almost equally hostile Peace Assembly, the Mukunashi government lost control of the daimyo, who now veered in the direction of the more or less moderate Peace Assembly. On 21/1/65 the daimyo sent a personal messenger, Mōri Motosumi, to the *shotai* camp at Sasanami, asking them to withdraw to Yamaguchi until a new program could be established. The *shotai* refused, aware that only continuous pressure would enable them to achieve their ends, but they did agree to a truce until 28/1/65. The week of truce was a week of inaction. Two members of the Mukunashi group were removed from their posts and Mukunashi himself was relegated to a position of relatively small importance; however, this was merely a façade, for underneath they retained control of the government. In spite of a stream of memorials from the Peace Assembly urging a complete change of officials, the effect of the truce was to put off the dismissal of the conservatives. On the twenty-eighth the *shotai* once again began to advance, and, on the same day, a warship under the control of the *shotai* fired at the Hagi castle as a reminder to those within of the growing *shotai* strength. (Since the daimyo was in the castle, the cannon were, of course, unloaded.)

On the same day, the daimyo and his Elders set themselves to the task of forming a new government. Mukunashi, Miyake, and all other members of their clique were dismissed, and in their place were appointed Kaneshige Jōzō and Yamada Uemon, both men associated with the former Sufu government. Lesser positions formerly held by the conservatives were also newly filled with supporters of the former moderate-extremist government. This change in government was very similar in nature to the change which had brought Mukunashi to power several months before. At that time the approaching Bakufu armies had caused the rise of the conservatives in the han; now the approaching *shotai* army had caused the fall of the same group within Hagi. In both cases the daimyo and a few Elder advisors who could be swayed in either direction by sufficient pressure dismissed one government and appointed another and so avoided the necessity of a battle in which the outcome was apparent beforehand. The mechanism by which the change was produced seems essentially the same as in 1853, 1855, or 1858. Political controversy had given way to military struggles, military power rather than public opinion became the force that moved the daimyo, and yet, as the focus of samurai loyalty, the daimyo never became the pawn or the captive of any one group. He could move in advance to recognize a new distribution of power even before it had been realized, and, in spite of having been a nonentity personally, in a certain sense he remained an independent variable in Chōshū politics even in the late Bakumatsu period.

Informed of the changes in the government, the *shotai* halted their advance on 2/2/65 and sent a letter to the daimyo explaining the reasons for their actions. On 9/2/65 the new government met before the daimyo and reaffirmed the 1858 han policy of "loyalty to the Court, trust to the Bakufu, and filial duty to the ancestors [of the house of Mōri]." It was observed that, "although from time to time changes of officials may take place for tactical reasons, the basic policy of the han remains unchanged."[12] Such a banal declaration can only be viewed as an attempt to mask the daimyo's facile changes between policies which in fact were basically antithetical. The Kaneshige-Yamada government ordered the Sempōtai to disband and removed the troops which had been guarding the approaches to Hagi. On 10/2/65 they sent four prominent members of the Peace Assembly to Yamaguchi to exchange opinions with the leaders of the *shotai*. The situation in Hagi was still very fluid: the relations between the *shotai* and the new moderate government were as yet undefined, and this exchange of opinions was expected to lead to a

[12] *Ishin shi* (Tokyo, 1939), IV, 435.

rapprochement between the two groups, which were basically alike in their aims.

On 11/2/65 after having conferred with the *shotai,* the four Peace Assembly samurai left Yamaguchi to report back to Hagi; three, however, were assassinated at Akiragi by members of the Sempōtai. This was part of a plot by which the conservative military hoped to exacerbate relations between the *shotai* and the Peace Assembly and then to persuade the latter to reinstate the government of Mukunashi. No sooner had the assassinations taken place than rumors of violence by the *shotai* were carefully spread, and the Sempōtai samurai gathered to go to the castle to "inquire into the guilt of the new government for trusting the *shotai.*" [13] But their plot miscarried. Even as they marched toward the castle, samurai from the Peace Assembly barred the gates and refused to admit them. At the same time messengers were sent to Yamaguchi urging the *shotai* to send troops to the rescue. The *shotai* responded, and by 15/2/65 they were in possession of Hagi. On 27/2/65 the daimyo and the government returned to the extremist stronghold of Yamaguchi. On 23/3 the *shotai* policy of "armed submission" to the Bakufu became the official policy of the han. Less than three months after the withdrawal of the Bakufu army, all that it appeared to have accomplished had been undone. Chōshū had once again become the protagonist in the movement against the Bakufu.[14]

The government newly formed in Yamaguchi is worth considering in greater detail, since it continued without further major shifts in personnel until the Restoration. If the former Sufu government can be thought of as resting on a moderate-extremist alliance, one in which bureaucratic positions were almost completely dominated by the moderate Sufu clique, the government formed after the civil war can be called an extremist-moderate alliance in which the decision making positions in the government were divided between the two groups. This did not, however, result in a split policy, since by this time both groups were committed to the extremist pro-Court

[13] *Ibid.*, p. 435.

[14] Undoubtedly the best description of the events during the Chōshū civil war is given in Suematsu, *Bōchō kaiten shi,* VII, 15–108.

doctrine. And yet one can ask where power did actually rest at this time. Was it held by the government or by the *shotai*? Did the extremists in the government merely represent the collective will of the *shotai* or, as former leaders of the *shotai* who had now advanced to obtain bureaucratic position, were they able to utilize their new authority to exercise an even more effective control of the *shotai*? The latter was almost certainly the case. Using their newly found authority and carrying out the program for which they had led the *shotai* against the Mukunashi government, they were able to re-impose a strict discipline and sense of hierarchic order on the hitherto autonomous *shotai* and to reform them within the cadre of a newly organized han army.

The most important moderate members of the new government were Kaneshige Jōzō of the former Sufu clique, Yamada Uemon, who had directed military reforms under Sufu, and Tamaki Bunno-shin, a career bureaucrat and the uncle of Yoshida Shōin.[15] The most important extremist officials were Kido and Takasugi, and of lesser stature were Maebara Issei, Ishikawa Kogorō, and Ōmura Masujirō. Kido was perhaps the most prominent figure in the han, almost if not quite as important as Sufu had been in the years past. Kido combined the career bureaucrat character of the moderates with the extremist character of the *shotai* leaders. Like the *shotai* leaders he had been a student at the school of Shōin and had participated in the countercoup in Kyoto. In a sense the two dominant governmental groups met in the person of Kido. Moreover, as a relatively high-ranking samurai (a fourth stratum *shi* with a stipend of one hundred and fifty *koku*) and a onetime secretary of the daimyo, he was more acceptable to the daimyo and the Elders than most of the other extremist leaders. Kido had gone into hiding in Tamba han after the unsuccessful countercoup; he returned to Chōshū after the civil war on 26/4/65 and was given official position on 27/5.

Takasugi's rise was based on his military abilities and on his position as the leader of the Chōshū loyalists. Put in charge of military reform by Sufu in 1863, he had founded the *shotai*; he was

[15] Tamaki, like Sufu, died the death of a samurai. In 1876, having learned that certain students of his had participated in the revolt of Maebara, he committed suicide.

the initiator of the han civil war; and he was (especially after the death of Kusaka in 1864) the unofficial leader of the Shōin group. But as we have seen, his leadership was not always recognized by the autonomous *shotai,* which, in time of war, often could not be controlled even by their own commanders. With Takasugi rose Ishikawa Kogorō (note that of the two *shotai* leaders who joined in Takasugi's initial attack it is Ishikawa and not Itō who is given position) and Ōmura Masujirō.

Ōmura is a particularly interesting figure: neither a member of Shōin's group nor an outstanding patriot, he seems to have risen solely because of his exceptional ability. Born a low-ranking samurai (I have not been able to discover the exact rank), he was adopted into a family who were the hereditary physicians to a district office of the han. Medicine led him to Dutch studies (*rangaku*), and because of his abilities he was made a samurai (*shi*) of Uwajima han in 1853 and given a stipend of one hundred *koku*. His duties there led him to translate works on Western military science; he became famous, and in 1860 he was invited to return to Chōshū with the rank and emoluments of a *shi*. It was chiefly Ōmura's influence that led Sufu to send Itō, Inoue, and others abroad, just when antiforeignism was at its peak in Chōshū. After the civil war Ōmura was appointed, probably by Takasugi, to reorganize the Chōshū military. It is clear that Ōmura's rise was predicated on his ability and his familiarity with Western science. My impression, while superficial, is that Ōmura was unlike the other extremist samurai of Chōshū and even unlike the Western-oriented extremists such as Itō or Inoue; perhaps, in spite of his military role, he may fruitfully be thought of as the Fukuzawa Yukichi of Chōshū.

The third of those appointed to lesser positions at this time was Maebara Issei. Like most of the others he advanced to the bureaucracy from the leadership of one of the *shotai.* Within the framework of Bakumatsu Chōshū his career appears to have been quite ordinary: after attending one or two Confucian schools he went to the Shōka Sonjuku of Shōin; he participated in the Chōshū defenses of the Hyōgo coast; he joined Kusaka and others in their impeachment of Nagai Uta, and then went on to be a member of one of the *shotai.* And yet he was not completely ordinary. Maebara was to be to Chōshū what Saigō was to Satsuma: an extremist leader who could not make the transition in the period following the Restoration. He became the leader of the Chōshū samurai who later rose against the new government. Captured while planning to sweep the "traitors" from the side of the Emperor, he was executed in 1876.

Finally, one may note in passing that Itō Hirobumi, Yamagata Aritomo, Inoue Kaoru, and Shinagawa Yajirō had all become *shotai* commanders by the end of the civil war. It is clear that this position was the springboard from which they would leap to eminence in the Meiji government. In many ways the leadership of one of the *shotai* provided a better channel for future eminence than did minor bureaucratic position in the han; many who rose through ordinary channels to become bureaucrats at this time remained, after the Restoration, in the obscurity of the officialdom of Yamaguchi-*ken*.[16] The *shotai* commanders, on the other hand, had opportunities to meet the leaders of the other important han and to develop contacts among the nobles in Kyoto. Also, in the next few years of war and plans for war, the influence of the *shotai* commanders on the extremist han leaders such as Takasugi and Kido, in unofficial if not official conferences, undoubtedly served to reinforce the personal ties formed at the school of Shōin or in the earlier phases of the movement. These ties may have been the determining factor in their later rise to power. Another factor which in its obvious way was of absolute importance was survival. For the *shotai* leaders survival alone was almost a sufficient condition for later success. Ōmura became Minister of War in the Meiji government because of the premature death of Takasugi, and after him, Yamagata, because of the assassination of Ōmura. Kusaka, Kijima, and many others would undoubtedly have stood among the Meiji leaders if they had not died in the course of the Restoration movement (assuming, of course, that they were of the Yamagata type and not of the Maebara type, and this is dubious in the case of Kijima).

[16] Of course, as we have already noted, the anti-Western Western orientation of Itō and Inoue was also vital to their rise. Yet, was this alone enough? One wonders what became of the other three samurai who were sent abroad with them.

It is also interesting to speculate what might have happened if Kido and Takasugi had died in 1864 and Kusaka had lived. Assuming that this would not have impaired Chōshū's role in the Restoration, might not the favorites of Kusaka rather than those of Takasugi (Itō, Inoue, Yamagata) have been chosen for position in the Meiji government? Or were personal relations so important during this period? It may be that the religio-political bond of the Bakumatsu pro-Court movement created a group solidarity in which such *oyabun-kobun* relations were of less importance than during the later secular maneuverings within the Meiji bureaucracy.

Finally, as Tōyama Shigeki has suggested in his brilliant work on the Meiji Restoration, the ups and downs, the wars and tensions, of the Bakumatsu period may have functioned as a shaking up or sorting out process in which only men of outstanding ability could rise, and during the process those who succeeded in rising were forged by that very process into a new type—that of the Meiji oligarch.[17]

The Problem of the Chōshū Civil War

Whether any event in the main stream of history can be singled out as more important than others is questionable, but intrinsic importance aside, some are undoubtedly more popular. If points were given according to the attention given to them by historians, the Chōshū civil war (along with the Tempō Reform and the sword hunt and cadastral survey of Hideyoshi) would certainly be among the winners in Japanese historiography after the Second World War. The reason for this is that the civil war is apparently tailor-made to fit the new conception of the Meiji Restoration as the outcome of a lower samurai-peasant alliance. In Chapter III the economic aspects of this interpretation were examined in the context of the Tempō Reform; here, the political, social, and military aspects of the same interpretation will be considered in some detail as they apply to the Chōshū civil war.

[17] Tōyama Shigeki, *Meiji ishin* (Tokyo, 1951), pp. 139–159.

There are many variations in the manner in which this theory is applied to the civil war, but they all concur on certain points. The *tōbaku* (overthrow the Bakufu) movement began with the civil war in Chōshū, emerging as it were from the ashes of the *sonnō jōi* movement. It began with a revolution of a sort by the mixed militia, the *shotai,* the embodiment of the samurai-peasant alliance. Accordingly, since the same alliance later went on to carry out the Restoration, the "Chōshū Restoration" is viewed as a scale model of what took place nationally three years later. Viewed in this light, the chief problem of Restoration history is to specify the nature and form of the alliance as it occurred in Chōshū and elsewhere.[18]

The pioneer application of this theory to Chōshū was made by Naramoto Tatsuya in his *Kinsei hōken shakai shiron* (A Treatise on Tokugawa Feudal Society). The civil war in Chōshū has subsequently been treated by Tōyama Shigeki, Inoue Kiyoshi, Horie Hideichi, and other leading historians of the Marxist school whose influence in Japan is strong today. The most influential and probably the most flexible proponent of this theory is Tōyama. Following Naramoto, he begins his history of the Meiji Restoration with a consideration of the Tempō Reform; this he views as the earliest confrontation of those classes whose struggles constituted, for the Marxist school, the political history of the Bakumatsu period.[19] As has been mentioned in Chapter III, the winning alliance in this struggle was that of lower samurai with well-to-do peasants (*gōnō*) and rural merchants; the losing alliance (representing the Tokugawa system) was that of upper samurai with merchant capital. As we have seen earlier, the rich peasants and rural merchants are viewed as domestic producers typical of the first stage of industrial capitalism, with interests opposed to commercial capitalism. Viewed politically, they are seen as having represented the first wave of the bourgeois democratic revolution.

For Tōyama, the first overt actions by sections of this alliance were the anti-Bakufu uprisings in Yamato in 8/63 and in Tajima

[18] Horie Hideichi, *Hansei kaikaku no kenkyū* (Tokyo, 1955), p. 1.
[19] Tōyama, *Meiji ishin,* pp. 21–22.

in 10/63. In both cases pro-Court samurai joined with well-to-do peasants who, as peasant officials or landlords, could mobilize the poorer peasants in their areas; in both cases, rebellions were begun and land seized within Bakufu domains, and, in both cases, after initial successes the rebellions were crushed by Bakufu forces. Tōyama argues that the weakness of these rebellions lay in contradictions between the interests of the landlords and those of the tenant peasants. In contrast to this, the Chōshū *shotai* uprisings were successful since the initial gains could be consolidated, using the power of the han.[20]

What was the nature of the alliance between lower samurai and well-to-do peasants? According to Tōyama, it was one in which the anti-feudal energy of the rich peasants was cut off from its source, twisted, and used for the political ends of the lower samurai. Thus, the bourgeois democratic character of the rich peasants (early industrial capital) was distorted. The Restoration does not represent a stage antithetical to feudalism but rather a compromise stage, absolutism, somewhere on the continuum between feudalism and bourgeois capitalism. Only after the Restoration, in the form of the Free People's Rights Movement (*jiyū minken undō*), did the pure anti-feudal energies of the rich peasants, which had been distorted throughout the Bakumatsu period, once again emerge.

How satisfactory is this theory? How well does it fit the facts of the Chōshū civil war? Using the evidence that has been uncovered in recent years, the following topics will be examined insofar as they apply to the above theory: the nature of the *shotai,* peasant participation in the civil war, the character of peasant officials, and the reasons for peasant aid to the *shotai* during the civil war.

The Shotai

Shotai were formed in several ways. Some were formed by direct government action; examples of these are the Kiheitai, Yūgekitai, and others formed by Takasugi in 1863 after the first foreign bombard-

[20] *Ibid.,* pp. 139–150.

ment of Shimonoseki. Others such as the Giyūtai or the Shūgitai were formed unofficially by extremist samurai living away from the castle town, in cooperation with rich peasants or merchants from whom they received their initial financial support. In the formation of such units, the peasant officials were acting in response to government directives; it would be misleading to view these as examples of completely independent action by well-to-do peasants. The Shūgitai, for example, was formed by Sakurai Shimpei and Satō Shinuemon in 8/63 with the support of a local peasant official. The memorial concerning its formation indicated that "its various expenses shall for the present be borne by the well-to-do peasants." [21] Two months later this unit was recognized by the han and given official support. Still other *shotai,* such as the Kōjōtai led by Inoue during the han civil war, were formed during the war to strengthen the extremist forces. In all, there were about fifty-eight *shotai* by the end of the civil war.[22]

Many early writers treated the *shotai* as peasant militia but this view has since been disproved. The composition of the *shotai* varied from unit to unit but in every case they were mixed units of both samurai and commoners. The Kiheitai, for example, had 292 members in 4/64. Of these 43 per cent were samurai, 33 per cent commoners, and 24 per cent of unknown origin. And of the 97 commoners, 36 were hunters incorporated into the subordinate Sogekitai, 50 were *yamabushi,* priests of a sect of nature worshippers with Shintoist overtones, 3 were townsmen, 1 was a fisherman, and only 7 were peasants. Of the 125 samurai, 12 were *shi,* 21 were *junshi* (acting *shi*), 44 were *sotsu,* 35 were *baishin,* and 12 were *rōnin* from other han.[23] Table 4 shows the composition of two other of the more important units, the Second Kiheitai and the Yōchōtai.

[21] Tanaka, "Shotai," p. 15.

[22] Tōyama speaks of there having been, in all, 156 *shotai* but many of these were undoubtedly formed during the period immediately preceding the Bakufu-Chōshū war. Even some of the 58 mentioned here were no doubt neutral *nōheitai* who did not join in the civil war. See Tōyama, *Meiji* ishin, p. 151.

[23] Umetani Noboru, "Meiji ishin shi ni okeru kiheitai no mondai," *Jimbun gakuhō,* 3:34 (Mar., 1953).

TABLE 4. An analysis of the Second Kiheitai and the Yōchōtai

	Second Kiheitai		Yōchōtai	
Class or occupation	Membership	Per cent	Membership	Per cent
Peasants (fishermen, townsmen)	78	57	123	54
Samurai	34	25	75	33
Shinto priests	21	15	7	3
Commoner physicians	0	0	1	1
Unknown	4	3	21	9
	137	100	227	100

Source: Tanaka Akira, "Chōshū han kaikakuha no kiban: shotai no bunseki o tōshite mitaru," *Shichō,* 51:12 (Mar., 1954).

These two units were probably more typical of the average *shotai* than the Kiheitai, which because of its early formation no doubt collected more than its share of the available extremist samurai. In terms of Tokugawa history, of course, the most remarkable fact about the *shotai* is that over half of their membership was made up of peasants. Yet, they were clearly not pure peasant militia as some seem to have suggested; even in these two units the samurai component was sizable.[24]

Of the samurai members of the *shotai,* some joined to participate in the *sonnō jōi* movement: this was certainly true of *rōnin* who came from other han to Chōshū. Others, *sotsu* or *baishin* who were unable to compete with the hereditary prerogatives of high-ranking *shi* in the regular han army, entered the *shotai* hoping to find a channel for advancement.

The nature of the *shotai* commoners is more difficult to ascertain: there are fewer records of their origins and even these are often contradictory. This would seem to suggest that the commoner component was heterogeneous, recruited from the bottom as well as from the top of commoner society. For example, in 12/63 the Chōshū government issued the following regulations: "Peasants and mer-

[24] Tanaka, "Shotai," p. 12.

chants may be permitted to enter the *shotai* if they have someone to take their place in farming or commerce. However, those who run away, leaving their families to perish shall in every instance be expelled [from the *shotai*]."[25] This regulation was enforced by requiring each commoner recruit to obtain a letter of clearance from his local official. As an example of how this worked, twenty-seven applications for entrance into the Kiheitai were received from Isamura, a village in the Yoshida district of Chōshū. Nakamura Seiichi was sent to investigate the applications, and, of the twenty-seven, seven were accepted, three were put under consideration, and seventeen were rejected—fifteen were ineligible since they could not be spared at home. This indicates that the regulations were observed and also suggests that commoner recruits were likely to be the second or third sons of fairly well-to-do families. This, of course, would be in line with an order to recruit commoners of "known origins."[26] These regulations, however, were instituted six months after the original order; at least twelve different units had been formed by that time and so the regulations were probably intended to check what in fact had already begun to take place. The regulations continued: "But in the cases of those who have already entered the *shotai* and are both courageous and trained in the use of the rifle, decisions will be made after [individual] investigations."[27] Since the pay of a *shotai* soldier compared favorably with what he could earn elsewhere, it seems probable that many poor peasants enlisted during the latter part of 1863 or during the civil war when regulations were not scrupulously enforced. On the other hand, some well-to-do peasant families discouraged their sons from entering the *shotai* because of their partisan character. Better to be a "rich peasant" than to hold a dangerous job with uncertain tenure at the bottom of the samurai class.

Other accounts speak of the *shotai* commoners as if they were the undesirable elements of their class. Commoner members are called

[25] *Ibid.*

[26] Seki Junya, "Bakumatsu ni okeru nōmin ikki: Chōshū han no baai," *Shakai keizai shigaku,* 21:33 (no. 4, 1955).

[27] Umetani, "Kiheitai," p. 58.

"unruly types" (*gontakure*),[28] and cases are cited of men who would "use their position to threaten people, to seize money or property, and to do other evil acts."[29] Another observed that "the Kiheitai is a gathering of crows . . . and if not governed with despotism, military discipline cannot be established."[30] The constant exhortations by the leaders that the *shotai* must be "the model for the people,"[31] the emphasis on the precepts of *bushidō,* the instruction in the Confucian classics given to samurai and commoner alike, all these were intended to mold the various class components into a unified disciplined military. On the other hand, disparaging comments on the *shotai* may have been merely the reflection of regular army resentment against the new units. A partial comparison of the two types of units, written by a member of the *shotai,* contains the following: "The *hachi-gumi* is the old regular army, but there now exist the Kiheitai and other *shotai,* all made up of volunteers . . . and as for *nōhei,* these are villagers. Today, the regular army is inferior to the Kiheitai and is laughed at even by the peasants."[32]

The motivations of the commoner recruits seem to have been relatively uniform. A few such as the Shinto priests mentioned in Table 4 may have been influenced by the *sonnō* ideology, and others, as we have already indicated, may have been attracted by the pay; the majority, however, joined to obtain the coveted symbols of the feudal class. A peasant member of the Seibutai, Kodama Jochū, related his thoughts at the time of his enlistment: "If I join the militia, in time of war I may travel to distant and interesting places; why should I not join up and swagger about?"[33] In time of war even the commoner members wore swords, received names, and could view themselves as fighting men. The uncertainty surrounding the *shotai* may have discouraged the conservative commoner, but for

[28] Seki Junya, "Bakumatsu ni okeru hansei kaikaku (Chōshū han): Meiji ishin seiritsuki no kiso kōzō (II)", *Yamaguchi keizaigaku zasshi,* 6:75 (May, 1955).

[29] *Ibid.*

[30] *Ibid.*

[31] Tanaka, "Shotai," p. 21.

[32] Seki, "Nōmin ikki," p. 33.

[33] Umetani, "Kiheitai," p. 48.

the ambitious or adventurous it offered the glory of the military profession. It was clearly this possibility of a sudden and brilliant rise in status rather than any concern or even awareness of its political or social implications that led the commoners to join the *shotai*. The same Kodama reminisces elsewhere: "At the time the people in general were not permitted to wear swords . . . but it was said that entrance into the militia would give one samurai status, and therefore the young all wished to enter and strut about." [34]

Yet even while receiving a name and sword, they remained commoners vis-à-vis their fellow members who were samurai. Perhaps a part of their pride in their new swords and names was the belief, or the desire to believe, that as fighting men they were better samurai than the samurai, in spite of being commoners. This sort of latent identification with their commoner background can be seen in their songs. The Shiyūtai made up of townsmen betrayed its origins with, "throw down the abacus and divide the enemy in two." [35] The Sogekitai, the unit of hunters subordinate to the Kiheitai, sang, "Don't miss the boars ravaging the country, you Sogekitai." [36] The Yōchōtai, whose composition was analyzed before in Table 4, hits a rather doctrinaire note with "don't look down on the Yōchōtai, the peasants are the base of the country." [37] In the past some have spoken of the Kiheitai (and by implication the other *shotai* as well) as "a mixture of free and unfree elements," [38] one that "welcomed to a certain extent run-away, that is to say, self-emancipated peasants." [39] The *shotai* leaders have been spoken of as controlling "from above a popular movement, never allowing it to get out of hand and to become a mass movement both propelled and led from below." [40] It seems clear today that the *shotai* never welcomed run-away peasants (except perhaps during the two crucial months of

[34] *Ibid.*, pp. 52–53.
[35] Tanaka, "Shotai," p. 21.
[36] *Ibid.*
[37] *Ibid.*
[38] E. Herbert Norman, *Soldier and Peasant in Japan: the Origins of Conscription* (New York, 1943), p. 31.
[39] *Ibid.*
[40] *Ibid.*, p. 32.

the civil war) and that *shotai* commoners were in no sense the conscious representatives of a movement from below.

If Japan in the Bakumatsu period were in fact poised on the brink of a series of peasant wars, then perhaps the *shotai* in Chōshū siphoned off just those ambitious and unruly types who would have been the most active during the peasant uprising. Even if the *shotai* were "feudal" in purpose, perhaps in effect they seized and twisted the revolutionary energy of the peasantry to their own ends. This sort of argument is difficult to refute since, passing beyond the *shotai* themselves, it ultimately rests with one's evaluation of the revolutionary potential of the Japanese peasantry during the Bakumatsu period. It is my contention that this potential was very small. The peasant uprisings in Chōshū during the 1830's were fairly small in scale; in the main they were protests against particular abuses of the system rather than against the system itself; they occurred only after successive years of poor crops, and they lacked a unified leadership or program other than the correction of specific abuses. In short, they did not appear to be the harbingers of a rising tide of revolution.

There are also two further arguments which would seem to vitiate the thesis of a revolutionary peasant movement. First, there were formed in Chōshū at this time two types of auxiliary militia containing commoners. One, the *shotai,* we have already treated; the other, the *nōheitai,* was a form of local guard unit, composed entirely of peasants. The natures of the two militia were quite different. The *shotai* were made up of full-time soldiers and were formed to supplement the regular han army; the *nōheitai,* on the other hand, were local guard units made up of full-time peasants who drilled only in their spare time. These local units were organized by local peasant officials; they were trained by samurai instructors hired by the local han offices and were to constitute a local defense corps for use in emergencies; their members were not given names and were not permitted to carry swords except when on duty. In contrast to the *shotai,* which could accept recruits from any area of the han, these local guard units could accept volunteers only from their own areas.

The side by side existence of these two types of military units, the pure commoner guards and the mixed commoner-samurai militia, suggests an experimental test under almost controlled laboratory conditions. If the strength or energy of the *shotai* were that of a "movement from below," that is to say, of a revolutionary peasant movement, then the *nōheitai,* composed almost entirely of peasants, should have been even stronger, even more revolutionary, and should have played an even more active role in the civil war than did the *shotai*—which, after all, were diluted with samurai. In fact, with a few exceptions, the *nōheitai* remained neutral. Commoners fought only when, as in the *shotai,* they were catalyzed by the firm control of extremist samurai leaders.

The *nōheitai* were controlled by their local peasant officials—in contrast to the *shotai* under their samurai commanders. The thought of these peasant officials, at least as reflected in their official statements, was remarkably like that of their samurai counterparts. The following petition sent to the han government in 7/63 is representative: "Although we are peasants, our ancestors have lived in Japan for generations. . . . Now that the ugly barbarians have appeared, it would be unworthy to flee before them. Therefore, if permission is granted, we will apply ourselves to this task [of forming *nōheitai*]. Preparations will be made in every village in this area [*saiban*], and if the time comes in which we are needed, we may be of some use and repay our debt to the han."[41] Yet their "debt to the han" did not include taking sides in the han civil war. When one considers that it was just this class of peasant officials—in its character as local entrepreneurs—which embodied in its purest form local "industrial capital," the apolitical nature of the *nōheitai* becomes even more significant.

A second point that can be made is that the willingness of the Chōshū bureaucrats to arm the peasants and to train them in the use of modern weapons suggests that they were totally unaware of any revolutionary potential among the Chōshū peasantry. Such a lack of awareness is not conclusive. The rigid separation of classes and the concentration of the samurai in the castle town (in normal

[41] Tanaka Akira, "Bakumatsuki Chōshū han ni okeru hansei kaikaku to kaikakuha dōmei," unpublished thesis submitted to the Bungakubu Kenkyūka of Tokyo Kyōiku Daigaku, pp. 434–435.

times) may, in some measure, have kept the han officials in ignorance of the feelings and conditions among the peasantry. Yet, along with the other available evidence, this tends to suggest that the "advanced nature" of the late Tokugawa agrarian class has been grossly overrated.

Yet, if it is easy to describe the *shotai,* it is more difficult to say what the action of the *shotai* meant historically. The military reform which the *shotai* represented was certainly an innovation in terms of the practice of the han in Tokugawa times. Yet even the conservative Mukunashi government, relatively uninfluenced by the West, early advocated the use of peasant troops. Precedents for the conscription and training of peasant troops existed from the period of Warring States, but the Chōshū *shotai* were certainly not a simple reversion or throwback to that earlier period. The rifle was traditionally the weapon of the lower *sotsu,* therefore when the demand arose for the formation of new military units, a new group of "lower *sotsu*" was recruited from among the commoners. The influence of the West was apparent in the new emphasis given to the rifle. Yet we must notice that not all of the *shotai* were given rifles. Some soldiers of the *shotai* were given swords, pikes, and other traditional weapons, and were trained in their use according to the traditional canons of individual combat. The question can be raised whether or not the *shotai* were technologically superior to the regular han army. Probably most of the *shotai* were—in organization if not always in their weapons—but no doubt a few were not. It was not until after the civil war, late in 1865 and in 1866, that the real reorganization and re-equipment of the *shotai* along Western lines was carried out. In a narrow military sense, the performance of the *shotai* undoubtedly did convince Ōmura, Yamagata, and the others of the practicality of a commoner army; this is extremely important. But it is also important to note that at the time of the han civil war, the Chōshū military leaders did not think of their *shotai* as the beginning of a more extensive reform.

The *shotai* were not considered as vehicles for social reform in Chōshū. An early contention: "The keen minds of the age, notably Takasugi Shinsaku, Ōmura Masujirō, and their companions of

Chōshū saw the need for drastic changes in the cultural and social life of the nation if lasting reform in the military system was to be effected,"[42] can no longer be held. It would be more accurate to say that the keen minds of the age were faced with the problem of creating a larger and more effective military without adversely affecting the traditional social and economic structure of the han.

We have already noted that admission to the *shotai* was restricted to those who were not needed for farming, business, or the support of their families. The care taken by the han leaders not to disturb the economic or social base of the han was equally reflected in the regulations concerning the preservation of traditional distinctions of class and rank within the *shotai.* The sleeve insignia of the *shi* were different from those of the *sotsu;* the latter, in turn, were different from those of the *baishin,* which differed from those of the commoners. Even in the ceremonies for those who died in battle a sharp distinction was maintained between samurai and commoners, and the commoner who, by proving his ability, had risen to a position of leadership and had been accorded the honors due a samurai was a rare exception.[43] In 11/64 the *shotai* leaders admonished their troops: "Let courtesy be your basic principle. It is important not to go against popular sentiments; courtesy does not confuse the degrees of highness or lowness. Therefore, let [each] act according to his position, and do not give yourself airs."[44] Punishments for those who broke the *shotai* regulations also varied according to class: allowances were often made for commoners, who were let off with lighter punishments than were samurai who had committed the same offense. It is not unnatural for military organizations to preserve sharp distinction in rank, but it is significant that the distinctions described here were based on feudal status.

These distinctions were preserved, in part at least, because the *shotai* were viewed as an emergency measure. The late Bakumatsu history of Chōshū can be viewed as a series of emergencies interspersed by periods of relative calm. The *shotai* were encouraged in

[42] Norman, *Soldier and Peasant in Japan: the Origins of Conscription,* p. 31.
[43] Tanaka, "Shotai," p. 21.
[44] Tōyama, *Meiji ishin,* p. 158.

times of crisis and restricted in periods of calm. They were formed to help protect Chōshū from foreign attack in 1863. Once the immediate danger had passed, orders were issued restricting their size and composition. The Kyoto countercoup, the attack by the four-nation fleet, the first Chōshū expedition, the civil war, the second Chōshū expedition—each in its own way spurred the growth of the *shotai*. Yet, after each crisis controls were reimposed: spontaneously organized units were disbanded, commoners untrained in the use of rifles were discharged, new leaders were assigned, and so on. *Shotai* autonomy, to the extent it existed at all, was at best a crisis phenomenon.

That the *shotai* were organized as an emergency measure, and not as a new type of unit favored by semirevolutionary leaders, can be seen even in the statements by their commanders. Takasugi spoke of their formation as a "hard-time policy for which there is no help." [45] He realized, of course, that it would serve his ends to use men of ability, but the will to use even commoners of ability does not prove a basic disenchantment with feudal society or its works. Toward the end of 1864 Takasugi wrote of Akane Taketo who had opposed his Shimonoseki uprising: "Is this Akane Taketo not a country bumpkin from Ōshima? And am I not a samurai with the hereditary patronage of the house of Mōri? I am not to be compared with such a baseborn fellow as Taketo!" [46] As a samurai with a stipend of one hundred and fifty *koku* and as the son of a direct inspector (the same position which had been held by Nagai Uta), Takasugi was proud and acutely conscious of his feudal position. Even to his wife he wrote, "It is important never to forget that the wife of a samurai is different from the wife of a merchant or peasant." [47]

In many ways one can think of the *shotai* commoners as a new stratum of samurai placed below the *ashigaru* and *chūgen* to strengthen the Chōshū military. And, from this viewpoint the commoners cut off from their peasant origins and permitted to enter

[45] *Tōkō sensei ibun* (*Shokan-hen*) (Tokyo, 1916), p. 107.
[46] Tōyama, *Meiji ishin,* pp. 158–159.
[47] *Tōkō sensei ibun* (*Shokan-hen*), p. 138.

were just those whose transformation would cause the least disturbance to the society as a whole.

Peasant Aid to the Shotai during the Civil War

Another facet of the problem of the Chōshū civil war is that of the aid given to the *shotai* by peasants outside the *shotai* organization. Proponents of the theory that the Restoration was accomplished by the energy of a "movement from below" cite this aid as further evidence of an alliance between "lower samurai" and certain groups in the peasantry. To establish a framework within which this problem can be treated, we may begin by stating three points. First, during the civil war both the *shotai* and the Hagi Mukunashi forces received a certain amount of aid from the peasants. Second, with one or two exceptions, neither side received aid from areas not under their direct control; those areas peripheral to the fighting maintained a scrupulous neutrality. Third, the evidence collected to date indicates that the *shotai* were far more successful than the Hagi government in mobilizing the peasants in the areas under their control. But before going on to consider who gave the aid and why, let us first examine its scope.

In the papers of the house of Tsubaki whose head at this time was the *ōjōya* of the Toshima region, its highest-ranking peasant official, it is recorded that the conservative government was able to recruit twenty-five or thirty peasant soldiers from the area immediately surrounding the castle town of Hagi. In these papers there are also preserved orders (*satagaki*) issued by the Mukunashi government for the mobilization of peasants, *chōnin,* and even *semmin,* as peasant militia.[48] It is significant that during the civil war, when both sides were fighting for the control of the government, neither side hesitated to arm even the *semmin,* the lowly or "unclean" class despised even by the peasants. The Mukunashi government, however, in spite of unflagging efforts to mobilize peasants in Toshima, had little success.

In the following year, when preparing for war with the forces of

[48] Seki Junya, personal communication to author.

the Bakufu, the restored extremist government permitted the formation of the Ishindan, a militia unit made up of butchers (*semmin,* of course) from the Kumage region of Chōshū. That both sides were willing to use commoners, even *semmin,* in militia units suggests that this in itself did not call for particularly advanced or enlightened views. One can also recall that it was the Tsuboi-Mukunashi government which in 1855 first called for the establishment of peasant militia. Perhaps Western historians have tended to overestimate the immediate short-run significance of commoner troops or to view them in isolation from the circumstances in which they were formed.

The Hagi government also sent four samurai to the Tokuji region "to quiet down the people in the area and to recruit peasant militia and prepare them [to fight] against the *shotai* uprising." [49] The four samurai were unsuccessful and they had to report back to Hagi as follows:

> Those in this area are afraid [to enter] the peasant militia; [therefore] only a few can be recruited. Only twenty hunters have been obtained, and their hearts have been bound only by giving each a *to* [one-half bushel] of rice. Because of this [failure] we were forced to send messengers to the neighboring Yamashiro region asking for help. But they too answered that in the present circumstances only the defenses of the *saiban* [district] could be preserved.[50]

This, of course, was an attempt to recruit soldiers in an area over which they had no direct military control.

We earlier quoted the partisan words of the Peace Assembly which stressed that the *shotai* were not only popular among the samurai and growing stronger day by day, but that they were supported even by the peasants. Elsewhere, a staff officer of the conservative forces, Yamagata Yoichihei, wrote in the same vein: "All are losing their vigor. I am extremely doubtful about the sympathies of the commoners. The *shotai* send out spies and by taking advantage of the fatigue of the Sempōtai [conservative army] and the unrest of the people; [they] come out before our eyes with this policy of night attack." [51]

[49] Tanaka, "Dōmei," p. 461.
[50] *Ibid.,* pp. 461–462.
[51] Tanaka, "Dōmei," p. 463.

In contrast to this, peasant aid to the *shotai* was of greater proportions and played a more crucial role in the war. Even Takasugi's initial attack of Shimonoseki was partly financed by a loan from a rich peasant official, Yoshitomi Tōbei of Yuda. Before the attack Takasugi wrote to Yoshitomi: "Observing that this is a time of shame for the han, your younger brother [i.e., Takasugi, a polite first person] and others of like resolve are planning a righteous uprising. . . . In connection with this a little money is needed, and I am in difficult straits since few will furnish it. My elder brother [i.e., Yoshitomi] is loyal and sincere. If you could give us four or five hundred [*kan*], this will be a great blessing for the han." [52] Yoshitomi sent a sum of two hundred *kan* to Takasugi, who used it to finance his attack.[53]

Almost all the fighting in the civil war centered on the area between Yamaguchi and Hagi. Most of the aid received by the *shotai* came from the regions of Ogōri and Yamaguchi (Yoshitomi, for example, lived in Yuda, a "suburb" of the city of Yamaguchi). Most of it was obtained from or through the Union of Peasant Officials (*shōya dōmei*), which was formed on 8/1/65 by the peasant officials of Ogōri under orders from the *shotai*. It was formed only after the *shotai* troops occupied the area. The *shōya* were summoned to a tea house; they were addressed by Nomura Yasushi (Wasaku at this time), who instructed them to cooperate with the *shotai* cause, and there they formed the union. The *shōya* in this area had already signed a pledge to the Hagi government, agreeing not to help the *shotai* in any way; but now under the influence of the *shotai* forces they pledged to cooperate with the *shotai* and to furnish them with porters, provisions, and militia. Hayashi Yūzō, the *ōjōya* at the Ogōri district office (*saiban*), was too busy to attend to this personally, but he appointed five of the *shōya* who had been present at the

[52] *Tōkō sensei ibun* (*Shokan-hen*), p. 149.

[53] There may well have been an element of coercion present in the request. Opinion on this subject seems to be divided. Still another source of support during the period of the civil war was that of funds already allocated to the *shotai* before the rise of the Mukunashi government. Roger Hackett writes, "before the Zokurontō swept aside all opposing sentiment from official positions, he [Yamagata] succeeded in securing, through a sympathetic official, sufficient funds to cover the expenses of the Kiheitai for a year." See Roger Hackett, "Yamagata Aritomo 1832–1922: A Political Biography" (diss. Harvard University, 1955), p. 59.

meeting to handle the aid; of the five, Akimoto Shinzō was the most prominent. This group of five *shōya* placed a quota for porters on each village in the area, paying each porter so much for his services. One account tells us that 1200 porters were furnished to the *shotai* at the time of the Ōda battle.[54] Three hundred and eighty-six *koku* of rice and two hundred and five *kan* of silver were also collected and given to the *shotai* in the form of a loan. A small incident occurring when the *shotai* first entered Ogōri is perhaps indicative of the attitude of the peasant officials. The newly arrived *shotai* went to the district office and demanded money. Hayashi Yūzō gave them thirty-five *kan* of government money but demanded a receipt in return. He was given an unsigned receipt; he returned it asking that it be signed, and they had no choice but to oblige.

Three military groups were also formed in this area: the Kōjōtai, organized and led by Inoue, the Tōshintai, organized and led by Akimoto Shinzō, and the Ogōri *nōheitai* of five hundred men formed of village peasant militia. The last unit is the notable exception to the earlier generalization that peasant militia remained neutral during the civil war. A somewhat over-enthusiastic description of the relations between the *shotai* and the commoners is given by Yamagata: "The people of Ōda were overjoyed at the arrival of the *shotai* and vied among themselves to greet them and to furnish them with men and equipment. When the money was returned to be used as alms for the poor in that area, the people became more and more joyous and greatly aided our *shotai* during the civil war."[55]

Yet it must be noted that the people did not aid the *shotai* directly; the people obeyed the *shōya* and the *shōya* supported the *shotai*. Effective control of the peasants depended on the cooperation of the peasant officials. Cooperation was obtainable only in areas controlled by one or the other of the military forces, but even then the *shotai* were able to win more help than the Hagi government. Moreover, in a few cases, such as those of Yoshitomi, Hayashi, or Akimoto, there apparently was a positive willingness to aid the extremists. Akimoto's famous statement to Kaneshige Jōzō of the former Sufu group,

[54] Tanaka, "Dōmei," pp. 459–460.

[55] Kano Masanao, *Nihon kindai shisō no keisei* (Tokyo, 1956), p. 115.

although recorded only after the Meiji Restoration and then by Akimoto himself, is telling:

Truly, today is an urgent time for the han. If the members of the loyalist group fail in this situation, it can be thought to be the end of the han. In this critical time it will not do for vassals of three-hundred-years standing to be inactive; yet until now they wait, and none have risen up. If the vassals will not do it [overthrow the Mukunashi government] with their hands, then I, Shinzō, with my own hands will cause a peasant uprising and restore the country. Therefore, Master Kaneshige, no matter what happens, do not become discouraged. If Hagi becomes dangerous, then we will do it with a peasant uprising.[56]

Contrast this with the letter sent by Hayashi and Akimoto to the Hagi government before the occupation of Ogōri by the *shotai*.

In regard to the matter of putting down the uprising [of the *shotai*], at a time when violence is being used it is natural [for the government] to turn to force if attempts to persuade [fail]. [However], since we are not yet trained, if they [the *shotai*] come, we will be unable to defend ourselves. . . . However, if the samurai of the high [ranking] houses of Hagi come here to defend us, then we will follow until our body is broken even unto dust and bones.[57]

The truth undoubtedly lies somewhere between the two statements. The latter quotation was a typical neutralist statement by peasant officials whose area was not under the military control of either group. It probably can be thought of as an anchor to windward to insure their own safety in case the conservatives should win the civil war. The former quotation, on the other hand, might well have suffered from the idealising distortion of memory.

The Chōshū Peasant Official

Who, then, were these peasant officials and what was the nature and mode of their control over the peasants in their areas? Several of them, such as Yoshitomi of Yuda or Tsubaki, were landlords.

[56] Tanaka, "Dōmei," pp. 447–448.
[57] Seki, "Meiji ishin," p. 76.

Tsubaki was an *ōjōya* and at the same time a landlord with an income of eighty *koku* from tenants. In private family documents he speaks of himself as the equivalent of a samurai with a stipend or fief of two hundred *koku*.[58] (The prevailing official tax rate was 40 per cent, therefore, a samurai with a stipend or fief of two hundred *koku* would have had in theory an income of eighty *koku*.) Landlordism, such as that represented by Tsubaki, we must note, was relatively rare in Chōshū. The farm village economy was comparatively stable, and the Kinki agrarian pattern, with very rich landlords and impoverished tenants, had not yet appeared. Most of the village officials in Chōshū, while more comfortable than the average peasant, were not landlords.[59]

The class of peasant officials formed a small social world of its own. The Hayashi house and the Akimoto house had for generations given each other daughters in marriage and extra sons for adoption when needed. (Hayashi Yūzō, however, was adopted from the Yoshitomi house.) The following letter sent to Hayashi Yūzō by his uncle Yoshitomi Tōbei is indicative of the way of life of the local official. "Though the Hayashi house is of a status that has for centuries produced *shōya,* this position of *shōya* is very deadly to descendants and we must be prepared for being employed today and having to change tomorrow. It will do great harm to our house if you stick a sword in your belt and put on airs. Being born a peasant, do not forget the way of the peasant." [60] This suggests that many did stick swords in their belts and "forget the way of the peasant."

The history of the Hayashi house and the career of Hayashi Yūzō illustrate yet another dimension of the Chōshū *shōya.* According to a stele found in Ogōri, the Hayashi family had existed in that area as a house of *gōshi* status since the time of the Ōuchi rule, which preceded the rise of the house of Mōri. Then, from 1644 on, for generation after generation the Hayashi house furnished village headmen in that area. Hayashi Yūzō, from 1828 to 1840, was the *shōya* of a samurai fief, the lowest rank of *shōya.* From 1841 to 1855 he was the

[58] Seki Junya, personal communication to author.

[59] Tanaka Akira, "Chōshū han ni okeru hansei kaikaku to Meiji ishin," *Shakai keizai shigaku,* 22:147–152 (nos. 5 & 6, 1956).

[60] Seki, "Meiji ishin," p. 59.

ōjōya of Ogōri, the highest peasant official at the Ogōri district office in charge of all the *shōya* in the area. His landholdings, however, were relatively small, much smaller than those of most peasant officials of comparable status in the Kinki area. During the 1820's he had three *chō,* eight *tan* of land producing seventy-nine *koku;* by the early Meiji era, this had increased to five *chō,* seven *tan* of land producing one hundred and two *koku.*

In contrast to the above rich local-peasant official type, Akimoto Shinzō, who personally organized, financed, and led the Tōshintai in the civil war, was a *shōya,* a rich peasant, and a dry goods merchant. Okamoto Sanuemon, a *shotai* sympathizer and friend of Takasugi and Kusaka, was a cotton merchant of Mitajiri "who indulged a taste for *kokugaku* and, emphasizing the righteousness of the *sonnō jōi* teachings, aided the patriots and dispersed his family fortune." [61] Another of this type was Shiraishi Seiichirō of Shimonoseki, who knew Takasugi, Kusaka, Kido, Nomura, Yamagata, and many others; he received extremist samurai from other han, and his home was used later as a meeting place for negotiations with samurai from Satsuma. In later years he was made a samurai for his services during the Bakumatsu period.

Nor were these the only contacts between samurai and peasant officials. Many of the high-ranking *shi* who later became prominent in the han bureaucracy first spent some time as the *daikan* in a local district office, where they were in frequent contact with the *ōjōya.* Yoshitomi, for example, was born in Yuda and from an early age he seems to have been on fairly close relations with Inoue, who was also from Yuda; there is even some evidence that from time to time he gave financial aid to the house of Inoue. We have already mentioned Yoshitomi's relation to Takasugi. Sufu was also well acquainted with Yoshitomi (had he been the *daikan* in the district in which Yoshitomi was *ōjōya*?) and when in 1864 he decided to commit suicide, he chose to do it in Yoshitomi's house.

The following incident which was recounted by Yoshitomi after the Restoration seems to have occurred during the civil war at the time of the formation of the Kōjōtai. A rumor reached the

[61] *Ibid.*, p. 78.

Kōjōtai that the daimyo and his son would come out to lead the Hagi forces against the *shotai.* Faced with this possibility, a conference was called at which the commander of the unit, Inoue, recommended that all "commit harakiri before their lord [literally, before his horse] as a reproof [for having come out against the *shotai*]."[62] Yoshitomi admonished him: "If on hearing that the two lords are to appear you wish to throw away your weapons and cut open your bellies, then it would have been better never to have risen up in the first place. Takasugi and the others at Ōda would never come forth with such a policy."[63] And instead he proposed an ambush in which empty cannon would be fired, confusion would ensue, and the son of the daimyo would be "greeted" and conducted to the *shotai* stronghold at Yamaguchi. To justify such an action, Yoshitomi invoked the authority of the founder of the han. In this context Tōyama speaks of the "commoner Yoshitomi" and "Yamagata of *ashigaru* origin" as "not being bound by the pure feudal ethic" and therefore "not refusing to fight with the daimyo."[64] This may explain the difference of opinion expressed above, but at the same time one should remember that *ōjōya* such as Yoshitomi were probably just as "feudal" or "Confucian" in the Tokugawa sense as most samurai. Also, why is it that Takasugi, who was neither commoner nor *ashigaru,* was also able to put aside the pure feudal ethic? Still another example of contacts between village officials and samurai was the relation between Yamagata and Hayashi Yūzō; even after the Restoration when Yamagata, Minister of the Interior, returned to Yamaguchi, he went to visit Hayashi whom, it is recorded, he addressed as *sensei* (teacher).[65]

Whatever facet of their complex character may have induced these men to aid the *shotai,* it was their position as village officials which enabled them to control and mobilize the peasants as they did. Hayashi explains this control as follows: "The peasants have no learning but if you explain they will understand. Therefore they

[62] Suematsu, *Bōchō kaiten shi,* VII, 40–41.

[63] *Ibid.*

[64] Tōyama, *Meiji ishin,* pp. 156–157. Actually, the successes of the *shotai* were so great that the daimyo never consented to appear at the head of the conservative troops.

[65] Tokutomi Iichirō, *Kōshaku Yamagata Aritomo den* (Tokyo, 1933), I, 108.

can be made to take the side of the *shotai.*"[66] This seems too simple, for it assumes as an implicit premise an understanding of the position of the *shōya* in the framework of the local government.

The Tokugawa village was an economic unit from which a fixed amount of tax rice had to be collected, a unit which in fact was often actually composed of several villages. Within this unit the *shōya* governed all aspects of the peasants' life; his decisions were final; and since samurai officials rarely went to the villages, he was the total representation of han authority to the inhabitants of the village.[67] The *shōya* of the various villages in each district were subordinate to the *ōjōya,* the highest peasant official who under the *daikan* ran the affairs of the entire district. Chōshū was broken up into fourteen such districts (*saiban*), although what is conventionally referred to as Chōshū also included the four branch han, each of which was about the size of a district. Ogōri and Yamaguchi were two of the fourteen districts.

Although the *ōjōya* was subordinate to the *daikan* and other samurai officials in the district office, his actual power if not authority was often much greater in the local areas than his subordinate position would indicate. Not only was he in close and constant contact with the *shōya* but he was also familiar with the conditions in the local areas and his post tended to be permanent. There was a far greater turnover of those in the position of *daikan,* and accordingly they had less opportunity to become intimate with the conditions and problems of the district. In some respects this *ōjōya* class might perhaps be compared to the permanent underofficial (*li*) class in the *hsien* office in Imperial China. Moreover, the *daikan* and his officials spent a large portion of their time in the castle town. One contemporary source sums up the situation: "The present *daikan* was born in the castle town and knows nothing of the people even in his dreams. Therefore his underlings can deceive him and do evil acts as they wish. . . ."[68] The confidence of

[66] Seki, "Meiji ishin," p. 78.

[67] The *shōya* was not, of course, the only peasant official in the village, but he was the most important.

[68] Naramoto Tatsuya, *Kinsei hōken shakai shiron* (Tokyo, 1952), p. 59.

the peasant officials in their own power over the peasants can be seen in the following words of Akimoto Shinzō, uttered when someone conveniently asked him whether a bag of his wheat would be stolen if left by the roadside. "If there were anyone in the neighboring villages or in the district who would take one *kin* of Shinzō's wheat or one *gō* of my rice, then my voice would have lost its authority. No, this sort of thing could never happen!"[69]

Another consideration affecting the power of the Chōshū *shōya* was that the village class structure at whose apex they stood was fairly stable. Studies made so far seem to indicate that although the well-to-do peasants were gaining ground in terms of their land holdings, they were doing so at the expense of the poor peasants; the stable middle stratum of peasants was in a comparatively good position. This is important since it was the relation between the *shōya* and this middle stratum which owned land, worked its own soil, and participated in village meetings, that determined whether the *shōya* could exercise effective control. In the Kinki region, where the rich peasants grew richer at the expense of the middle group, the declining middle tended to oppose the wealthy *shōya* or even to join with the poor peasants in uprisings against the *shōya*. In Chōshū the comparatively contented middle layer followed the lead of the *shōya*.[70]

The class of peasant officials, if one is to judge from the outstanding examples about whom materials are available, was a curious and interesting group. They were peasants or commoners, but in terms of real income they were higher than most samurai,

[69] Tanaka, "Dōmei," p. 380.

[70] Tanaka Akira, "Tōbakuha no keisei katei: Chōshū han bakumatsu hansei kaikaku to kaikakuha dōmei," *Rekishigaku kenkyū,* 205:8–12 (Dec., 1957). At the time I wrote this section I had not yet read Thomas Smith's analysis of peasant class structure during the Tokugawa period. In terms of his analysis, it is my impression, based on the writings of Tanaka Akira and Seki Junya, that the agrarian base in Chōshū had evolved from the "original" post-Hideyoshi *fudai-tezukuri* pattern to the *nago-* (in Chōshū called *kado*) *tezukuri* pattern, but had not advanced, with a few exceptions, to the *kosaku* pattern. This suggests, in terms of Thomas Smith's analysis, that the dominance exercised by the *shōya* class occurred in Chōshū within a more or less traditional village cooperative structure. See Thomas C. Smith, *The Agrarian Origins of Modern Japan* (Stanford, 1959).

and they probably had an education at least equal to that of the *sotsu*. In contact with samurai and attending to the business of the han, even their ideas, as we shall see, were like those of the samurai. And yet they were not samurai. They can be viewed as having constituted a potentially radical group, since they knew the frustrations of having reached a position from which no further advancement was realizable and since they possessed power without its accompanying prestige; this might account for certain strains in their thought such as the idiosyncratic interest of a few in *kokugaku* or for the enthusiasms of an Akimoto Shinzō. At the same time they may just as well be viewed as having been an essentially conservative group, for, having once risen to the upper limit prescribed for their class there was nowhere to go but down. And not infrequently they did go down: they were responsible for the success of tax collections, they were blamed when a reform miscarried, and they were usually made the scapegoats when *ikki* broke out. As peasants they had a power and a standard of living for which there was no logical justification in the Tokugawa ideology; therefore, they were uneasy. Children growing up in the household of a *shōya* partook within peasant society of their parents' prestige and, therefore, were likely to develop an attitude of self-importance. Yet such a self-important peasant would not be tolerated, much less appointed to the ranks of the peasant officials. Hence, "this position of *shōya* is very deadly to descendants." As a result, the children of *shōya* were brought up to honor the "way of the peasant" and to avoid entanglement in the factionalism of samurai. Thus, the potential radicalism inherent in the *shōya's* position was offset by their conservative philosophy, and became apparent only when reinforced by the teachings of *kokugaku,* which as we have noted were not widespread in Chōshū. In this perspective, it is not hard to understand why the majority remained neutral during the civil war in the han.

What does this discussion of the peasant official imply concerning the theory of a movement from below? It would appear that there existed no such movement. The vast bulk of the peasants were apolitical. They rioted in time of drought or famine or against various specific grievances; individuals among them were stirred, when the opportunity arose, by a blind desire for samurai status; but they were not swept by the sort of revolutionary class fervor that a rapid disintegration of traditional village structure might conceivably produce. Or, if one must speak of a movement from below, it was a movement sparked by the peasant official class. Why, within the limits described above, did it aid and support the *shotai?*

The Motivations of the Peasant Officials

The following is taken from an address given to the Union of Peasant Officials at the time of its formation:

> Since the action in Kyoto in the seventh month of last year [the countercoup], the *zokuron* party has risen. . . . They have interfered with the production of the people and finally, they have willfully brought immeasurable troubles to the two lords [the daimyo and his son]. Now, twisting the intentions [of the daimyo], this wicked group slay innocent men. . . . If things continue in this manner, it will not be long before Chōshū is destroyed. . . . Having trod the soil of Chōshū and eaten its rice, today, in this time of crisis, let those who cannot endure to stand still and watch hasten to help [the *shotai*]. The *shotai* do not engage in violent actions, and peasants and merchants need not fear them. Rather, all ought to join together for the sake of the han. If anyone has any other intentions, he had better watch out.[71]

The last sentence contains the teeth of the argument: the *shotai,* inspired by a righteous zeal, were less tolerant, or more coercive, toward the *shōya* in their areas than was the Mukunashi government in the Hagi area.

The theory has been advanced by Naramoto Tatsuya that the peasant class structure in the mountainous areas near Hagi was different from that in the coastal area close to the Inland Sea. He suggests that the former is a "Tōhoku type" area with a relatively greater number of poor peasants who would not cooperate with their *shōya* as did their counterparts in the Inland Sea area.[72] This

[71] Tanaka, "Dōmei," pp. 457–458.
[72] Naramoto, *Kinsei hōken shakai shiron,* p. 18.

is plausible, but judging from the evidence presented so far it seems more likely that the *shōya* in the Ogōri-Yamaguchi area were more willing to cooperate with the *shotai* than the Hagi *shōya* were to cooperate with the Mukunashi government.

Another obvious reason why peasants might have been disposed to favor the *shotai* was that the *shotai* contained many peasants, and, since they were recruited from all over the han, every district if not every village was represented. And so wherever the *shotai* went they had ready-made contacts. (This was undoubtedly a factor contributing to their military efficiency as well.) Still another type of personal tie which may have predisposed the *shōya* to aid the *shotai* was the tie between samurai officials and peasant officials. We have already mentioned the personal relations between Inoue and Yoshitomi, Yamagata and Hayashi, and Sufu and Yoshitomi. Similar bonds also tied Tamaki Bunnoshin, a moderate samurai who sympathized with the *shotai,* and *shōya* in the Ogōri and Yoshida districts where Tamaki had been *daikan* in the past. As the Sufu clique had been in power for six years between 1858 and 1864 it is probable that most of the men who served as *daikan* during this period were men sympathetic to the policies of the extremists.

The following letter sent by the Peace Assembly to the Hagi government during the civil war suggests another reason why the *shōya* may have aided the *shotai:*

> At this time the peasants of the various districts are not submitting to the *shotai* for gold or silver, [rather, they are submitting to them] because they feel that the daimyo's policy of loyalty to the Court and trust to the Bakufu is just. Moreover, up to the present the virtuous policy [*sonnō jōi* policy] [spread] by exhortation and proclamation has penetrated above and below and won the hearts of the people. In the present situation the [people] feel that virtue is on the side of the *shotai* [and because of this] the peoples' hearts yield to them.[73]

"Peoples' hearts" probably referred to the hearts of the *shōya.* Most of these *shōya,* to the extent that one can judge from their memorials, were swayed by the same "Confucian" ethic held by

[73] Suematsu, *Bōchō kaiten shi,* VII, 57.

the samurai. A few of them, however, may have also been inclined toward the pro-Court movement because of the teachings of *kokugaku*. Earlier we noted that politically-oriented *kokugaku* as an organized movement contributed almost nothing to the rise of the *sonnō jōi* movement in Chōshū; only a few disciples of the Hirata school were to be found in the han and these were not samurai but peasants. Now we must note that these peasant disciples were just those who were active in support of the *shotai* during the civil war. Suzuki Shigetane, one of Hirata's outstanding disciples and a prominent teacher in his own right, had come to Chōshū several times and his followers included Yoshitomi, Hayashi, Akimoto, and Shiraishi. Okamoto Sanuemon, the merchant of Mitajiri who "indulged in a taste for *kokugaku*," was a disciple of Suzuki Naomichi, a teacher of *kokugaku* and a Shinto priest of Chōshū. This attachment to the teachings of *kokugaku* may have been the essential difference marking the *shōya* who were active from those who were not. One must not assume, however, that their thought was radically different from other *shōya*. At best *kokugaku* complemented the Confucian teachings which seem to have dominated samurai and peasant official alike. One *shōya* from the Ogōri district, writing a letter in 4/63 to accompany money collected in his village for the work of expelling the barbarian, spoke of their contribution as "the repayment of our debt to the han." [74] A document quoted earlier regarding the formation of peasant militia began by speaking of "ancestors who, having lived in the divine land, have incurred an obligation to it," and ends with talk of "each fulfilling his debt to the han." [75] Nomura addressed the *shōya,* first appealing for action with the words, "We have trod the soil of Chōshū and have eaten its rice." [76] Samurai and peasant official both spoke the same Tokugawa language of duty and obligation.

None of the reasons covered so far suggest that the *shōya* were a part of a revolutionary movement from below. Those who hold

[74] Tanaka, "Dōmei," pp. 431–433.
[75] *Ibid.*, pp. 434–435.
[76] *Ibid.*, pp. 457–458.

this theory explain peasant or *shōya* aid to the *shotai* in the following manner: The *shotai* by 1865 had merged or become identified with the Sufu party and the Sufu tradition of reforms (that of Tempō and that of Ansei). This type of reform was favorable to early industrial capital, therefore, peasant producers in this category support the *shotai* out of self-interest. The *shōya* themselves were (often, but not always) local producers and at the same time they were the interpreters of han policy to other such producers among the peasantry. This is indeed a plausible argument but the evidence has not yet been presented that would make it wholly convincing. First, as we noted in the section on the Tempō Reform, it is not yet clear how much local producers stood to gain by the abolition of the merchant-controlled monopolies, which was advocated by the loyalist-Sufu clique. Second, no evidence has been presented to show that the *shōya* succored the *shotai* in the hope of selling more *shōyu,* wax, or indigo with less han interference. One might suggest that the economic self-interest of the *shōya* was more closely tied up with their political position than with the particular form of the han monopoly system and that all those *shōya* who could remained neutral, because they did not wish to jeopardize their special position in the han. This neutrality, it must be kept in mind, was the central fact concerning the role of the *shōya* in the civil war.

Negative Forces in the Civil War

It would be inadequate, however, to describe the Chōshū civil war solely in terms of the *shotai,* the winning forces. Ideally, one should also present a more extended analysis of the Peace Assembly, of the groups supporting the Mukunashi government, and of the conservative Sempōtai. But this is difficult since relevant materials or monographs are scarce. Few in the past cared to record the glories of those who fought and lost, and there are few historians today who are not exclusively preoccupied with the nature of the winning team. Some materials, however, are available concerning the actions of the Chōshū branch houses and certain groups of rear vassals during the civil war. These, I maintain, are just as important as the actions of the peasants discussed above, for, had these groups joined the Mukunashi camp, it almost certainly would have won in spite of the peasant aid given to the *shotai*.

We will first consider the *baishin* of Mōri Chikuzen, most of whom lived on his fief in Migita. During the civil war, Ōkura Genuemon had been sent to Migita by the Mukunashi government to mobilize these *baishin* for service with the conservative troops. He found that "the majority of them sympathized with the *shotai*."[77] He transmitted the order of the Hagi government, but as they did not respond he had to report: "They say they will guard only the residence of Chikuzen. . . . Even Chikuzen cannot control these *baishin;* since we have nothing to gain in this place, I request to be speedily returned."[78] Considering that Mōri Chikuzen became the leading Elder in the new extremist government formed after the war, it may well have been that his vassals had been secretly instructed not to respond to the above order; whatever the case, had they participated in the war, the government forces would have been undeniably stronger.

Even more striking is the case of the *baishin* of Masuda Danjō, one of the three Elders executed for his part in the Kyoto countercoup. Masuda had 538 retainers who, according to the Register of Retainers Fiefs of 1870, were graded as follows:[79]

263 *shi*	5 House Elders (*rōshin*)
	53 upper *shi*
	96 middle *shi*
	109 lower *shi*
275 *sotsu*	68 *ashigaru*
	207 *chūgen*

The majority of these retainers lived on the Masuda fief in the Okuabu district of Chōshū. This district, which fronted on the

[77] *Ibid.*, pp. 465–466.
[78] *Ibid.*
[79] Kimura Motoi, personal communication to author.

Sea of Japan in the mountainous area to the east of Hagi, was nominally under the control of the Hagi government. But the Masuda retainers were thrown into conflict among themselves when they received an order to join the forces of the conservative army; and consequently, a microcosmic civil war, one which reproduced most of the features of the han civil war, broke out in the Masuda fief.

In 1858 Masuda had sent Ōtani Bokusuke, the head of the Ikueikan, a private school maintained by Masuda for the education of his vassals, to study under Yoshida Shōin. Oguni Yūzō and Kawakami Hanzō, two other teachers in the same school, were also close to Shōin and together with Ōtani were instrumental in spreading his teachings among the Masuda vassals. When the civil war broke out in Chōshū, the extremist minded among the Masuda *baishin* formed in 1/65 the Kaitengun to join with the *shotai* and avenge the execution of their lord. This was opposed by other conservative *baishin,* who on 25/2/65 formed the Hokkyōdan to fight against the Kaitengun. Here as elsewhere, the conservatives used peasants to strengthen their force. We described earlier how the Mukunashi government in desperation had offered to restore the fiefs of the three Elder houses if their vassals would join the conservative army. The conservative group among the Masuda retainers now argued that it was their duty to preserve the house of Masuda (that is to say, the fief) and that they should serve the son rather than avenge the father. There is a remarkable parallel between this and the arguments of the Mukunashi clique who also argued that it was its duty to preserve the han through a policy of submission rather than to fight a war that might end with the destruction of the house of Mōri. The outcome of the battle between the two factions, the Masuda civil war (Susa *naikō,* Susamura was the headquarters of the Masuda fief), fought on 1/3/65, was contrary to that of the han civil war. The extremist Kaitengun was defeated and its leaders were ordered by the victorious conservatives to commit suicide. The victory proved to be short-lived, for, soon afterward, the *shotai* who had won the han civil war reversed by fiat the situation among the Masuda retainers:

the surviving extremists were placed in control of the Masuda house government.

Yet another small-scale civil war was fought among the vassals of Tokuyama han, one of the Mōri branch han within the Chōshū domains. After the attempted countercoup in Kyoto two factions appeared among the Tokuyama retainers. The extremists joined to seize the Tokuyama government, which was in the hands of the Tokuyama conservatives. Their attack was a failure, and they themselves were seized and imprisoned. Some were executed on 24/11/64 after the rise to power of the Mukunashi clique in Chōshū, and others were put to death on 14/1/65 following the outbreak of the Chōshū civil war. And, as an index of the degree of autonomy possessed by the branch han, it should be noted that Tokuyama remained under the control of its conservative or *zokuron* government until 6/65, when a mission from Chōshū persuaded Tokuyama to dismiss its conservative clique headed by Fujita and to establish an extremist government.

These last examples point up the difficulties involved in an overly rigid application of a theory of conflicting classes representing different stages of economic development. To say that the Hagi forces lost the war because they could not mobilize the peasantry would be too simple. They were also unable to muster enough samurai, even among the high-ranking *shi*. And, in the same Hagi area the conservatives of the house of Masuda could and did utilize peasants in their Hokkyōdan to win in their small-scale war. Again, if the struggle within Chōshū is to be seen as a contest of strength between Hagi commercial capitalism and the burgeoning early industrial capitalism of the Inland Sea area, one would be hard put to explain the conservative victory in the branch han of Tokuyama, which was entirely in the Inland Sea coastal area.

The idea of economic classes bound by certain common interests based on their relation to the productive process is obviously an important one which it would be folly to ignore; yet this too is only one among many threads in a complex fabric. And, not only was this a single thread among other threads, but it was woven, along with other threads, on a particular loom. Chōshū was different in many respects from other han, and within Chōshū there existed a variety of smaller units. The total configuration of forces within the han cannot be seen apart from its internal structure any more than Japan as a whole can be understood apart from the individual han, for, like the han, smaller units in the han were functional even in the Bakumatsu period.

Finally, as we noted earlier in the chapter, the han civil war was a struggle between two groups, each possessing a certain claim to legitimacy. Each side tried to persuade the uncommitted groups in the han that its claim was better than that of the other. In the struggle for the minds of the inhabitants of Chōshū, the loyalist-moderate clique was the more successful. Some of the reasons for their success have already been considered; one or two others we will take up now.

Not the least important reason for the success of the loyalists in this battle of opinion was that the Sufu clique, with which the *sonnō jōi* party had merged, had been in power within the han for all but two years since 1853. Prior to 1853, public opinion in the Western sense of the term hardly existed in Chōshū. Not only were commoners completely excluded from the affairs of the han, but even most samurai were expected to obey without understanding the decisions made by the han's highest councils. Thus, a student at the school of Yoshida Shōin was warned by his father that it was unpardonable to discuss the political affairs of the han. Or, the very act of informing the samurai of the han budget at the time of the Tempō Reform was considered as a startling innovation. An even more stringent prohibition existed against discussion of the affairs of the nation, particularly concerning those areas under the control of the Bakufu.

That discussion was banned did not mean, of course, that public opinion of a sort did not exist. When in Tempō times the han government asked for opinions on reform, a number of opinions were submitted, though mostly by high-ranking *shi*. But almost all of the opinions consisted of very ordinary Confucian platitudes. Then, Perry came to Japan; the Bakufu drew the han into national politics; and in 1858 the Court-Bakufu split shattered the former single climate of Bakufu authority. And parallel to these events, within the han a more developed form of political opinion began to emerge within the school of Shōin and about the programs of the two beaureaucratic cliques. Yet these early manifestations of opinions were feeble, and were subsequently silenced during the Purge of Ansei. They reappeared in a stronger form in 1861 and 1862, but even the mediations during this time cannot be spoken of as popular programs. Rather, the samurai supported the han, and the han government supported these programs. Only during the period between 1862 and 1864 did the majority of samurai come to identify with a particular program; and since the dominant position of the han government during these years was that of the Sufu-loyalist alliance, the majority of the samurai in Chōshū seem to have accepted this position as their own—or at least, they hesitated to actively oppose this position by giving support to the Mukunashi government during the civil war.

Any study of nationalism during this period, while qualitatively dependent on an analysis of the nature of the society, will measure its quantitative rise in terms of this emergent public opinion. An important phase in the transition from han nationalism to Japanese nationalism also occurred at this time. The Sufu-loyalist alliance with its doctrine and program for action on the national scene provided a bridge leading from the Tokugawa identification with the feudal lord to the Meiji identification with Emperor and nation. In contrast to this, the Mukunashi position, reflecting in its own way an equally strong identification with the han, lacked an emphasis on the nation.[80]

The claims of the *shotai* called for the attainment of an independent polity (on both the han and national level.) The *shotai*

[80] Obviously, this is a summary statement of a very complex question. There is one strain (at least) of "Japanese nationalism" which existed alongside the "han nationalism" of the Tokugawa period, and of which traces can be found even in earlier periods. Certain elements of this—the sense of what is Japanese in Japanese culture and so on—were picked up and intensified during the modern period. Yet I would strongly maintain that the nationalism which provided the foundation for the accomplishments of the unified Meiji state was, in good part, a transmuted form of Tokugawa loyalty (han nationalism) rather than this earlier strain.

were even willing to oppose the daimyo—or to sweep the evil ones from his side—in order to achieve these ends. The Mukunashi government, on the other hand, interpreted the basic Tokugawa concept of loyalty in a different light. It was willing to renounce all hopes for future glory in order to placate the Bakufu. To pose this contradiction even more sharply, the *shotai* were willing to risk the loss of the han in the name of the founder of the han, the symbol of what the han stood for, whereas the Hagi government was willing to risk the destruction of the daimyo—the negation of what the han stood for—in order to preserve the han and the house of Mōri. In a sense it is a commentary on the vitality of the system that, when forced to choose, a majority in the han should have preferred to preserve those values in terms of which the system was justified, rather than those supporting the forms of the system. That they were forced to choose suggests that life in Chōshū, as it had existed earlier, was already beginning to disintegrate under external political pressures.

In conclusion we can agree that peasant participation in the *shotai* and peasant aid to the *shotai* were crucial to the victory of the loyalists, even though such aid was largely obtained under duress. This is not, however, to say that the Chōshū civil war was a small-scale replica of the struggle which would later take place within the nation as a whole. Peasant aid was more important during the civil war than at any subsequent time, and the winning team in the han civil war was not representative of the combination that carried out the Restoration. One important difference is that the superiority of the Chōshū-Satsuma forces in the wars against the Bakufu was technological, whereas in the han civil war the technological advantage of the *shotai* was very small. Just as important as peasant aid to the *shotai* victory in the civil war was the neutrality of other groups in the han; a neutrality resulting from a paralysis of will in the face of legitimate yet conflicting claims, from self-interest, and from abhorrence of a fratricidal war —a neutrality extending even to groups of high-ranking *shi,* whose neutrality conflicted with the conservative interests of their social and economic position.

CHAPTER IX

The March to Power

The Background of the Second Chōshū Expedition

The opportunity provided by Chōshū's attack on Kyoto in 1864 for an expedition in the name of the Court proved to be of great benefit to the Bakufu. In this expedition all the great han were united once more under the military leadership of a representative of the shogun; the antagonist *sonnō jōi* movement was pursued to its last stronghold and silenced; even the rival *kōbugattai* (union of Court and Camp) movement, faced with the immediate common purpose of reducing Chōshū, lost its force, as the separate political identities of its components were submerged within the larger formation of the joint Bakufu-han army. And, the apparently successful conclusion of the expedition had had the same effect on the Edo *fudai* bureaucrats who ran the Bakufu as had the Satsuma-Aizu coup in 1863. In each case the Bakufu bureaucrats took the defeat of their opposition as a signal to reassert the supremacy of the Bakufu throughout the land.

To accomplish this the Bakufu took action in three areas. First, it decreed on 25/1/65 that the old *sankin kōtai* system, which had been done away with in 1862 and re-established by decree if not in fact in 1864, would once again be strictly enforced. Secondly, the Bakufu sought to regain control of Kyoto, on the one hand by obtaining Tokugawa Keiki's return to Edo, and on the other by moving Bakufu troops into Kyoto. Keiki was the legal guardian of the shogun, and, to the extent that he was under the influence of the Court and of the daimyo in Kyoto, he was an obstacle to the plans of the Edo Bakufu. By sending troops to Kyoto, the Bakufu could weaken the position of the various han and could force the Court to follow its lead. Thirdly, dissatisfied with the lenient terms accorded Chōshū by the Bakufu commander, Tokugawa Keishō, it also sought to have the Chōshū daimyo brought to Edo as further evidence of the Bakufu subjugation of that han.

All three measures were separate items but they were to become intermeshed in the early months of 1865 after the end of the first expedition against Chōshū. In the eyes of the han or of the Court, they were seen as a single program, and opposition to any one was viewed as opposition to all. This must be borne in mind to understand the attitudes of the various han toward the Bakufu's subsequent relations with Chōshū.

When Tokugawa Keishō disbanded the Bakufu army on 27/12/64, he felt that his job as the commander of the army was finished. Chōshū had submitted; whatever its punishment was to be, it was for Edo and not himself to administer it. He recommended that the two Mōri (the daimyo and his son) be forced to retire to a monastery and that the Chōshū domains be cut by 100,000 *koku*. Some han favored a harsher disposition and others a more moderate one. The Bakufu was unhappy with Keishō's action, and, in spite of the fact that he had been given complete authority over the expedition, it sent a messenger instructing him to escort the two Mōri to Edo. The message arrived on 4/1/65, after Keishō had already set out for Kyoto. His answer was that he had done what was best to improve relations between the Court and the Bakufu, and, having so replied, he continued on his way.

Determined to bring the two Mōri to Edo, the Bakufu on 5/2/65 ordered Komai Chōon, a Bakufu *ōmetsuke,* and Mitarai Kanichirō, a Bakufu *metsuke,* to conduct the two Mōri to Edo; at the same time it ordered Tokugawa Keishō to send a contingent of troops to Osaka to be placed under the command of Komai. Keishō flatly refused to send troops on the grounds that "it would be hard to tell what sort of incident" might arise.[1] Komai also refused the assignment. The Bakufu then dismissed and punished Komai and appointed in his place another *ōmetsuke,* Jimbō Sōtoku. Pleading illness, Jimbō also declined. The Bakufu next assigned Tsukahara Masayoshi to the task, once more ordering Keishō to send troops to Osaka. Keishō refused again. The Bakufu now ordered the Tatsuno, Uwajima, and Ōsu han to furnish troops to escort the two Mōri to Edo. Tsukahara and Mitarai left Edo on 6/3/65 and arrived in Osaka a few days later. Even before their arrival, however, the three han mentioned above had refused to send their troops. As a result, the Bakufu envoys sat waiting in Osaka.

Meanwhile, the Bakufu had met with a similar failure in its efforts to extend control over the daimyo and the Court. On 2/65 two *rōjū* entered Kyoto with 3000 troops. They hoped both to gain control of the Court and to persuade Keiki to return to Edo. Keiki refused to return saying that as the representative of the shogun his duty was to guard the Court. We may note here that Keiki's actions at this time, as those of Keishō, must be interpreted as the continuation of the earlier 1855–1857 struggle between the *fudai* and *shimpan* factions within the Bakufu. If one were to view the decline of Bakufu power throughout the Bakumatsu period as stemming from a gradual loss of support by its traditional adherents, such dissident *shimpan* leaders as these, who withheld their aid just when the Bakufu needed it the most, must certainly be counted among those who brought about its destruction.

The attempts of the Bakufu envoys to gain control of the Court were to be equally unrewarding. On 2/3/65, the Court summoned the Bakufu representative in Kyoto and issued an order stating that for the present the two Mōri should not be taken to Edo and that the old *sankin kōtai* system should not be re-established. The order was issued under the influence of Satsuma, and it is said to have been written by Ōkubo himself. The Bakufu officials in Kyoto were aware of the influence of Satsuma, and, fearing the reaction of Edo, they asked that its transmission be temporarily deferred; at the

[1] *Ishin shi* (Tokyo, 1939), IV, 383.

same time they privately informed the Bakufu of the Court's intentions. The Bakufu was determined to override the opposition which had risen against each of the three measures, and it reacted to this new evidence of resistance by announcing to the various daimyo on 29/3/65 that if the two Mōri should refuse the summons to Edo, the shogun himself would lead another Chōshū expedition and that the daimyo should make their preparations in advance. The wording of the announcement, "we hear that the situation in Chōshū is not completely settled," and "it seems that the violent ones [the extremists] have again broken out," [2] suggests that the Bakufu was already aware of the continuing existence of the pro-Court party in Chōshū but had not yet learned that the party had come to dominate the han government.

Just at this time the news of the actual situation in Chōshū arrived. This was an affront and a potential threat which the Bakufu could ill afford to ignore. It was the negation of all that the Bakufu had believed it had accomplished by the first expedition against Chōshū, and, coupled with the resistance of the daimyo and the Court, it constituted a denial of the Bakufu's entire program to reassert its supremacy. Some in the Bakufu argued that the shogun should first subjugate Chōshū and then, having again exercised his military leadership over the han, bring Kyoto under control on his way back to Edo. Other Edo officials disagreed, saying that the Bakufu should first gain control of Kyoto and then proceed from this base against Chōshū. The debate ended in a compromise: on the surface, the shogun was to set out against Chōshū; in reality, the plan was to halt in Osaka and to first extend Bakufu control over the Court. This plan had many advantages. The Bakufu felt that the announcement alone of a second expedition would suffice to obtain Chōshū's submission. It was also felt that the Court would be unable to oppose an expedition to which the Bakufu had so openly committed itself. And, by making Chōshū and not Kyoto their objective, the Bakufu hoped to avoid embarrassing questions from the Court as to why they had not closed the port of Yokohama as they had previously

[2] Tokutomi Iichirō, *Kinsei Nihon kokumin shi: Baku-Chō kōsen* (Tokyo, 1937), p. 213.

promised. On 19/4/65 it was finally announced that the shogun would depart from Edo by the sixteenth of the following month. He departed on schedule and arrived in Osaka on 22/5.

Over a year was to elapse from the time of the shogun's arrival in Osaka until 7/6/66, when the Chōshū-Bakufu War began. It was a year of waiting, of conferences and petty decisions, of tentative positions, of feints, and seemingly endless and tortuous maneuvers. It was a year in which opinions were the chief commodity: messengers came and went, traveling between the shogun in Osaka, the Court, the Edo officials, and the various han. The Bakufu was utterly committed to the destruction, or at least, to the subjugation of Chōshū, and yet it was loath to employ military force, and so it tried to subdue Chōshū with a barrage of edicts, proclamations, and negotiations. Only the small *fudai* han, whose destinies were intertwined with those of the Bakufu, gave it their support. Other small or medium-sized han were, as usual, passive, and a few of the larger han, or those close to Chōshū, who were unable to oppose the Bakufu directly, contributed polite comments on the unwisdom of the whole expedition. Since it is hardly worthwhile to trace the events of the year in detail, we will try to single out the main events and characteristic attitudes which formed steps in the progression toward the Chōshū-Bakufu War.

During the first two months after his arrival in Osaka, the shogun did nothing; it was hoped that Chōshū would submit without even an invitation. Chōshū, however, responded not at all. It was then decided to summon and question the daimyo of the Chōshū branch han. The order was transmitted, but both of the designated daimyo declined, pleading sickness. Again the order was sent for them to appear, and once again the response was the same. Unable to obtain even a conference with the representatives of Chōshū and increasingly harassed by criticism from the various han, the Bakufu moved to strengthen its hand by requesting the Court for an Imperial edict authorizing the subjugation of Chōshū. One faction of *kōbugattai* nobles who argued that the Chōshū problem could be solved only by a council of the daimyo opposed the edict. A pro-Bakufu faction supported by Keiki replied that, since Chōshū was the "enemy of the Court," the Court alone could arbitrate the matter. The pro-Bakufu faction proved the stronger, and on 20/9/65, the day before the shogun was to appear at Court, it was decided to issue an Imperial edict for the subjugation of Chōshū.

In the meantime in Chōshū, the han government had begun to fear that its intransigence toward the Bakufu might be viewed by the various other han as an unwillingness to negotiate. Therefore, when the Bakufu next requested a conference, it was decided to send a representative of the han to meet with Bakufu officials and to state the position of Chōshū. *Shotai* leaders in Chōshū objected to this on the grounds that it signified a softening of Chōshū's opposition to the Bakufu, but they were quieted by assurances from the han government that this was merely a tactical move to influence opinion and to gain time for the completion of the military reforms being carried out within the han. The Chōshū mission arrived in Hiroshima on 22/10/65; its chief representative during the talks was Shishido Tamaki. The Bakufu mission headed by the *ōmetsuke* Nagai Naomune arrived in Hiroshima on 16/11/65; the meetings began on 20/11/65. The Chōshū representatives were questioned on the fighting within Chōshū, the movement of the daimyo from Hagi to Yamaguchi, contacts with foreigners, the purchase of foreign arms, and so on. In each case, Chōshū's answer was to humbly deny the charge and to protest against the second Bakufu expedition. Nine days later, representatives of the *shotai,* who had also been asked to the meetings, arrived in Hiroshima to meet with the Bakufu representatives on the last day of the month. They chose an even more uncompromising stand than had Shishido; the Bakufu representatives were put on the defensive and found themselves trying to explain the actions taken by the Bakufu. Consequently, on 17/12/65 the Bakufu party returned to Osaka and reported on the following day that Chōshū's submission was a pretense and that, while dissembling, it, in fact, was preparing for war. It was therefore recommended that the Bakufu act speedily to subjugate Chōshū.

The next issue which was to delay the expedition (for yet another month) was a controversy as to what should be done with Chōshū after it had been subjugated. It was not a new problem; a conference on the same subject had been held seven months earlier when the shogun first entered Kyoto. At that time the Edo *rōjū* had held the harshest opinion, arguing that the two Mōri should be executed and their domains confiscated. Matsudaira Katamori of Aizu had been almost equally severe, advocating death for the two Mōri for having fired cannon at the Imperial palace (during the 1864 countercoup) and a reduction of the Chōshū domains. But since the most powerful han held that the lives of the two Mōri should be spared and the domains cut by only one-third or one-half, and since the support of these han was needed, no decision was arrived at during the early meetings.

During these early months there was a widespread opinion among the daimyo in Kyoto and Osaka that a conference of daimyo should be called to settle the Chōshū question. This was the proposal which we mentioned earlier as having been supported by the *kōbugattai* faction in the Court when the question arose as to whether an Imperial edict should be issued for the subjugation of Chōshū. The Bakufu, however, insisted that the question would be settled by its sole authority. And as the months passed, in opposition to this Bakufu authority, the opinion of the han concerning the question of Chōshū tended to become more and more lenient: in many cases the opinions of the han at the end of 1865 were the reverse of what they had been at the time of the first expedition. Even for the *kōbugattai* han, Chōshū was no longer a dangerous rival; rather it had become a fellow han oppressed by the Bakufu.

Matsudaira Keiei criticized the second expedition saying it "would cause confusion within the country and more and more disturb the Emperor."[3] Matsudaira Mochiaki, the daimyo of Fukui, sent a petition to the Bakufu in which he stated: "The entire country felt relieved when the first expedition against Chōshū was concluded without resort to arms. If great numbers of troops are again mobilized, the various daimyo will be impoverished, the people will dislike the Bakufu, the country will fall into disorder, and who knows what will happen [under such circumstances]. . . . Truly this affair is

[3] *Ishin shi,* IV, 417.

grave, one which may affect even the destinies of the Tokugawa house."[4]

Ikeda Shigemasa, daimyo of Bizen, petitioned the Bakufu *rōjū,* Mizuno Tadakiyo, saying that a clement government would win the hearts of the people but a cruel government would lose them. He later wrote to Ikeda Yoshinori, the daimyo of Inaba, predicting that the fighting of the second expedition against Chōshū could not be confined to Chōshū alone but would give rise to disturbances throughout the country. He further stated that, as a member of a side branch of the Tokugawa house, he could not sanction this state of affairs. The daimyo of Inaba replied that he too, was opposed to the second expedition but, since petitions would not influence the present leaders within the Bakufu, the best thing to do was to wait and watch. The daimyo of Bizen also sent similar letters to the daimyo of Aki and Tokushima. Tōdō Takayuki of Tsu han petitioned the Bakufu officials in Osaka as follows: "According to the laws of military science, to push one's enemy to the wall is extremely bad tactics. If Chōshū is pushed to the wall, one man will resist with the strength of one hundred, and one hundred with that of ten thousand. Therefore, even if there were no doubt about the Bakufu's prospects for victory, the war would be long and who can tell what disaster might occur in the interim."[5]

Furthermore, since the only charge contained in the announcement of the second expedition against Chōshū was that of "a grave plot"[6] on the part of Chōshū, many han directed their criticism at the vagueness of purpose of the expedition. Tokugawa Mochitsugu, daimyo of Kii, wrote to the *rōjū* Abe Masatō that the charges against Chōshū were unclear. The daimyo of Tokushima wrote that the matter could not be given serious attention until the charges against Chōshū were better defined. The daimyo of Kumamoto, Hosokawa Yoshiyuki, cautioned that "even when commoners are punished, if the nature of the offense is not made clear, the people become uneasy. [There-

[4] *Ibid.,* p. 416.
[5] *Ibid.,* pp. 419–420.
[6] *Ibid.,* pp. 389–390.

fore] even more in the serious matter of subjugating a great han, it is natural for the various han to feel uneasy, and who can tell what upheavals may be caused."[7]

Within this very negative climate of opinion, the earlier emphasis on a severe punishment could not be maintained, and so, following the return of the Bakufu representatives from Hiroshima in 12/65, the Bakufu *rōjū* meeting in Osaka decided that the Chōshū daimyo must retire in favor of his son and that the Chōshū domains would be reduced by 100,000 *koku*. To this decision Keiki replied from Kyoto that if indeed all the troops of the various han were once more to be mobilized, this was too light a punishment. A further series of meetings took place; the final decision rendered on 19/1/66 was that the two Mōri should both step aside in favor of the daimyo's grandson and that their domains should be reduced by 100,000 *koku*. The Bakufu officials felt that the Bakufu was not prepared to wage war and hoped that these comparatively lenient terms would persuade Chōshū to submit as it had done at the time of the first expedition. On 12/1/66 the Court approved of these terms.

A *rōjū,* Ogasawara Nagamichi, was then sent to Hiroshima to inform Chōshū of the above terms. He arrived on 7/2/66, and on 22/2/66, using the daimyo of Aki as a go-between, he summoned the daimyo of the Chōshū branch han and two Chōshū Elders, Shishido Tamaki and Mōri Chikuzen, to Hiroshima. Chōshū replied on 27/2/66 that, with the exception of Shishido, who was already in Hiroshima, none of those summoned could appear because of sickness. The Bakufu representatives again summoned the same persons, saying that, if they were unable to appear, representatives should be sent in their place; this second summons was cast in the form of an ultimatum: if the Chōshū representatives did not appear in Hiroshima by 21/4/66, the Bakufu armies would advance. Chōshū's response to this was made in essentially the same spirit in which it had been parrying with the Bakufu for many months. Within the han Chōshū hastened its total mobilization for war; formally, it continued to protest the injustice of the second expedition and at the same time appointed representatives to proceed to

[7] *Ibid.*, p. 421.

Hiroshima where, with a mixture of reason, moderation, and compromise, they could delay the actual fighting. By 23/4/66 the various representatives from Chōshū had all arrived in Hiroshima. On 1/5/66 the representatives were summoned to meet with the *rōjū* Ogasawara. Though present in Hiroshima, Shishido, the representative of the Chōshū daimyo, did not attend the meeting because of "pains in his leg." Those present were instructed to take the Bakufu terms to Chōshū. They objected, saying that this could be done only by the representative of the Chōshū daimyo, but these objections were overruled. On 9/5/66 Shishido was again ordered to appear, but he again declined on the same grounds. The Bakufu *rōjū* thereupon sent troops to seize Shishido and handed him over to the custody of Aki han. The Chōshū branch han representatives hurried back to Chōshū bearing news of a new ultimatum: Chōshū must submit by 20/5/66. Chōshū requested that this be postponed until 29/5. This was granted but the appointed day came and passed without a word from Chōshū, and so on 7/6/66 the army of the second Bakufu expedition against Chōshū finally began its attack on the han. Before taking up the war, however, we will consider first, the relations developing between Chōshū and Satsuma, and second, the developments within Chōshū itself at the time.

The Satsuma-Chōshū Alliance

The most important development in the period between the first and second Bakufu expeditions against Chōshū was, undoubtedly, the establishment of the Satsuma-Chōshū alliance. In the years between 1861 and 1865 the antagonism between Chōshū and Satsuma was, like the antipodality of the Court and the Bakufu or the rivalry between *shimpan* and *fudai* factions within the Bakufu, one of several horizontal axes determining the movements of politics in Japan. Both han were leaders of national movements: Chōshū, of the *sonnō jōi* movement, and Satsuma, of the *kōbugattai* movement. Each offered a solution to the problems confronting the country, each hoped for power as a reward for success, and each directed its energies toward the defeat of the other.

It was within this context that the Satsuma leaders welcomed with jubilation the Bakufu announcement of the first expedition against Chōshū. The annihilation of Chōshū, they thought, would open up a clear path for the success of their own movement. Thus Saigō Takamori held views on the disposition of Chōshū as harsh and uncompromising as those held by Aizu: the Chōshū leaders should be executed and the han dismembered. Similar views were common among other Satsuma leaders.

The animosity toward Chōshū reflected in these views was a consequence of Satsuma's hopes for its own *kōbugattai* movement. One must note, however, that this did not necessarily imply a high estimation of the Bakufu on the part of Satsuma. The *kōbugattai* program was one of "union of Court and Camp," but in the eyes of Satsuma this was to be a union watched over and directed by Satsuma itself. The Bakufu was not expected to play an active role; it need only cooperate. Saigō, for example, had become aware of the Bakufu's inability to act as early as 9/64. At that time Katsu Awa, the commander of the Bakufu naval forces, had confided to Saigō his desire to attack the foreign ships and to enter into negotiations with the foreigners that would not humiliate the Imperial country. But, he had lamented, he could not, for the Bakufu knew only procrastination. A few days later Saigō wrote to a friend: "Of late the Bakufu has entered the period of its decline, and it is no longer able to function as hegemon. Now, when the barbarians are on all sides of us, if the great han, at least, do not join together, it will mean the decline of the Imperial country." [8] The great han led by Satsuma were to join and act for a senile Bakufu and the infantile Court.

But would the Bakufu cooperate, and should Chōshū be destroyed after all? By the end of 1864 Satsuma had become less certain. Saigō, for example, as a staff officer of the Bakufu army surrounding Chōshū, contended that Chōshū should be purged of its extremism by Chōshū's own conservatives and that the Bakufu army should disband after attaining its minimal objectives. This comparatively moderate view reflected the first glimmerings of apprehension by

[8] *Ibid.*, p. 445.

the Satsuma leaders concerning the reliability of the Bakufu. Chōshū extremism must be crushed, but the han itself should be maintained as additional ballast in the west against the uncertain Bakufu in the east.

The doubts concerning the Bakufu were strengthened during the early months of 1865 when the Bakufu renewed its efforts to reimpose its authority over all of Japan. In the face of such efforts, Satsuma's *kōbugattai* program was reduced to an absurdity. Therefore, little by little, almost by default, Satsuma was forced into opposition against the Bakufu until finally it found that its position was not too different from that of Chōshū. And each increment of change in Satsuma's political position produced an increment of change in Satsuma's attitude toward Chōshū. Thus, when the Chōshū civil war broke out, Saigō held that this was an internal affair concerning Chōshū alone. When the Hagi government executed Maeda Magoemon and others of the former Sufu-extremist coalition, Saigō deplored the loss of these men who had worked virtuously on behalf of the Imperial country.

By the spring of 1865 when the Bakufu annouced the second expedition against Chōshū, Satsuma had swung into complete opposition to the Bakufu, and its views were the exact reverse of what they had been less than a year before. Saigō spoke of the expedition as a "private war" [9] and vowed that Satsuma would not send troops to join the Bakufu army against Chōshū. Ōkubo wrote that there were no examples in history, whether "in the past or in the present, in Japan or in China, of victory in a war that was contrary to the principles of honor [*meigi*]." [10] Ōkubo also traveled to Fukui to urge Matsudaira Keiei to go to Kyoto to oppose the expedition. Keiei agreed to this and set off for Kyoto on 1/10/65, but he changed his plans en route and turned back when a retainer suggested that the Bakufu would suspect him of collusion with Satsuma. (Again and again in Bakumatsu history the gulf between *fudai* and *shimpan* power appears to have been crucial.)

Convinced finally that nothing could be done with the Bakufu,

[9] Tōyama Shigeki, *Meiji ishin* (Tokyo, 1951), p. 182.
[10] *Ibid.*

Saigō and Ōkubo petitioned Hisamitsu to abandon the *kōbugattai* position held by Satsuma since 1862 in favor of a *sonnō* policy based on the union of the great han. Hisamitsu accepted the petition and privately expressed his agreement to Saigō and Ōkubo. The Satsuma officials in the han, however, were not told of the policy change at this time, since the officials surrounding Hisamitsu felt that it would only arouse conservative opposition which could be better dealt with at another time. This decision, in a sense, resembled the 1862 Chōshū decision to adopt the *sonnō* policy. In both cases bureaucratic cliques headed by men who were not in any simple sense *sonnō* ideologues decided to adopt, while away from their respective han, the *sonnō* policy for reasons of political expediency. The consequences of this decision were almost as momentous for Satsuma as they had been for Chōshū; the most important of these was Satsuma's changed relation to the Bakufu. However irresolute, the shogun was the hegemon; however dilapidated, the Bakufu apparatus still functioned in all but the most vital of its former spheres. Opposition to the Bakufu in the name of the Court (which indeed had little more to offer than the sanction of its name) put Satsuma in a position almost identical with that of Chōshū, and yet the antagonism existing between the two han proved strong enough to hold them apart. That this was eventually overcome was in part due to the efforts of two young samurai from Tosa.

As a han, Tosa played a very insignificant role in Bakumatsu politics. Except for a span of several months in 1862, when under the control of its extremists Tosa had joined with Chōshū in promoting the *sonnō jōi* movement in Kyoto, Tosa's position was typical of the pro-Bakufu *tozama* han: it was willing to share in the political authority which had previously been the monopoly of the Bakufu but it hesitated to attack the Bakufu as such. And, many of those within Tosa who wanted to join with Chōshū had been executed or imprisoned by the government of Yamanouchi Yōdō. Two members of the Tosa extremist party, however, left their han to become free agents in the gathering confusion of national politics, and it was in this capacity, and not as representatives of Tosa, that they were able to bring about the alliance between Chōshū and Satsuma.

Nakaoka Shintarō was an intimate of the Chōshū extremists. Leaving Tosa in 1863, he joined the entourage of the extremist nobles who had withdrawn to Chōshū. He fought with the Chōshū forces in the attempted countercoup of 1864 as the commander of a *rōnin* militia unit. Nakaoka also fought with the Chōshū forces against the foreigners at the time of the four-nation bombardment of Shimonoseki. Sakamoto Ryūma, on the other hand, was the contact man on the Satsuma side. As a former student of Katsu Awa, he had been introduced to Saigō by Katsu. At one time when Katsu was in difficulties with the leading Bakufu officials, he asked Komatsu Tatewaki to care for Sakamoto. Komatsu was an Elder of Satsuma who with Ōkubo and Saigō formed the triumvirate which led Satsuma during the last years of the Bakumatsu period.

In 5/65 Sakamoto accompanied Saigō to Satsuma; they arrived just when the policies of the han were shifting in the direction of the *sonnō* position of Chōshū. As soon as he learned of the changes, Sakamoto set off for Chōshū in the hope of bringing about an alliance between the two han. He arrived in Shimonoseki where he was met by Kido, anxious for news from Satsuma. In the meantime, Nakaoka Shintarō had gone from Chōshū to Kyoto with the same end in mind. There he met with another Tosa *rōnin,* Hijikata Hisamoto, and together they planned a meeting between Kido and Saigō. On 24/5/65 Hijikata set out for Chōshū and Nakaoka for Satsuma. In Satsuma Saigō agreed to Nakaoka's proposal to visit Shimonoseki en route to Osaka. Kido was apprised of this, and he also agreed to meet with the Satsuma leader. But while en route to Osaka, Saigō suddenly changed his mind and passed on without stopping at Shimonoseki. This did not enhance relations between the two han; the Tosa mediators felt betrayed, and Kido returned to Yamaguchi impressed by the insincerity of Satsuma.

Fearing that this incident would further estrange the two han, Sakamoto and Nakaoka asked the Chōshū leaders not to abandon the idea of a reconciliation with Satsuma and proposed a test of Satsuma's sincerity. Chōshū at this time was busy training its troops against the day that the second expedition would arrive, but because of Bakufu control of trading it had been unable to purchase the foreign arms needed to complete its military reforms. Itō and Inoue now suggested that the needed arms be bought in the name of Satsuma. Sakamoto and Nakaoka agreed and left for Kyoto where they obtained Saigō's approval. Just at this time, a Chōshū samurai sent to Nagasaki to buy arms had returned empty-handed, and, having heard from Sakamoto, Kido sent Itō and Inoue to Nagasaki. The two arrived on 21/7/65, and, staying at the Satsuma residence, they were able by the end of the eighth month to arrange for the purchase of a warship and 7300 rifles.[11] Itō remained in Nagasaki to complete the transaction, and Inoue returned with Komatsu to Kagoshima to plan for a reconciliation of the two han. Moved by this tangible proof of good will, the Chōshū daimyo sent a letter to Satsuma expressing his desire for better relations in the future. Orders were also issued to give good treatment to Satsuma vessels visiting the shores of Chōshū.

This was the first step toward an alliance of the two han. At the same time, however, there arose a strong reaction within Chōshū to both the purchase of foreign arms and to the proposed alliance. Of antiforeign feeling we will speak later; as for anti-Satsuma feeling it was very wide-spread in Chōshū among the conservatives as well as the extremists. Some in the *shotai* threatened that any Satsuma man who planned to come to Chōshū should regard the Shimonoseki Straits as the River Styx (*sanzu no kawa*).[12] Others swore that even though they should be forced to make peace with the barbarians, they would never be friends with Satsuma. It was said that Shinagawa Yajirō, at this time a *shotai* commander, had written on his clogs, "Satsuma bandits, Aizu villains," in order to enjoy the pleasure of walking on them with each step.[13]

That such hostile feelings were eventually overcome was due in part to the exertions of the Chōshū leaders. Inoue, for example, wrote

[11] Of the 7300 rifles, 4300 were Minies, the weapon with which the American Civil War, which had ended only a few months earlier, had been fought. These cost 77,400 gold *ryō* (about 6569 silver *kan*). The other 3000 were older weapons costing only 15,000 gold *ryō* (about 1275 silver *kan*). Thomas Glover, the English merchant at Nagasaki who arranged for the sale, wrote later: "Of all of those in rebellion against the Tokugawa government, I felt that I was the greatest rebel." See Tanaka Akira, "Chōshū han ni okeru Keiō gunsei kaikaku," *Shirin,* 42:107 (Jan., 1959).

[12] *Ibid.,* p. 182.

[13] *Ishin shi,* IV, 450.

to Kido that many opposed a union with Satsuma out of respect for the memory of those Chōshū samurai who had died fighting in the *shotai*. This, he continued, is not a valid argument. Those who fell had fought for the han. Today it is necessary for the han to ally itself with Satsuma, and since the end is the same, we need not feel that the dead are abused.[14] On 9/65 Satsuma, aware of the mounting tension in Kyoto, decided to send troops to that city. At this time Saigō sent Sakamoto Ryūma to Chōshū to ask whether provisions for the Satsuma troops could be obtained at Shimonoseki. This move seems to have been prompted less by economic considerations than by political motivations. The Chōshū response was affirmative, and it was also decided that although Chōshū would fight alone if necessary, an alliance should be formed, if possible, with Satsuma, Bizen, Aki, Chikuzen, and other great han against the Bakufu.

On 25/10/65 Komatsu and Saigō led troops from Kagoshima to Kyoto. Aware of the tensions in the city and convinced that war was imminent, they sent a Satsuma samurai to Chōshū to invite Kido to Kyoto for a meeting with the Satsuma leaders. Due to the opposition within Chōshū, Kido at first demurred, but, urged on by Takasugi and Inoue, he finally decided to go. On 8/1/66 he arrived at the Satsuma Fushimi residence accompanied by Shinagawa Yajirō and other *shotai* leaders. They were met by Saigō, who accompanied them to the Satsuma residence in Kyoto. They were treated cordially, and frequent meetings over a period of ten days with Ōkubo, Komatsu, Saigō, and other important Satsuma leaders were spent in general discussions of national affairs. Neither side, however, mentioned what was uppermost in their minds, the question of an alliance, for each felt that to broach the subject would entail a loss of face.

Kido became discouraged and was preparing to return to Chōshū when Sakamoto arrived in Kyoto from Shimonoseki. When Sakamoto heard of the situation, he was indignant and reproached Kido for having left Chōshū at this crucial time only to squander days in Kyoto in empty discussions. To this Kido rejoined:

At present the position of Satsuma is different from that of Chōshū. Speaking frankly, Satsuma openly serves the Emperor, openly meets with

[14] Suematsu Kenchō, *Bōchō kaiten shi* (Tokyo, 1921)), VII, 495–497.

the Bakufu, and openly deals with the various daimyo. Therefore, openly and without hindrance it can participate in national affairs. In contrast, all are the enemies of Chōshū and even now press on us from all sides. Our samurai are at ease solely in their hearts as they prepare to [fight until they] die. One can say that Chōshū's danger is extreme; it has no way out. In this condition, can samurai of Chōshū propose an alliance with Satsuma? From their point of view, even if it did not appear that we were inviting them to share our danger, we would seem to be asking for help. This is not in the hearts of Chōshū samurai.[15]

Recognizing the delicacy of the Chōshū position, Sakamoto directed his arguments toward the Satsuma leaders, who finally agreed to begin negotiations. Once opened, discussions progressed rapidly and a formal agreement on the terms of the alliance was reached on 21/1/66.

The actual agreement took the form of a six-point pact designed to cover all possible contingencies. The points outlined were as follows: 1) If the Bakufu goes to war against Chōshū, then Satsuma will send troops to secure Kyoto (this was a commitment to exert military pressure rather than to take military action). 2) If Chōshū should win the war, Satsuma would mediate on its behalf at the Court. 3) If Chōshū should appear to be losing the war, since this would not come about quickly, during this time Satsuma would do all that was possible on its behalf. 4) If there is no war, Satsuma will mediate for Chōshū at the Court. 5) If Satsuma is absolutely unable to mediate because of interference by the Bakufu, Aizu, or Kuwana, then it will go to war against the Bakufu. 6) Pardoned by the Court, Chōshū will join Satsuma in working for the glory of the Imperial country.[16]

That this alliance had been achieved was, of course, a closely kept secret. After returning to Chōshū, Kido and other leaders of the han government decided that it should be kept a secret even within the han on the grounds that the fighting spirit of the Chōshū samurai would be higher if they were unaware of the alliance with Satsuma. And it goes without saying that the Bakufu, although increasingly

[15] *Ibid.*, VIII, 140–141.
[16] *Ishin shi*, IV, 468–469.

suspicious of Satsuma, pushed forward its plans for the subjugation of Chōshū, little dreaming of the newborn coalition. This alliance was the second great step in the movement to overthrow the Bakufu, second in importance only to the Chōshū civil war. By the end of 1865 Nakaoka Shintarō could write to a friend in Tosa: "I see, as in a mirror, that in the near future the entire country will be following the commands of Satsuma and Chōshū."[17]

Developments within Chōshū

In the previous chapter we characterized the Chōshū government formed after the civil war as an extremist-moderate coalition within which the extremists were dominant. Their political influence was based on their former positions as *shotai* commanders, and at the same time it included the new dimension of bureaucratic power which had enabled them, for example, to bring the *shotai* under tighter governmental control. The underlying policy of this new government was embodied in a proclamation made by the daimyo on 17/3/65: externally, Chōshū was to pursue a policy of penitent submission; within the han all classes were to unite in efforts to strengthen Chōshū militarily. Yet it should be pointed out that all classes within Chōshū were not united during the period immediately following the civil war; the wounds of the civil war were slow in healing. The defeated conservative forces were still antagonistic to the new government. The branch han of Iwakuni remained a stronghold of conservative thought. Many neutralist samurai were no more anxious to support the new extremist government than they had been to support the government of Mukunashi. Peasant officials and peasant militia which had been neutral during the han civil war remained essentially passive toward the new government. This postwar political multipolarity was further aggravated when in 3/65 Takasugi, Itō, and Inoue were forced to flee from Chōshū to escape assassination for having proposed that Shimonoseki be opened to foreign trade. The loss of these men, and particularly of Takasugi, created a hiatus between the han government and the *shotai* com-

[17] *Ibid.*, p. 473.

manders just when cooperation between the two groups was most urgent. To say the least, there certainly did not exist in the han at this time a social base from which a new program of anti-Bakufu activity could be launched. Had the Bakufu chosen to ignore the changes produced in Chōshū by the han civil war, Chōshū might have remained neutralized for some time by the unsettled conditions and divided sentiment within the han.

Instead the Bakufu announced on 19/4/65 the formation of a second expedition against Chōshū. This announcement forced Chōshū to look to the future rather than to the past, thus diverting its attention from the wounds left by the civil war. The first step toward the creation of a new sense of unity within the han was achieved only a few weeks later when Kido Kōin returned on 26/4/65. On 13/5 he met with the daimyo and was appointed to a leading position in the han government, thereby effectively bridging the hiatus between the *shotai* and the government caused by the loss of Takasugi two months earlier. As we noted earlier, Kido could fill this pivotal role since he combined in one person the character of the extremist disciple of Shōin with that of the career bureaucrat close to the daimyo.

On 22/5/65, the same day on which the shogun arrived in Kyoto to carry out the expedition against Chōshū, a meeting of the leading figures in Chōshū was held. It was decided that Chōshū would fight should the Bakufu actually mount a second expedition against Chōshū. Why did the government of Chōshū make this decision when barely eight months ago at the time of the first expedition it had submitted? Part of the answer is obvious: the Sufu government had been dependent on the will of the daimyo and Elders, who were influenced by the conservative criticisms of the Mukunashi clique; the post-civil war government was dependent primarily on the military power and intransigent opinion of the *shotai*. And, not only was the nature of the dependence different, but the nature of the dominant groups also differed significantly. The Sufu clique had been primarily composed of bureaucrats, or samurai bureaucrats. By the autumn of 1864 they realized that their policies had brought misfortune to the han. Right or wrong, they had clearly failed. Sufu himself atoned for his failure by suicide. The *shotai* leaders, on the other hand, judged the same course of events from the religio-political framework of the *sonnō* doctrine. From this standpoint, Chōshū had not failed in its struggle for power; rather, the forces of darkness had triumphed over the forces of light. In 1864 the *shotai* leaders had not possessed sufficient influence within the government to resist the rise of the conservatives (had they, Chōshū would undoubtedly have been crushed by the first Bakufu army). They rebelled, and achieving a dominant position, they confidently resolved to resist the second Bakufu expedition against Chōshū. Moreover, as we have seen, there were some differences between the first and second expeditions in the external scene: the second expedition had far less popular support than the first, and the terms of surrender it presented to Chōshū were much more severe.

Having obtained the decision to resist the new Bakufu expedition, Kido acted in three areas to increase the strength of the han: he effected a reconciliation with the conservative Kikkawa of the branch han of Iwakuni, he extended the power of the extremists in the han bureaucracy, and he appointed Ōmura to reform the han army.

The problem of Kikkawa is fascinating not only because it illustrates the semiautonomous position enjoyed by the branch houses within Chōshū, but also because it points up how the shadows of the pre-Sekigahara past lingered to complicate political relations over two centuries later. In 1600 Kikkawa Hiroie, the grandson of Mōri Motonari who had established the fortunes of the house of Mōri, followed Mōri Terumoto to Sekigahara as a member of the Western camp. Perceiving, however, that the Western camp could not win, he first tried to persuade Terumoto of the folly involved, and when this failed, he came to an understanding with Tokugawa Ieyasu. After Sekigahara, Kikkawa Hiroie, in spite of having deserted Mōri Terumoto in his hour of need, interceded on his behalf and pleaded before Ieyasu that the house of Mōri should be continued. In a sense, since he was the head of a branch line, it was only proper that he should try to preserve the continuity of his main house. On the other hand, feudal might not infrequently neglected such a concern for ancestral right. In any case, Kikkawa Hiroie and his vassals felt that it was due primarily to their efforts that the lands of Mōri were not totally confiscated. But Mōri Terumoto and the Mōri vassals felt that the defection of Kikkawa at Sekigahara was one cause of the defeat of the Western camp, and, cramped on their drastically reduced domain, they gave little credit to his intercession.

As a consequence, an antagonism arose between the two houses that continued into the Bakumatsu period. The initial antagonism was furthered by the ambiguous position of the Kikkawa house vis-à-vis the main house of Mōri and the Bakufu. In spite of the fact that the Kikkawa was as much a branch house as the other three, it was called a vassal (*kerai*) by the Mōri, and in this context "vassal" was pejorative.[18] The Tokugawa, on the other hand, recognized the Kikkawa as an independent daimyo house; Kikkawa Hiroie was accorded an audience with Tokugawa Ieyasu in 1604, and three years later he was told to proceed to Edo under the *sankin kōtai* system. Out of deference to Mōri, Kikkawa declined to participate as an independent daimyo in this system; nevertheless, a Kikkawa residence was established in Edo, and the custom arose that once every generation the new Kikkawa lord would receive an audience with the shogun. It appears that the favorable treatment meted the Kikkawa by the Bakufu was a deliberate policy stemming from the time of Ieyasu, for in a manual of instructions left by Ieyasu to future shoguns on the subject of controlling the daimyo, there is a passage directing that the Kikkawa of Chōshū, the Inada of Tokushima, and the Isahaya of Saga be treated as daimyo in order to stir up dissension between them and their respective main houses.[19] This was but another small component in the elaborate system of devices and controls by which the Tokugawa ruled in spite of their weakness of numbers. The hostility existing between the Mōri and the Kikkawa was overcome in 1863 when, for the first time since 1600, the Chōshū daimyo visited the Kikkawa domain.[20] But in spite of this, the difference in the traditions of the two families was certainly one element leading Kikkawa Tsunemoto

[18] Suematsu, *Bōchō kaiten shi,* I, 20–24.
[19] *Ibid.,* p. 23.
[20] *Ibid.,* II, 225.

to support the conservative clique of Mukunashi in the autumn of 1864 and to fight against the *shotai* in the spring of 1865.

When the *shotai* forces proved victorious, Kikkawa Tsunemoto withdrew to his Iwakuni domain. This created an impossible situation for Kido and the extremist government: in the face of the Bakufu expedition they could not tolerate a neutral or unsympathetic power within the han, and yet they were disinclined to use force against the semiautonomous branch han daimyo. It was also a difficult position for Kikkawa Tsunemoto: it was humiliating to drop his conservatism and join the extremists whom he had been fighting only a few months earlier, but, if the Bakufu armies should come and defeat Chōshū, his domain would no doubt be confiscated along with the other Chōshū holdings, and, after all, Chōshū did represent the main line of the house of Mōri.

On 6/i5/65 a conference to determine the official han policy was held. To this conference Kikkawa Tsunemoto was invited along with the daimyo of the other branch han. Kikkawa attended the meeting, having decided beforehand that he had no other choice than to throw in his lot with Chōshū. He thereupon seized the bull by the horns: in his opening speech he announced that the rumor of a second Bakufu expedition against Chōshū was so unreasonable as to be incredible, but that, should it in fact be true, Chōshū would have no choice but to fight to the death.[21] Since this conference had been called primarily to obtain the cooperation of Kikkawa his opinion was readily accepted.

The second point in Kido's program to strengthen the han was the extension of extremist power within the han government. This was accomplished, first, by appointing to official position persons sympathetic to the extremist cause, such as Maebara Issei or Hirosawa Saneomi, and second, by calling back to the han and reappointing Takasugi, Itō, and Inoue. Hirosawa was an interesting figure in that his career pattern is atypical of those who later rose to position in the Meiji government. He rose not through the extremist *shotai* but through the han bureaucracy. He did not attend the school of Yoshida Shōin, he was too young to be a member of

[21] *Ibid.*, VII, 170, 171, 220.

the Sufu clique, yet in the years between 1858 and 1863, he advanced through various subordinate positions in the han bureaucracy. In 9/63 he was appointed to the position of treasury official (*kuramoto yaku*), an office of considerable importance. Hirosawa was put in jail by the Mukunashi clique during his brief ascendancy for having extremist sympathies, and finally, after the han civil war, he was given one of the most important positions in the han by Kido.

The third and perhaps the most important measure taken by Kido was the appointment of Ōmura Masujirō to reform the han army. Ōmura was also made a member of the *ōgumi,* the fourth highest stratum of *shi* so that his rank would be commensurate with his function. In the period immediately following the civil war some measures had been taken to limit the number of *shotai,* but they had had little effect. The *shotai* were now brought under the firm control of the han government, and discipline was strictly enforced. They were standardized in weapons and training. Units which had previously employed a variety of weapons were converted into rifle companies. Peasant members of the *shotai* whose training had been inadequate were discharged. The *shotai* were assigned to various local areas to cooperate with the local defense corps or peasant militia. This rationalization of the han military was not limited to the *shotai*. Perhaps even more important was the organization of the other han samurai and rear vassals into military units which could easily be mobilized in time of crisis.[22] *Shotai,* vassals, and rear

[22] Tanaka Akira, in a recent article, has analyzed in some detail this period of military reform. The gist of his argument is that during this period the locus of power in the han was moved from the *shotai* (representing the movement from below) back to the han government (where there rapidly formed a nucleus of absolutist bureaucratic officials of the Meiji type). See Tanaka Akira, "Chōshū han ni okeru Keiō gunsei kaikaku," *Shirin,* 42:104–124 (Jan., 1959). If the above is put in a slightly changed theoretical framework I feel that it is in essential agreement with my own analysis: 1) that the non-*shotai* power of the moderate *chinsei kaigi* (Peace Assembly) still had some voice in the han government after the civil war; 2) that effective authority shifted in the process of the reform of the military from the *shotai* back to the han government.

Mr. Tanaka, of course, means much more than this when he speaks of absolutist Meiji-type bureaucrats. Absolutism in the sense in which Japanese historians use the term is derived from Karl Kautsky's categorization of the term: "The political form in which the power of the state is not directly a tool of class domination, but appears

vassals all engaged in periodic maneuvers. Many sanguine accounts exist, chronicling the zeal and intensity with which they prepared for war. The following description from a letter by Nakaoka Shintarō is not exceptional:

Han policy has been stabilized, the government has been reformed, and the people have resolved to fight unto death. In this state the samurai spirit has grown ever more steady, the preparation of weapons increases day by day, and words have been replaced by deeds. In every way the forces of the han have been renewed; only companies of rifle and cannon exist, and the rifles are Minies, the cannon breech loaders using shells; in every respect the military system is being reformed. Cavalry units also flourish; within the han great maneuvers are carried out; in one day [as many as] forty-six units may practice gunnery without cease; truly, Chōshū's forces are unsurpassed. All this is wholly due to the war, and it cannot be equaled by other han, no matter how hard they try; even Satsuma cannot compare with this.[23]

The morale or military spirit of the Chōshū samurai can be seen in memorials sent by the *shotai* to the han government. Many were critical of the han's external policy of submission, arguing that, since its cause was just, Chōshū should begin the hostilities. As early as 26/5/65 the daimyo summoned the *shotai* commanders to admonish them as follows: "At this time there is a rumor that a second military force has been sent to subdue us. Even if by chance this force should arrive at our borders . . . the *shotai* are strictly forbidden to engage in skirmishes or violence until orders have been

to have an independent existence transcending parties and classes. This can only be established when all those classes that have weight in society hold one another in equilibrium in such a way that no one class among them is strong enough to seize the power of the state for itself." See Karl Kautsky, *Die Klassengegensätze im Zeitalter der Franzosischen Revolution* (Stuttgart, 1919), p. 10. For Mr. Tanaka, the absolutist equilibrium existed between the *shotai* (representing the bourgeois energy of their peasant and townsmen components) and the regular feudal vassal groups which were at this time reformed into more or less rationalized rifle units. This balance of bourgeois and feudal elements was put under the Kanjōtai, a military unit (of higher echelon) composed of Peace Assembly members. This in turn was controlled by the han bureaucrats whose character, by the nature of the balance of elements which they straddled, was absolutist.

[23] Osatake Takeshi, *Meiji ishin* (Tokyo, 1943), II, 671–672.

issued."[24] In spite of several such warnings, as hostilities became imminent, the restless *shotai* became more and more bellicose, until finally, on 4/4/66, one hundred samurai of the Second Kiheitai seized weapons and fled the han. They attacked and burned a Bakufu *daikansho* at Kurashiki in Bitchū, and were dispersed only after the arrival of Bakufu troops and troops from the surrounding han.

The Chōshū government was disconcerted by this action since its central policy at this time was to win the sympathies of other han by making the Bakufu appear the aggressor. It also feared other such incidents, since many young samurai in other *shotai* praised the Second Kiheitai for its daring. It therefore took an extremely stern attitude toward the incident, and, by the end of the following month, forty-eight of the Second Kiheitai members who had made their way back to Chōshū had been executed, and those in other *shotai* who advocated similar uprisings were also punished. In many respects the attitude of the 1866 extremist government toward the over-zealous Second Kiheitai was similar to that of the 1858 Sufu government toward Yoshida Shōin. Each was firmly in control of the han, each faced the threat of Bakufu action, each had its own plan for national action, and neither was willing to have its plan jeopardized by irresponsible elements. The 1866 government was dominated by the extremists who in 1858 had broken with Shōin on the grounds that his plots would bring disaster to the han. In 1866 the same men criticized the Second Kiheitai for the same reasons. Yamagata wrote to Kido that the Second Kiheitai had acted "without regard for the han,"[25] and Inoue wrote that it had "shamed Chōshū."[26] This strongly suggests that, not only in their goals but also in their means, the 1866 leaders were much closer to Sufu than to Yoshida Shōin.[27]

[24] Suematsu, *Bōchō kaiten shi,* VII, 165.
[25] *Ibid.,* VIII, 360.
[26] *Ibid.,* pp. 359–360.
[27] Marius Jansen has suggested, in the context of Tosa history, much the same thing: "As with the extremists of the Shōwa years, Takechi and his followers [the Tosa equivalent of Yoshida Shōin or the Second Kiheitai] affected their times more by the creation of unrest than by the establishment of new patterns. Indeed, it could be said

Finally, it should be stressed that the great military power created in Chōshū at this time was only in part a result of the resources discussed in earlier chapters. The *buikukyoku* enabled Chōshū to purchase arms, the Tempō Reform improved the condition of its samurai class, the size of the han furnished a base of men and productive power; and yet, to a large extent the strength of Chōshū in 1866 stemmed from two emergent factors. One of these was the technological superiority of the Chōshū military forces that had been achieved since the time of the civil war. Certain aspects of this have already been mentioned: reorganization, strict discipline, modern weapons, and so on were all obtained in response to the crisis facing Chōshū, and all these in turn contributed to the strength and morale of the han military forces. The second emergent factor stemmed from Chōshū's desperate position. A Bakufu army now threatened the han. Should Chōshū be defeated, its daimyo would be punished, its samurai would lose their fiefs and stipends, the stores of its merchants would be looted, and even the lands and crops of its peasants devastated. Faced with this grim prospect, all found an immediate common interest. The preservation of the han and the attainment of the values for which it stood now merged into a single program, and extremists, conservatives, moderates, and peasant militia alike made common cause. "The two provinces of Chōshū, silently as in the dead of night" awaited the Bakufu attack." [28] Even in this condition it is doubtful whether "one man had the strength of one hundred, and one hundred of ten thousand," [29] but in contrast to the Bakufu army, the strength of Chōshū's forces was overwhelming.

The Bakufu army first and foremost put a severe strain on the Tokugawa finances. Matsudaira Mochiaki, daimyo of Fukui, wrote that, "if troops are moved, it will inevitably bring chaos to the land,

---

that many of his victims, and certainly Yoshida Tōyō [the Tosa equivalent of Sufu], were closer to modern Japan than was this martyred loyalist." See Marius B. Jansen, "Takechi Zuizan and the Tosa Loyalist Party," *The Journal of Asian Studies,* 18:199 (Feb., 1959).

[28] Seki Junya, "Bakumatsu ni okeru hansei kaikaku (Chōshū han): Meiji ishin seiritsuki no kiso kōzō," *Yamaguchi keizaigaku zasshi,* 6:78 (May, 1955).

[29] *Ishin shi,* IV, 419–420.

it will impoverish the daimyo and arouse the resentment of the people."[30] One Bakufu *rōjū* revealed that payments for the army alone came to 180,000 gold *ryō* a month and that 3,000,000 *ryō* had been spent since the expedition was first announced. Matsudaira Keiei wrote that at one time the Bakufu possessed only 20,000 gold *ryō* in cash. It had ordered 3,000,000 collected from Osaka merchants but as this had not yet been obtained, it was thinking of borrowing from the foreigners.[31] Money to pay for the Bakufu's expenses during the expedition was eventually collected from the merchants of Osaka in return for a mortgage on the revenues of three provinces, but the daimyo participating in the expedition continued to complain of their expenses.

The Bakufu was also plagued with a wave of rice riots, which broke out about a month before the beginning of the war and which seem to have been caused in part at least by the burden of the Bakufu troops. The riots began in Hyōgo and soon spread to Osaka where the shogun was in residence. Later in the month they also broke out in Edo. Shimazu Hisamitsu of Satsuma sent the following petition to the shogun at this time:

> Since last year signs of a great civil disturbance have appeared. On several occasions armies have been moved, commoners have been killed, and now . . . peasant uprisings break out in Tamba and Yamato, and riots occur in Hyōgo, Osaka, and Edo. As the shogun is presently in Osaka . . . his august command and military might should be shining forth in all directions; on the contrary, the merchants and lowly people at his very feet break the law without regard for his authority. This is the result of what is called "a misery so great that life is unbearable" and it is an intolerable situation.[32]

Even more important than finances and riots were the state of the Bakufu troops and the nature of their weapons. With the single exception of one unit from Kii han, the troops of the Bakufu army were equipped with old-fashioned weapons. And, from accounts that remain, the morale of the troops was inferior even to their weapons. The Bakufu *hatamoto* had been arriving in Aki han since

[30] *Ibid.*, p. 416.
[31] Osatake, *Meiji ishin*, II, 664.
[32] *Ishin shi*, IV, 499.

the end of 1865, but their behavior was most unmilitary. Kondō Isamu of Aizu on his return from Hiroshima reported that the *hatamoto* passed their time buying presents to take back to the East and thought only of the day of their departure. He therefore recommended that, since there was little hope for victory if war should begin, any sign of submission from Chōshū should be accepted without further investigation.[33] A Mito samurai, Hara Ichinoshin, who at this time was on the staff of Tokugawa Keiki, wrote: "If the government acted with decision, the morale of the Bakufu troops would improve. Since it is weak and indecisive, they tend in general to think only of returning East. If an order were now issued saying those who wish to go West may go West [to fight against Chōshū], and those who wish to go East may go East, not one would choose to go West."[34] In the words of Keiki, the Bakufu was "riding a tiger,"[35] and, unwilling to get off for fear of the consequences, it was carried forward by the inertia of its own position.

The Chōshū-Bakufu War

The war began on 7/6/66. In the annals of Chōshū it was called the "war on four sides,"[36] since the Bakufu troops and the troops of the han that joined the Bakufu forces attacked Chōshū on four different fronts. Bakufu naval forces attacked Ōshima in the southeast, and the Bakufu armies moved against Chōshū from Aki han in the east, Tsuwano han, in the north, and Kokura han, in the southwest. In many respects, accounts of the actions of the fragmented Bakufu army in this war remind us of the wars fought in fourteenth-century Europe, when troops might be withdrawn in the middle of a battle and army commanders, who were themselves independent feudal lords, would conclude a peace without regard for any higher authority. And it is because of this that one can view the war as if it were in fact four separate wars.

The first of these began on 7/6/66 with an attack on the island of

[33] *Ibid.*, p. 482.
[34] Osatake, *Meiji ishin*, II, 666.
[35] *Ishin shi*, IV, 424–425.
[36] Suematsu, *Bōchō kaiten shi*, VIII, 408.

Ōshima by Bakufu naval units. Troops of Matsuyama han and the Bakufu, who landed there, held the island for a week; by the fifteenth, however, Chōshū *shotai* under Takasugi had retaken the island. This early one-week occupation of Ōshima was to be the only Bakufu success and the only fighting to take place on Chōshū soil during the entire war. And this was possible only because the defense force on Ōshima had been small, since it was not considered to be a strategic area. The naval forces of the Bakufu were superior to those of Chōshū. Had Great Britain not barred them from operations in the Shimonoseki Straits, on the grounds that this might interfere with foreign shipping, the Bakufu might have been able, at least, to harass more effectively the advancing Chōshū armies.

The fighting along the border between Chōshū and Aki han began on 13/6/66 in the area along the Inland Sea. By the end of the first day Chōshū had advanced several miles, occupying two coastal villages. Two further attacks were launched by Chōshū on the nineteenth and the twenty-fifth, but they were checked by the best equipped and strongest of the Bakufu troops. On 25/6/66 the Bakufu *rōjū,* Honjō Munehide, who at this time was in Hiroshima, privately decided that the Bakufu army could not win the war. In a letter to the *rōjū* Itakura Katsukiyo, he wrote that Chōshū had the backing of Satsuma and Great Britain and that the entire han was burning with the conviction of victory. In contrast to this, he continued, the troops of the various han in the Bakufu army had no desire to fight, their equipment was poor, and the war was already lost.[37] He therefore released Shishido Tamaki, the Chōshū representative who had been taken captive during the course of earlier negotiations, and ordered the Bakufu army in Aki to withdraw. Honjō hoped that Chōshū would follow suit and soon propose to end the war. Chōshū, however, answered that, since the Bakufu had begun the war, it was not for Chōshū to propose a truce, and, instead, it ordered its troops to advance. Just at this time the Bakufu front line commander, feeling that he had been betrayed by the Hiroshima headquarters, resigned in disgust. After some negotiations, this resignation was refused, the Bakufu commander in Hiroshima, Honjō Munehide,

[37] *Ishin shi,* IV, 507–508.

was replaced, and an unofficial truce with Chōshū was obtained, which lasted almost until the end of the seventh month. On 26/7/66 the Bakufu forces launched against the Chōshū positions a second all-out attack which continued for several days but no gains were made. Finally, on 7/8/66 representatives of Aki han, whose troops had remained neutral throughout the above fighting, met with Chōshū representatives and promised to seal off the Chōshū-Aki border if Chōshū troops would withdraw from the territory they had gained within Aki. This was done, putting an end to the second "side" of the war.

The third border of Chōshū along which fighting took place lay to the north and the northeast. Immediately adjacent to Chōshū was the small han of Tsuwano and beyond that, Hamada. Tsuwano withdrew all its troops to its castle town to avoid a conflict with Chōshū. As a result, the Chōshū troops passed through Tsuwano and on 17/6 attacked the troops of Fukuyama and Hamada on Hamada territory. By 5/7 the Chōshū troops had advanced through a number of small towns and stood ready to attack the castle town of Hamada itself. Hamada sent word to Hiroshima asking for aid and also requested Bizen to send troops to help it fight Chōshū. The Bakufu leader in Hiroshima, thereupon, ordered the daimyo of Inaba to the rescue. The daimyo of Bizen refused, pleading sickness, and the daimyo of Inaba not only refused, but also sent a messenger to Hamada suggesting that it would be prudent to make peace with Chōshū. Negotiations begun on 15/7/66 were broken off three days later when the Hamada samurai, who were unwilling as samurai of a Tokugawa branch han to submit to Chōshū, burned down their castle town and fled with their daimyo to Matsue han. Hamada and the neighboring Ōmori *daikansho* fell to the Chōshū troops.

The fourth front in this multipartite war lay in Kokura han across the Shimonoseki Straits, where the troops of several han had gathered. Chōshū forces attacked on 17/6, 3/7, and 27/7. On the last day of the seventh month the commander of the Kumamoto troops decided to go home, and his example was subsequently followed by the troops of Karatsu, Kurume, and Yanagawa han. Kokura sent a letter demanding aid to the Bakufu headquarters and received the answer that Kokura had no alternative than to surrender. The Bakufu commander Ogasawara Nagamichi (the heir to the daimyo of Karatsu) had already secretly left his headquarters, slipping out of the rear gate at night to return to Osaka, and so the Kokura samurai burned their castle and retired to the hills to engage in guerilla warfare. This continued until the end of 1866 when a peace was achieved between Kokura and Chōshū by mediators from Satsuma and Kumamoto.

In the meantime on 20/7/66 the shogun Iemochi died in Osaka, leaving Tokugawa Keiki and Matsudaira Keiei as the two dominant figures in the Bakufu. The latter argued that the death of the shogun should be used as a pretext to put an end to the war and that Bakufu absolutism must now give way to rule by a council of the han. To continue the war without the support of the daimyo was folly.[38] Keiki opposed this plan, saying that it would cause the Bakufu to lose face completely; he proposed instead first to defeat Chōshū and then call a council of the daimyo.[39] He, Keiki, would personally lead the Bakufu forces against Chōshū. As the successor to the shogun, Keiki's views carried the greater weight and on 8/8/66 he received the permission of the Court to lead the Bakufu army against Chōshū; it was announced that he would leave on 12/8. The day before his scheduled departure, word reached Osaka that his army had been routed, that the troops of the various han had withdrawn, and that his commander had fled. Left without an army to lead, he was forced to adopt the plan of Keiei.

The degree to which the expedition had been a failure can best be seen in the negotiations which took place between Chōshū and the remnants of the Bakufu army in Aki and Hamada. Katsu Awa, the Bakufu naval commander who had opposed the second expedition against Chōshū, was sent to negotiate the peace. Meeting with Chōshū representatives on 2/9/66, he spoke of Keiki's plans to reform Bakufu politics and to call a council of daimyo to settle the Chōshū question. He requested that the Chōshū forces withdraw to the boundaries of Chōshū as the forces of the Bakufu withdrew to

[38] *Ibid.*, p. 520.
[39] *Ibid.*, p. 521.

Hiroshima. The Chōshū representatives replied that Chōshū would not disband its forces until the Bakufu formally announced the end of the expedition. Katsu finally had to be content with the Chōshū promise not to attack the Bakufu troops as they withdrew. Under these ignominious conditions, the troops of the Bakufu retreated to Hiroshima and to Osaka. The expedition was not officially ended until 25/12/66, when the Bakufu invoked as a pretext the death of the Emperor Kōmei.

The Political Aftermath of the Chōshū-Bakufu War

The defeat of the Bakufu at the hands of a single han made clear to all that the Bakufu hegemony was ended. Even the Tokugawa understood and, to a certain degree, accepted the fact. It was not clear, however, what political structure would emerge in its place. The first attempt to arrive at a new synthesis of power was made by Tokugawa Keiki, who, on 16/8/66 the same day that he asked the Court to call off the expedition against Chōshū, proposed a conference of the leading daimyo to decide on national policy. This was the same proposal that Keiei had made only a few weeks earlier, at which time it had been rejected by Keiki.

The proposal was an attempt to make the best of a bad situation. If the Tokugawa could no longer rule Japan, at least they could be the first among equals, presiding over a council of daimyo. As the leader of such a council Keiki hoped to rally by diplomatic means the other han against the extremism of Chōshū and Satsuma. The council was conceived of by Keiki as a policy-making body whose decisions would be executed by the existing Bakufu organization. Had it been realized, it would have ensured the preservation of the Tokugawa domains, the continued existence of Tokugawa military power (which Keiki planned to reform), and Tokugawa control over the actual administration of Japan.

In contrast to this view, Satsuma and the extremist faction of the Court nobles headed by Iwakura Tomomi, who also supported the idea of a council of daimyo, conceived of it as an opportunity to forward the interests of the han and the Court at the expense of the

Bakufu. The seemingly trivial point on which the two conceptions conflicted was the question as to who should issue the orders summoning the daimyo: the Court or the Bakufu. The decision rested with the Court, but in the Court at the time the balance of power teetered between a faction of pro-Bakufu nobles and of the moderate nobles who had previously supported the *kōbugattai* party.[40]

The pro-Bakufu faction appeared to have triumphed when on 28/8/66 the Court decided that Keiki should be instructed to summon the various daimyo. A few days later, however, there occurred at the Court a shake-up in which the moderates moved to the fore. In their petition, among other points they asked that "the various han be summoned directly" by the Court.[41] Consequently, on 7/9/66 it was the Court and not the Bakufu who sent orders to the daimyo to proceed to Kyoto to participate in a council on national affairs.

One must stress that this small victory for the Court was won not by extremist nobles but by moderates over the pro-Bakufu clique of nobles. As always, since the Court had little strength of its own, the balance of power within reflected the dominant power without. The rise of the moderate faction reflected the new power of the han, which was one result of the defeat by Chōshū of the Bakufu. Extremist nobles were still very weak in number, position, and influence. Iwakura Tomomi, the unofficial leader of the extremist nobles in Kyoto, was still in confinement outside of Kyoto. He had originally been a moderate but had turned subsequently to an anti-Bakufu position. How closely his change in views followed the change in Satsuma's policy is hard to say. In any case, from 1866 on, his contacts with Satsuma became wider and more important, and he soon gained a position vis-à-vis Satsuma that was similar to the position held by Sanjō Sanetomi in relation to Chōshū. But his role in politics at the time remained necessarily limited.

In spite of the machinations of the various groups in and about Kyoto, the plan for a council of daimyo was to be a failure, for of the twenty-four daimyo summoned to participate, only five had

[40] *Ibid.*, pp. 530–534.
[41] *Ibid.*, p. 531.

appeared in Kyoto by 10/66: Kaga, Okayama, Tsu, Matsue, and Awa. The majority excused themselves on the ground of illness. A token conference was called, but it soon collapsed as the daimyo, realizing what had happened, returned one by one to their han. Lacking monographic studies one may speculate that most of the han remained in the grip of the same inertia that had immobilized them since 1853. That the Bakufu was about to fall seemed obvious; to support it would be folly, yet to openly oppose it might still be dangerous. An air of incipient change hung over the country; better to wait until events were more clearly defined than to act rashly.

The months that followed the collapse of the plan were relatively quiet; the period from 10/66 until 5/67 was a time of ferment without dramatic change. These were months when most groups in Japan, either from necessity or choice, turned inward to attend to domestic affairs. Tokugawa Keiki was finally persuaded in 12/66 to become the fifteenth and, certainly, the most reluctant shogun on the condition that he would be given a free hand to reform the Bakufu. After the failure of the council of daimyo, he devoted himself to the task with striking success. Both a loan to buoy up Bakufu finances and military aid were obtained from France; new offices were created to coordinate vital areas of administration; men of ability were appointed without regard for rank; military, judicial, and fiscal reforms were put through, and changes were made in every part of the Bakufu apparatus. The results of these reforms were such that even the anti-Bakufu forces were impressed. Iwakura Tomomi wrote with negative concern that "the actions of the present shogun Keiki are resolute, courageous, and of great aspiration; he is a strong enemy and not to be despised." [42] In the same vein Kido Kōin wrote: "Now the Kantō government is being reformed and remarkable [changes are taking place] in its military system. The courage and resourcefulness of Hitotsubashi [Tokugawa Keiki] cannot be despised. If the opportunity to restore the Court government is now lost and the lead is taken [from us] by the Bakufu, then it will truly be as if one were seeing the rebirth of Ieyasu." [43]

[42] Oka Yoshitake, *Kindai Nihon no keisei* (Tokyo, 1947), p. 95.
[43] *Ibid.*

Kido went on to compare the political situation within Japan to "a ball balanced on a mountain peak" which, if moved "the breadth of a hair" in any direction, "might plunge downwards for thousands of feet."[44] Who, he asked, would be the one to move it?

The Court was equally engrossed in its own internal affairs. The Emperor Kōmei died in 12/66 and was succeeded by the child Emperor Meiji. Since Kōmei had favored the moderate nobles, this was a development favorable to the extremists, for not only was the negative force of Kōmei removed, but, in the amnesties of 1/67 and 3/67 following his death, many extremist nobles who had earlier been banished from the Court were permitted to return. Iwakura, for example, who had been in confinement since 10/62 in Iwakura-mura outside Kyoto, was given a partial pardon on 29/3/67.

Most daimyo were also primarily concerned with affairs in their own han. Some were engaged in last minute military reforms in the hope of participating in whatever changes were to come, but others merely remained in their habitual torpor. Chōshū, which might otherwise have been active, was still officially the "enemy of the Court" and as such it was barred from entering Kyoto. Chōshū samurai traveled on han orders, mostly to gather information, to Kyoto, Nagasaki, and Kagoshima, but activities on this level could not produce major political change.

Satsuma was the most active of the han at this time. Saigō and Komatsu Tatewaki in particular were busy in Kyoto developing contacts among extremist nobles, discussing policy with Chōshū samurai such as Shinagawa or Yamagata who lodged from time to time at the Satsuma residence in Kyoto, and in general laying the base for future action. But in the face of a Bakufu engrossed in internal reforms, a more or less quiescent Court, and the passive han, even Satsuma could do little more than to talk and plan for the future. Decisive change could only be produced by military action, and there was little to indicate that Satsuma was prepared for such a move. Hisamitsu was willing for his triumvirate of ministers to fish in the muddy waters of national politics, but opinion in Satsuma was still conservative and unready to support a military campaign against the Bakufu.

[44] *Ibid.*

Therefore, in order to break the impasse, the Satsuma ministers decided to call their own conference on national affairs to be attended by Shimazu Hisamitsu, Date Munenari of Uwajima, Yamanouchi Yōdō of Tosa, and Matsudaira Keiei of Fukui han. Hisamitsu agreed to this plan as a pacific means to forward the ambitions of Satsuma; the other three daimyo also agreed to participate, probably in anticipation of an inside position in some future political settlement, although Keiei may also have hoped to act as liaison between the conference and the Bakufu. By 1/5/67 all four daimyo had arrived in Kyoto.

The two questions which confronted the group, as well as the Bakufu and the Court at the time, were those of Chōshū and the opening of Hyōgo to foreign commerce. In 1862 it had been agreed that Hyōgo, Niigata, Edo, and Osaka were to be opened by 7/12/67. As the date drew closer, France and Great Britain alike counseled the Bakufu that it should open the ports. On 5/3/67, and again on 22/3/67 under heightened foreign pressure, the Bakufu asked the Court for permission to open the ports. Both times the requests were refused. Therefore, hoping to use the four daimyo as a lever against the Court, the shogun Keiki summoned them to Osaka on 14/5, where, after praising their efforts on behalf of the country, he asked their opinions concerning the Chōshū and Hyōgo problems.

Hisamitsu replied that these were indeed pressing problems but that, of the two, the problem of Chōshū was the most urgent and should therefore be taken up first. To this Keiki replied that the problem of Chōshū was, after all, only one small internal problem, whereas the question of Hyōgo, involving as it did the foreigners, concerned the very future of Japan. The question of whether Chōshū or Hyōgo was the more important problem was to be the focus of long and tedious conferences for several weeks. The Bakufu, unable to defeat Chōshū, was willing to forgive and to restore it to the good graces of the Court, but first, Keiki held, Chōshū must apologize for its past offenses.

Chōshū, of course, was in no mood to apologize. Rather, hoping to overthrow the Bakufu with force, it did not want to see the Chōshū problem solved. When Itō Hirobumi was informed that the Bakufu was prepared to deal leniently with Chōshū, he wrote to

Kido that "the clever plot of the Bakufu" to establish peace in Kyoto was the greatest obstacle to the pro-Court movement, and that, since "Keiki is an extremely fearsome enemy who is not to be scorned," the pro-Court han should quickly take action before the Bakufu reforms were completed.[45]

Satsuma was well aware that the Chōshū problem was insoluble on the Bakufu terms and that considerable pressure was being put on the Bakufu by the foreign powers over the Hyōgo question. Nevertheless, it continued to demand that the Chōshū problem be given primacy. Thus, the real issue at stake was the continuing struggle of the strong han against Bakufu authority. In this struggle Satsuma and Uwajima were adamant in their refusal to compromise, while Tosa and Fukui, one in debt to the Bakufu since Sekigahara and the other a *shimpan* daimyo, were more moderate in their stand. But the Bakufu failed to exploit these differences, and, consequently, it broke off negotiations with the four han and began to negotiate directly with the Court, applying pressure by the various means at its disposal.

On 24/5/67 the Court finally gave its permission to open the port of Hyōgo, and, at the same time, it instructed the Bakufu to deal mildly with Chōshū. Thwarted by the Bakufu, the four daimyo now retaliated by sending a scathing condemnation of Bakufu policy to the Court. This was to do them little good, however, and one by one they left Kyoto for their respective han. But the triumph of the Bakufu was more apparent than real; it had thrown away its last opportunity to compromise. Disgruntled by defeat, Tosa now began its movement for the peaceful restoration of power to the Court, and Satsuma joined with Chōshū to plot the overthrow of the Bakufu by military force.

The Two Restoration Movements

Of the two anti-Bakufu Restoration movements, that of Satsuma was the first to begin; it can be thought to have started on 29/5/67 when, dissatisfied with the Bakufu's response to the proposals of the

[45] Suematsu, *Bōchō kaiten shi,* IX, 298.

four daimyo, the Satsuma samurai in Kyoto decided to join Chōshū in a military coup against the Bakufu. The feelings of the Satsuma leaders at the time can be seen in a letter sent to Satsuma by Ōkubo. He argued in it that "the Bakufu has shown absolutely no intention of adopting the views of the four han, or of repentance, or of upholding the Imperial edicts." On the contrary, "by all means, it tries to suppress the righteous han." [46] Because of this, Ōkubo concluded that military force alone would avail; Satsuma troops must be sent to Kyoto and Saigō must go to Chōshū to inform it of the Satsuma decision and to draw up concrete plans.

Since the Satsuma-Chōshū alliance of 1866, Satsuma and Chōshū had maintained fairly close relations. At the time of the decision described earlier, Shinagawa Yajirō and Yamagata Aritomo were staying at the Satsuma residence in Kyoto collecting information for the han. They were about to return to the han bearing the news of the Court's decision to open Hyōgo and to accord lenient terms to Chōshū, when they were summoned before Hisamitsu, who told them that although he had come to Kyoto with the other three daimyo for the good of the country, his opinions had been rejected by the Court and that so far the Bakufu had shown no signs of change. He also told them that Saigō would probably be sent to Chōshū with a message in the near future. At the interview he personally presented revolvers to the two Chōshū samurai. The interview was followed by a conference between the Satsuma triumvirate and the two Chōshū men at which Komatsu further stated that ordinary methods would not suffice to destroy the evil plots of the Bakufu and that Chōshū and Satsuma must cooperate to this end. Shinagawa and Yamagata then inquired for further details of the Satsuma policy, but Komatsu only answered that its aim was to protect the Court and establish its authority, to carry out Imperial edicts, and to rectify the evil government of the Bakufu. On the following day, 17/6/67, the two Chōshū samurai left Kyoto for the han.

In the meantime the parallel Tosa movement had also begun. Yamanouchi Yōdō, the retired daimyo of Tosa, had left Kyoto on

[46] *Ishin shi,* IV, 663.

27/5/67 owing to illness, but before leaving he had sent word to Gotō Shōjirō, a high-ranking Tosa samurai, to proceed to Kyoto to make what he could of the situation. It is probably correct to assume that this was an attempt to forestall a Chōshū-Satsuma military coup that might exclude Tosa from the final Bakumatsu settlement. Gotō arrived in Kyoto on 13/6/67 with a plan to petition the Bakufu to restore to the Court the powers of government. If the Bakufu did not agree, it would be overthrown by force. If it accepted the Tosa proposal, then Keiki, as the greatest of the feudal lords, would head an assembly of daimyo that would govern under the Emperor. This Tosa plan may be viewed as a compromise measure which would at the same time enable Tosa to fulfill its loyalty to the Tokugawa and to play a prominent role in whatever new government would emerge. That some new government would be formed was by this time obvious to all. We may also note that much of the Tosa plan was suggested to Gotō by Sakamoto Ryūma who, having earlier left Tosa without permission, could not himself mediate in an official capacity.[47]

Gotō had hoped to obtain the approval of Yamanouchi Yōdō and then immediately to present his petition. But Yōdō had already left; therefore, a council of the Tosa officials in Kyoto was called, and, contingent upon the approval of Yōdō, Gotō's plan was adopted as the official Tosa policy on 23/6/67. For such a plan to succeed it was necessary to obtain the support of other powerful daimyo. Gotō presented his plan on 17/6 to Date Munenari, the former daimyo of Uwajima. Date approved of the plan but added that since the time was still too early, he should discuss the matter with Satsuma. Gotō subsequently obtained the approval of both Satsuma and Aki han. On 3/7 Gotō left for Tosa to obtain the approval of Yōdō, and, upon receiving it on 13/7, his policy became the official Tosa policy; Gotō was now free to present his petition to the Bakufu. Just at this point, however, there took place an incident involving Tosa in negotiations with foreigners which delayed Gotō's return to Kyoto until 4/9.

Before returning to the Satsuma-Chōshū movement, we may ask why Gotō was able to obtain Satsuma's approval of his plan. Was

[47] *Ibid.*, pp. 615–616.

not Satsuma already committed to joint military action with Chōshū? The answer to this question is not altogether clear. The original Satsuma-Chōshū alliance of 1866 certainly did not commit Satsuma to joint military action with Chōshū. The last item of the six-point pact stated merely that the two han would work together on behalf of the Court. The four-daimyo conference which followed the Chōshū-Bakufu war also illustrates that Satsuma, while embarrassing the Bakufu with the Chōshū issue, was still content to advance its own interests by peaceful means to the extent that it was possible. When it proved impossible, the 29/5 decision to stage a military coup had been made; it was made privately by Hisamitsu and his triumvirate in Kyoto, and it was not communicated to Satsuma, since it was felt that the opposition in the han would be too strong. Moreover, Komatsu would not commit himself beyond the rather evasive statement that usual methods would not suffice to destroy the evil Bakufu plots. This together with the statement that Saigō would probably be sent to Chōshū in the near future did imply some sort of joint action, but it was never made explicit.

And yet an understanding seems to have existed between the two han, for, even when a messenger was sent from the Satsuma residence in Kyoto to Chōshū informing it of Satsuma's approval of the Tosa plan, the Chōshū government continued to believe in the good will of Satsuma. The motives of the Satsuma leaders in approving the Tosa proposal are not clear, but it seems that they viewed it as a means and not as an end. At the time the approval was given, Saigō rebutted the arguments of the extremists who wished to reject it, saying that it was a first step which could be followed later by military action. In a letter to Shinagawa and Yamagata, which was sent through the same messenger, Saigō first apologized for not having come to Chōshū as was promised and went on to speak of the Tosa proposal as a "transitional device"[48] in the movement against the Bakufu.

Aware that changes were imminent, Chōshū sent Shinagawa and others to Kyoto on 17/7/67 to confer with the Satsuma leaders.

[48] Suematsu, *Bōchō kaiten shi,* IX, 340–341. Literally, "watari ni fune o esōrō kokochi."

For the first time Chōshū was informed explicitly of Satsuma's plans for an armed coup against the Bakufu. Messengers arrived in Chōshū on 24/8 informing the han of the Satsuma plan. On 17/9/67 Ōkubo traveled to Yamaguchi accompanied by Shinagawa and Itō who had been staying at the Satsuma residence in Kyoto. After stating Satsuma's intention to overthrow the Bakufu, Ōkubo formally requested the aid of Chōshū. As the Bakufu had previously ordered Chōshū to send one of its Elders to Osaka to negotiate a pardon, Ōkubo suggested that this be used as a pretext to send troops to the Kyoto area. Chōshū agreed to the proposal and a detailed program was drawn up: Satsuma troops were to arrive in Chōshū by 25/9/67 and the troops of the two han were then to march on Kyoto.

In the meantime, on 4/9 Gotō returned to Kyoto to find the situation vastly different from what it had been two months before. The Satsuma leaders who had previously agreed to support his plan now argued that a petition would not suffice in the present situation. Saigō told Gotō that Satsuma, while not opposed to Tosa's petition, was determined to use military force. Gotō was at a loss; on 24/9 he met with Nagai Naomune, a Bakufu official with whom he had discussed his plan before returning to Tosa. Nagai urged him quickly to present his petition to the Bakufu in order to preclude military action by other han.[49] Gotō feared that his plan would fail without the backing of Satsuma and he once more tried to persuade its leaders to support his petition. Finally, on 2/10/67 he was able to obtain the approval, if not the backing, of Komatsu, the most moderate of the Satsuma leaders; on the following day he presented his petition to the *rōjū* Itakura.

It was not at all clear at this time what form of government this petition was intended to establish. Some patriots of the pro-Emperor movement felt that the system established at the time of the abortive restoration of Godaigo should be taken as a model. Iwakura, with greater enthusiasm than practicality, argued in favor of the forms established by the Emperor Jimmu. Others such as Gotō Shōjirō and Sakamoto Ryūma felt that some form of representative government such as those existing in the West would be best. But whatever

[49] *Ishin shi,* IV, 732–733.

form of government was to follow, it was clearly better for the Bakufu to give up voluntarily its authority to the Court than to be stripped of both authority and lands by a coalition of the han. Moreover, the Tosa plan offered the Bakufu a chance to put aside its national responsibility while maintaining the sources of its power. Consequently, on 9/10/67 Gotō was informed that the shogun, after two hundred and sixty years as hegemon, would return his authority to the Emperor. On 13/10 this decision was formally announced to the representatives of forty han assembled for this purpose at Nijō castle. There was no opposition. On the following day the Bakufu sent a memorial to the Court requesting it to take over once again the government of the country. On 15/10/67 this was accepted by the Court: the first but illusory restoration had been accomplished.

On the same day orders were sent out to all daimyo with domains exceeding 100,000 *koku* to come to Kyoto for deliberation on national affairs. Within a week it became apparent that the Court now faced the same embarrassment that the Bakufu had suffered months earlier: it had issued invitations to a party and no one had come. Therefore, on 22/10/67 the Bakufu was once again entrusted with the affairs of the nation. On 25/10 a second set of orders was issued summoning the daimyo to Kyoto, but again no one came. Some daimyo felt that the situation in Kyoto was too uncertain, that it was wiser to mark time until the more powerful han such as Satsuma revealed their intentions. Others refrained out of respect for the Bakufu. The *shimpan* and *fudai* daimyo, especially, were opposed to this development and argued that the shogun should not have returned his authority to the Emperor. Some of them felt that the Court would be unable to rule; others went so far as to announce that they would rather lose their status as independent daimyo than to serve the new government. Matsudaira Keiei, who realized the inevitability of a new balance of power, was opposed to the Tosa proposal on the grounds that it "arbitrarily called for the establishment of an assembly on a Western model."[50] The elation which Gotō and the other Tosa officials had experienced at the success of their plan soon changed to chagrin: if his plan for an assembly of

[50] *Ibid.*, p. 747.

the daimyo did not materialize, Gotō thought, it would mean the failure of his entire venture. He therefore contacted the representatives of the various han in Kyoto to urge the presence of their daimyo in Kyoto. And on 3/11/67 Gotō left Kyoto to report to Yamanouchi Yōdō and to urge his presence in Kyoto. Yōdō promised to go to Kyoto in support of Gotō's plan, and Gotō returned to Kyoto.

Meanwhile the Chōshū-Satsuma plan for a second military restoration had been progressing steadily. On 20/9/67 Aki han agreed to join the Satsuma-Chōshū plan. In Kyoto Iwakura, who had been informed of the plan, began preparations for action within the Court. Some difficulties arose in Satsuma, when high-ranking conservatives raised their voices in opposition to the proposed military gamble in Kyoto. This may have been merely one form of protest against the usurpation by men such as Saigō or Ōkubo of what they considered their right to leadership in the han. In any case Hisamitsu finally quieted the opposition by explaining that the Satsuma troops sent to Kyoto would be used only for guard duty. By this time, however, the date set for the arrival of the troops in Chōshū had already passed.

The pact between Ōkubo and Kido called for the arrival of the Satsuma troops at Mitajiri, an Inland Sea port in Chōshū, by 26/9. Chōshū troops had been duly prepared and equipped, and on the prescribed day they waited for the Satsuma troops. The month ended and still the Satsuma troops did not appear. Many among the *shotai* in Chōshū argued that Chōshū should advance alone and stage a coup without the aid of Satsuma, for latent hostility toward Satsuma was still very strong among the *shotai*. The reason why Kido had sent so many Chōshū samurai to the Satsuma residence in Kyoto as observers in previous months was not because of a need to reconnoitre the situation in Kyoto, but rather to convince men such as Yamagata or Shinagawa, who were trusted by the *shotai,* of the sincerity of Satsuma's intentions toward the Court.[51] All who went came back convinced; nevertheless, there re-emerged among the *shotai* a considerable degree of anti-Satsuma sentiment. Conferences were held within Chōshū; by now even some of its highest officials

[51] Suematsu, *Bōchō kaiten shi,* IX, 312.

favored independent action, but the final decision on 3/10 was to wait and see. On 6/10 and 9/10 two groups of Satsuma troops finally arrived; all doubts about the intentions of Satsuma were now dispelled; it was decided, however, that these should remain in Chōshū before pushing on to Kyoto until Imperial edicts were obtained and more troops were readied.

In Kyoto, Ōkubo of Satsuma and Hirosawa of Chōshū were advancing their plans for a coup. On 8/10 it was decided at a meeting to ask the Court for edicts instructing the han to overthrow the Bakufu. The request was communicated to nobles favorable to Satsuma who thereupon signified their approval. The following day Ōkubo visited Iwakura and told him of the plans for the Satsuma-Chōshū coup. Iwakura drafted a Restoration petition which he submitted to the Court Councillor Nakayama Tadayasu. Whether the petition was actually submitted to the boy Emperor is uncertain; in any case, on 13/10 an edict was issued pardoning the two Mōri for their part in the Kyoto incident of 1864 and restoring their rank, and on 14/10/67, the same day on which the Bakufu petitioned the Court to once again take over the government of the country, Imperial edicts for the overthrow of the Bakufu were granted to Chōshū and Satsuma. Aki han was not trusted sufficiently to be given an edict.

Bearing the edicts, Hirosawa, Shinagawa, and the Satsuma triumvirate hastened back to their respective han. Komatsu and Saigō stopped off in Chōshū to arrange for the details of the troop movements. The triumvirate arrived in Kagoshima on 26/10; they presented the Imperial edict to Shimazu Hisamitsu and his son, the daimyo Shimazu Tadayoshi, and asked that the daimyo personally lead the troops of the han to Kyoto. This was accepted, and on 13/11/67 the daimyo of Satsuma set out at the head of his troops for Kyoto. He arrived in Kyoto on 23/11; there were said to have been almost 10,000 troops from Satsuma at the time, but judging from the number actually participating in the fighting which subsequently took place, this may have been an exaggeration. Chōshū troops left the han on 21/11 and 25/11, some by land and some by sea. They took up positions in the vicinity of Kyoto. The stage was now set for the coup and final plans were made, taking the 18/8/63 Satsuma-Aizu coup as a model.

On 2/12/67 Gotō learned of the Satsuma-Chōshū plan; on 8/12 he was told that the coup would take place on the following day and was asked to join in the plan. Taken aback and unable to decide what to do, Gotō asked that the plan be delayed for two days until the arrival of Yamanouchi Yōdō in Kyoto. This was refused, although a delay at the Court enabled Yōdō to arrive in time. Gotō then consented to act in concert with Chōshū and Satsuma in the formation of a new government. Had Tosa refused, then it, along with Aizu and Kuwana, might have been branded an enemy of the Court and would certainly have been excluded from the ensuing political settlement. Also on 8/12 orders were issued by the Court granting the Chōshū troops the right to enter Kyoto; they had been barred from the city since 1863. On the morning of 9/12/67 the samurai of the anti-Bakufu han sallied forth to "seize the jewel" (*tama o ubau*).[52] The palace gates were quickly secured, daimyo in Kyoto were summoned to appear before the Emperor, the offices of shogun and those of the Bakufu officials in Kyoto were abolished, and the Imperial Restoration was proclaimed.

The Political Settlement

The most immediate problem facing the new government was that of Keiki and the Tokugawa domains. A meeting was called at which Yamanouchi Yōdō, Matsudaira Keiei, and Tokugawa Keishō insisted that the Bakufu should be allowed to retain its domains and that Keiki should be permitted to participate in the new government on an equal footing with the other daimyo. In opposition to this, Iwakura and Ōkubo contended that the Bakufu was guilty and that the Restoration would become a hollow phrase if the Bakufu domains were not confiscated. Keiki could join in the new government only after giving up his lands. The argument continued back and forth until a recess was called. During the recess Iwakura

[52] Tōyama, *Meiji ishin*, p. 210.

informed Yamanouchi Yōdō that the fate of the Tokugawa had already been decided and that there was no room for further argument. The real power of the new government lay in the Satsuma-Chōshū troops and policy-making was in the hands of their leaders; the other han were guests and as such were on their good behavior. The following day Matsudaira Keiei and Tokugawa Keishō carried the terms of the new government to Keiki at the Bakufu headquarters in Nijō castle.

Keiki himself was willing to give up his position as shogun (he had already done so once) and perhaps even willing to surrender his domains to the Court. His retainers, however, were not. The *hatamoto, fudai* daimyo, and branch houses such as Aizu and Kuwana joined in heated opposition. They argued with sufficient reason if not complete impartiality that Satsuma and Chōshū, together with a few nobles, had staged a coup to achieve their own ends without even waiting for the daimyo to assemble in Kyoto according to the Tosa petition. Because of this opposition among his retainers, Keiki, while accepting the terms of the new government, asked for a period of time in which to quiet down his retainers before giving his answer regarding the Bakufu domains. Impressed by the turbulence of the Bakufu troops, Tokugawa Keishō agreed to this and suggested that Keiki withdraw to Osaka to prevent a clash between the Bakufu and Satsuma-Chōshū troops. On the night of 12/12/67 Keiki left by the rear gate of Nijō castle; the following morning the Bakufu troops also withdrew to Osaka as soon as they learned of his departure. They continued, however, to advocate war.

In the meantime, *rōnin* troops, recruited at the Satsuma residence in Edo for the purpose of provoking the Bakufu, were carrying out a campaign of lawless and unruly actions which so threatened the equilibrium of society in Edo that on 24/12/67 the Bakufu ordered the troops of Shōnai han to destroy this band at its source, the Satsuma residence. Two thousand troops attacked at dawn and completely destroyed the Satsuma compound. When news of the attack reached the Bakufu army at Osaka, the Bakufu retainers, already inflamed by rumors of the secret edicts for the subjugation of the Bakufu, became uncontrollable. Their arguments soon overcame the cautionary reticence of the *rōjū* Itakura and the reluctance of Keiki, and consequently on 1/1/68 Keiki, charging that Satsuma was guilty on five different counts, ordered his troops to advance on Kyoto.

The Bakufu troops were led by Aizu and Kuwana. The Imperial troops, as they now called themselves, were those of Chōshū and Satsuma. Yamanouchi Yōdō of Tosa, not wanting to commit himself to either side, called this a "private war" [53] and ordered his troops to remain neutral. The main battles took place at Fushimi and Toba, but in spite of the fact that the Bakufu troops were in high spirits and outnumbered their enemy by almost three to one, the fighting went against them. By the night of 3/1/68 the Bakufu army was defeated and two days later it was in full flight. As during the Chōshū-Bakufu war, the southwestern han had proved to be better equipped, better trained, more experienced, and more efficiently organized. The rout was completed on 6/1/68 when, after announcing that he would lead his army in person, Keiki again retreated by a rear door, this time for Edo. Western Japan was now in the hands of the Imperial forces.

But no one could tell what would follow. The military defeat of the Tokugawa forces had been so sudden that it had astonished even Chōshū and Satsuma who had been prepared to seize the Emperor and flee to the west had the outcome of the battle at Fushimi and Toba been different. The leaders of the Imperial forces therefore began the first of a series of makeshift attempts to consolidate their gains. An Imperial government with the offices of *sōsai* (Supreme Executive Office), *gijō* (Senior Councillor), and *sanyo* (Junior Councillor) was established: the infant prototype of what would in a few years emerge as a supra-han national government. And at the same time orders were issued for the formation of an army for the conquest of eastern Japan. In Edo, Keiki decided not to oppose the Imperial forces, and on 21/4/68 Imperial forces took over the Edo castle. A few of the Bakufu retainers decided to resist, sporadic battles were fought, and some even withdrew to Hokkaido in a

[53] *Ishin shi*, V, 137.

last-ditch effort to sustain the old regime, but all was in vain. On 28/3/69 the Emperor moved to the Edo castle, the former residence of the shogun, and the name of the city was changed to Tokyo, the Eastern Capital. The Restoration was now completed; the revolution was about to begin.

Conclusion

What then was the nature of the Meiji Restoration? The most common view, whether in Japan or the West, stresses the weakness of the Tokugawa polity. It depicts the Restoration as a revolution against the Tokugawa system carried out by lower samurai, and, in the eyes of some, by certain strata of commoners as well. It emphasizes the social and political frustrations which, supposedly, turned these groups not only against the Tokugawa rule but against the very system itself. It describes an impoverished military class suffering from both the demands of the governments of the han and the exigencies of a rising commercial economy to which its fixed incomes could not adjust. It stresses the breakdown of the traditional peasant village and the rising tide of peasant rebellions. It speaks of the lower samurai as a class which, by participating in the Restoration movements, was somehow striking out at its feudal fetters, a class which, finding no outlet for its ambitions within the existing society, was willing to turn tradition upside down to found a new order.

Many of the elements in this picture are certainly true. Many samurai were impoverished. Social status was largely determined by birth, and political position was limited to the top strata of *shi*. Traditional village structure had altered in some regions, land tenure was changing, and peasant rebellions did occur with increasing frequency throughout the period. But one cannot simply infer from this that the majority of samurai were disaffected. Sword, stipend, and status may have decreased in value over the Tokugawa period, but few samurai were willing to abandon them. Rather, they often clung even more tenaciously to their statuses, and to the patterns of action which these entailed, as their economic condition approached that of the commoners. This is not to say that there was no relation between samurai (and peasant) dissatisfactions and the Restoration movement. Consider the following passage from the writings of a samurai on the evils of the times:[1]

To cut in two the retainers' stipend is an action without a precedent. If a crime is committed, only then is a stipend cut, a fief confiscated, or suicide demanded. There is, however, no precedent for reducing the fief of one without guilt. In general, to take from the samurai was unheard of in the past. The retainer is one who is willing even to give up his life for his lord. In some circumstances he will serve even after his fief has been lost; but, when the lord is not worthy of his allegiance, when the lord is concerned only with his own wants and is without compassion, then the samurai cannot help feeling bitter.

How should the writings of samurai moralists such as the above be viewed? To begin with, it is obvious that the intent of the above writer is not seditious. He is not advocating rebellion, nor is he suggesting that the loyalty of the samurai to his lord be conditional. On the contrary, the samurai cannot help feeling bitter because they are loyal and their loyalty is not requited. It is the Confucian duty of a retainer to remonstrate when his lord deviates from the path of good government. Such remonstrations were, in fact, the proof of true loyalty.

What did such "bitterness" mean in terms of Tokugawa political process? Since feelings such as these were clearly subordinate to other strong positive values, and since there existed no ideology in late Tokugawa Japan in terms of which such feelings could be turned against the Tokugawa system, I feel that they had relatively little influence in their own right. At the same time, however, such feelings undoubtedly were one element contributing to reform sentiment within the han, and subsequent to 1858, to the Bakumatsu political movements. Chōshū samurai had had a portion of their stipends taken by the han government for over one hundred years, and many were in debt. The awareness of such frustrations, as well

[1] *Ishin shi,* I, 338.

as the desire to strengthen the han by strengthening its samurai class, may have been back of the emphasis by both reform cliques in 1840 on the problem of samurai debts. The publication of the Tempō budget was obviously an attempt to strengthen the reform by converting such bitterness into a feeling of righteous sacrifice for the sake of the han. To the extent that this conversion took place—and this was probably considerable, to the extent that such feelings actually led to reforms, to that extent they helped to preserve the han as a viable political unit. Only after 1858, feeding into the pro-Emperor movement, did such feelings help destroy the system. We noted in the Chōshū civil war that all samurai, even neutrals, were strongly committed to the han, but that some gave primacy to its preservation and others to the achievement of those goals in terms of which it was justified. It was to the latter activist loyalty that such samurai dissatisfactions probably contributed. Glimpses of this occur from time to time in materials dealing with the *shotai,* although even within these units this was by no means the most important source of energies.

Yet it is important to note that such dissatisfactions which, to a greater or lesser extent, were the common property of the samurai of every han never gave rise to class action or to a stronger class consciousness. When a samurai of the Tokugawa period considered the institutional integration of his society, he saw it in terms of the Confucian four-class formula. Samurai as the ruling class were more noble, less concerned with profits. At the same time, the units within which samurai acted were not horizontal classes, but rather, as we noted in Chapter IV, hierarchically ordered groups bound by ties that were both vertical and particular. In the Japan before Perry these two dimensions were fixed and little conflict occurred. After 1858, however, efforts by samurai to promote their own han at the expense of others became more intense and reinforced their particular identification with their own group at the expense of the horizontal order. Thus, to the extent that dissatisfactions contributed to action, they led away from a conception of self in class terms. (On the other hand, the content of the political movements as a supra-han, anti-foreign, pro-Emperor ideology undoubtedly strengthened their overall sense of identification with Japan, if not with a particular class.)

Dissatisfactions, however, even as transformed by the symbols of the Bakumatsu movements, were not the sole, or even the chief, internal factor determining the course of the Restoration. On the contrary, this study of Chōshū's history in the pre-Restoration period suggests that the Restoration stemmed more from the strength of the values and institutions of the old society than from their weaknesses. It suggests that the power of the Meiji state to respond successfully to the challenge of the West was to a considerable extent based just on the "feudal" elements in the Tokugawa system. To recapitulate the findings of the earlier chapters, the background factors which made it possible for Chōshū to engage actively in national politics during the Bakumatsu period were:

1) Chōshū was one of the largest han in Tokugawa Japan; its lands were extensive and its samurai were more numerous in proportion to those lands than were those of most other han. The latter point is of particular significance in relation to the Bakufu's "weakness of numbers."

2) Chōshū was a *tozama* han, one of the few large *tozama* han traditionally hostile to the Bakufu. Consequently, both its position in the Tokugawa state and its history encouraged Chōshū to chafe against the Tokugawa hegemony. When most han were paralyzed by the conflict between their desire to increase their own power (at the expense of the existing order) and their sense of obligation to the Tokugawa, Chōshū could act.

3) Chōshū's special relation to the Court provided an additional point about which its anti-Tokugawa animus could crystallize. The significance of this relation was reinforced by the emphasis on the Emperor in the climate of late Tokugawa thought.

4) Chōshū's institutional response to fiscal adversity, the *buiku* system, seems to have been a vital factor enabling Chōshū to act when most han were financially immobilized. More important than numbers of samurai alone were the 7000 rifles purchased from the West. Without these Chōshū would probably have been defeated in the Chōshū-Bakufu war.

5) Chōshū's Tempō Reform, while not an outstanding success in most ways, effected a substantial improvement in the condition of its samurai class. If one may view for a moment the reforms of Murata and Tsuboi as a single unit, during the seven years of reform from 1840 to 1847, the amount taken from the stipends of the Chōshū samurai was successively reduced, the debts of the samurai were either repudiated or taken over by the han, and special grants were made for the reconditioning of their arms. All these reforms enabled the samurai of Chōshū to enter the Bakumatsu period in better financial condition, and therefore with higher morale, than the samurai of most other han.

6) The strength of bureaucratic cliques composed of fairly high-ranking *shi,* and the competition between these cliques within the particular Chōshū government structure, appear to have given to the Chōshū government an able administration and, even more important, the flexibility needed to adapt successfully to the changing conditions and political opportunities of the Bakumatsu period.

Each of these six constitutional factors reflects areas of traditional strength within Chōshū. Traditional hostilities were kept alive by the vitality of traditional values and institutions. Chōshū samurai were not hostile to the Tokugawa because of their straitened circumstances (except insofar as they saw these as a consequence of the reduction of the Mōri domains after Sekigahara); they were hostile to the Tokugawa because they were loyal to the daimyo of Chōshū. It was the strength of this loyalty that sustained among them the memory that the Mōri fief had once been the second greatest fief in Japan. And, it was the strength of the Tokugawa control system that kept Chōshū, a *tozama* han, excluded from a role in national politics for over two hundred and fifty years.

Similarly, the *buiku* system succeeded against all probability because of the strength of "feudal" controls which the han government could exercise over the han economy, and because of the relative stability of Chōshū's peasant society. That it was these controls rather than its commercial wealth can be seen by comparing Chōshū's finances with those of bankrupt han in commercially advanced central Japan. The same strength and stability also explain the measure of success achieved by Chōshū's Tempō Reform as opposed to the dismal failure of the Bakufu Reform. Even the bureaucratic cliques—whose prominence derived from the personal weakness of the daimyo and Elders—reflected the vigor of Chōshū's Confucian intellectual tradition and of its hierarchical society. That anything so unbureaucratic as "feudal" loyalty should have been responsible for bureaucratic efficiency is anomalous; yet, to a considerable degree it was the sense of duty and obligation to the han that this engendered that led to the great emphasis on the attainment of collective ends. The almost complete lack of corruption—unusual within such an organization, the deep moral concern for the real problems of governance, and the "bureaucratic suicides" of top officials such as Nagai Uta or Sufu Masanosuke, all indicate that power and the emoluments of office did not blunt the self-sacrificing loyalty that was both expected and received during this period.

It is hard to document such abstract qualities as intellectual vigor or loyalty, but concrete evidence of the strength of Chōshū's traditional society is abundant. The samurai of Chōshū, in spite of having had a portion of their stipends taken for over one hundred years, were still in reasonably sound condition. Their loyalty was never open to question, and their morale was high. The class (or rank) base of power in the han had been broadened slightly since the early part of the Tokugawa period, but bureaucratic position was still restricted to the top four strata of the class of *shi*. The prestige of these men was high among the samurai class, and the prestige of the samurai class remained immense among those who did not belong to it. It was certainly the prestige of this status that made possible the recruitment of commoners for the *shotai*.

The peasant base in Chōshū was also firm: the early Tokugawa village class structure had not broken down as it had in some other regions—notably, in the Kinki and Kantō regions, the Tokugawa heartland—and independent landholders were still in the majority. Strictly speaking, of course, the Chōshū peasants were neither independent nor landholders, since peasants remained attached to the soil, and land tenure lay in that twilight zone between feudal tenure and private ownership; yet they were independent landholders in the sense that they were neither tenants nor landlords. (The strength of the feudal village in Satsuma was even greater than that of Chōshū since it preserved something very akin to the pre-Tokugawa type of direct peasant village control by samurai [*gōshi*].)

The same, essentially late feudal, vitality can be seen in the commercial relations of the han. Although the land tax, much of which was collected in kind, was sold as a cash crop and other commercial products were either taken as taxes or controlled by han monopolies, the sales in almost every case were handled outside of the han, and most of the profits went to the han government. Consequently, there occurred relatively little commercial erosion of traditional relations within the han. Maintaining samurai control over the sale of tax rice and, directly or indirectly, over the han monopoly system, the han never lost control of its own finances; it never became the creature of an Osaka merchant house.

Within Chōshū commerce expanded considerably during the Tokugawa period, but the inferior position of the merchants vis-à-vis the han government remained clear-cut. Although the number of merchants increased, the highest honor they could hope to obtain was an appointment as *yōnin* (commercial agent) for the han. A few bought the right to a name and sword—these were sold to augment the han revenues in spite of *buiku* successes—but they received the outer symbols without the inner grace of samurai status. A few other merchants who controlled official monopolies were given official positions roughly equivalent to those of the *shōya,* but all were members of a lower caste. Certainly their position was inferior to that of merchants in the Kinki han, where fragmented domains often made monopoly control impossible, and the propinquity of Osaka induced a far greater commercialization of agriculture.

The same vitality of traditional forms within Chōshū can be seen in the independence of action possessed by the four Chōshū branch han and, to a lesser extent, by the various Elder houses. Just as the han was sovereign over its own domains within the limits of the Tokugawa system, so within the limits of the han, these existed almost as administratively independent sovereign fiefs. The important qualification that must be made to the above is that in the case of the Elder fiefs the agrarian base, although not the body of retainers which it supported, had come under the control of the han government. The Bakufu support given to the Kikkawa fief in order to weaken Chōshū illustrates their degree of structural independence. This is an excellent example of surviving private power in the truest feudal sense of the term. Moreover, these internal fiefs were not merely formal structures from which all life had disappeared: even during the Bakumatsu struggles, and particularly during the han civil war, they actually functioned as autonomous political units.

Tōyama Shigeki has suggested in his work, *Meiji ishin* (The Meiji Restoration), that the contribution of traditional strength to the restoration movements was vital. He writes: "We cannot simply say that the han which demonstrated a positive power to act in the Bakumatsu period were those with a high degree of bourgeois development and advanced economic structure." Rather, political power rested on military power, and military power on economic power; "therefore, in order to rebuild straitened finances and to realize an accumulation of wealth, what was needed more than anything else was the strengthening of control over peasants and a thoroughgoing control over commerce . . . in short, a certain degree of commercial production . . . and feudal authority sufficiently strong to organize and bring this under han control." And, emphasizing the importance of the latter authority, he continues: "Especially in a situation where military power had to be created at any cost . . . it was the existence of old feudal power that contributed favorably to the leadership for historical advance." [2]

Yet, in spite of such penetrating comment, Tōyama remains basically committed to the view that the Restoration was accomplished by the energy of the movement from below, a movement whose class basis represented a fundamental breakthrough in the productive process. I feel that this view is difficult to reconcile with his earlier emphasis on feudal strength. To the extent that Tōyama speaks of only "a certain degree of commercial production," it is difficult to visualize this as having furnished the revolutionary energy for the Restoration. In Chōshū some commoners, usually those who had

[2] Tōyama Shigeki, *Meiji ishin*, pp. 35–37.

become adherents of the Hirata school of *kokugaku,* did contribute money to loyalist ventures, and under coercion the peasant officials of Ogōri gave considerable support to the *shotai.* But, neither the occasional support of the loyalist movement under such circumstances nor the desire of peasants to swagger about as samurai can be used to prove a class motivation in the Marxist sense.

Commoner participation in Bakumatsu politics was rarely crucial; perhaps it was only so during the Chōshū civil war. To the extent that it occurred at all it took place within a vertical structure under the control of samurai. And even then it occurred only in certain han. In Chōshū it was restricted to certain administrative districts and almost always was under the control of peasant officials who, as was suggested earlier, were in many ways the lowest echelon of the han bureaucracy. The use of peasants within such a context was not primarily an alliance based on mutual areas of economic interest (although a few such areas did exist). Rather, the fact that it was easier for samurai to join with peasants of their own han than with samurai of other han was merely another indication of the vertical cleavage in the society. The only other way in which peasants may have positively contributed to Bakumatsu politics was through peasant uprisings. How important these were is hard to say in the absence of monographic studies illustrating their effects on various han governments. Uprisings may have been one factor inhibiting political action by certain han. But, one cannot simply take at their face value statements by would-be reformers regarding peasant unrest; within the Confucian context of their thought such statements were almost obligatory.

Peasant conditions were bad, but what needs to be explained is why this did not lead in troubled Bakumatsu times to large-scale peasant rebellions. One can speak of this as passivity; yet it was not merely negative. It was characterized by a strong group identification, by a positive commitment to the group and to group action (if not necessarily to its leaders at any given moment), and by the hierarchic tendency for each subordinate group to fit into the next higher level of organization. In this work I have emphasized the active "han nationalism" of samurai. Equally important, however, for the workings of the total society is this passive commoner attachment to local political units; this is especially visible in the case of the Tokugawa peasant village, but it existed elsewhere as well. In its moral emphasis on individual selflessness, and in its stress on duties and personal obligations, this was very much like samurai loyalty, and although it was not nationalism in the sense of individual identification with the symbols of the national polity, it functioned like nationalism in providing the social cohesion by which the entire nation became joined.[3] Neither in the Bakumatsu nor in the Meiji period could Japanese society be referred to as "a sheet of loose sand."

To the extent that one can speak meaningfully of classes at all, it is primarily to the military classes that one must look for the energy responsible for the Restoration movement, since the positive role of nonfeudal classes as nonfeudal classes was relatively slight during the Bakumatsu period. Here too a vertical cleavage is very much in evidence, the energy was unevenly distributed, it belonged to certain han rather than to a class. Viewed in this light, the events of the Bakumatsu period appear at first sight as the redistribution of late feudal power. Political hegemony passed from the weakened authority of the Bakufu into the hands of Chōshū and Satsuma, han which

[3] Another way of viewing this incipient nationalism would be to play down the particular focus of loyalty (be it han or village) and to see both commoner and samurai as possessing a similar type of commitment to their respective polities. In both cases individuals tend to find themselves within, rely on, and work for, their groups. In both cases these tendencies can be thought of in terms of *on, giri,* and the situation-centered ethic of which Benedict has spoken. Whether this commitment became active or remained passive would then depend not on the character of the commitment itself, but on other factors: the position occupied by the particular group, its function, its economic health, and so on. The advantages of this view are: 1) One can view the peasant villages and inactive han as having had, at their respective levels, and for different reasons, the same sort of stability during the Bakumatsu period; this would also suggest that it was more than accidental that those han which were active during the Bakumatsu period (or other han like them) were also, in many cases, just those that rebelled against the new government in the early years of the Meiji period. 2) It would also help explain the apparently similar motivation of the early Meiji entrepreneurs, regardless of their class background. The drawbacks of this point of view stem mainly from the fact that, in the Bakumatsu context at least, the differences between a village and a han are so great that loyalties to them work in radically different fashions.

had maintained the strength necessary to adapt successfully at this time of crisis.

The Bakufu authority had been weakened, on the one hand, by the pressure of the foreign powers, and on the other, by the economic concomitants of its own control system. On the internal scene it was this commercial growth, this unplanned commercialization of its "substructure," that sapped the Bakufu of its earlier vigor and precipitated its downfall. As outlying han, Chōshū and Satsuma were less affected by the rise of commerce. Other outlying han which like Chōshū or Satsuma possessed relatively stable societies may have been unable to participate in this redistribution of power because of traditional obligations to the Bakufu, financial insolvency, a lack of bureaucratic flexibility, or other factors as yet unknown.

The Meiji Restoration was not a revolution, not a change in the name of new values—such as *liberté, égalité,* and *fraternité* in the French Revolution. Rather, it was what is far more common in history, a change carried out in the name of old values. It was a change brought about by men intent on fulfilling the goals of their inherited tradition. It was a change brought about unwittingly by men who before 1868 had no conception of its eventual social ramifications. The Tempō Reform with its essentially reactionary aims and its program resembling those of the reforms of the Kyōhō and Kansei periods, the decision to enter Chōshū into national politics taken by a traditional type bureaucratic clique in the name of the past greatness of the house of Mōri, even the *sonnō* doctrines which, in spite of their use of the prefeudal symbol of the Emperor, reflected in an idealized form the Tokugawa loyalty ethic, all illustrate the crucial importance of traditional motivation in Chōshū's Bakumatsu history.

Yet, it is not enough to say that Chōshū emerged pre-eminent because of its embodiment in the strongest form of the positive values of Tokugawa society. Having triumphed in the Bakumatsu struggles, Chōshū and Satsuma did not go on to build another Bakufu. Instead, intent on preserving or re-creating the virtues of their old society, they ended by destroying it. Why? My answer is threefold.

(1) In part the centralization of the Meiji Restoration was a

consequence of Japan's particular feudal heritage and of the changes which had occurred in it during the Tokugawa period. To explain why this was so, and to explain in the broadest terms possible the meaning of Chōshū's action during the period in which the Bakufu came to an end, I feel it is necessary first to make certain tentative suggestions concerning the character of the Tokugawa state as a feudal system.

Feudalism, conceived of as an abstract model, is generally considered to be a highly unstable political form. The European institutions which have been labeled feudal were subject to a very rapid process of evolution and decline. One author writes, "There is indeed a dramatic irony in the fact that the better feudalism works the more rapidly it generates a political structure which is no longer completely feudal."[4] In every known example feudal systems have shown a tendency to evolve into more highly centralized forms of government. While particular in many respects, Japan was no exception to this rule. The centralization which took place under Hideyoshi in the latter part of the sixteenth century was similar in many ways to that which occurred in the West after the European "period of Warring States."

In both cases it was the concern with power, the concern to maximize power, that led to this centralization. Warring States' society was composed of autonomous fiefs. In that society where each lord was the potential enemy of every other lord and every vassal a potential lord, power and its symbols, even more than money and its symbols in modern society, were the source of almost all security and prestige. In the struggle between fiefs only the fittest survived, and those domains which did not actively attempt to increase their strength were eliminated by a military process of natural selection. It was this process that created the tight unit-fief, confirmed the rule of a military aristocracy, and led the daimyo to reward loyalty above all other virtures. Peasant armies were formed, productivity was encouraged, efforts were made to increase the productivity of cities—all in the name of military necessity. And, under these circumstances,

[4] Joseph R. Strayer and Rushton Coulborn, "The Idea of Feudalism," in *Feudalism in History,* Rushton Coulborn, ed. (Princeton, 1956), p. 9.

military necessity is merely another name for power. That the loyalty of the period of Warring States was reciprocal, conditional, and functional—as we saw in Chapter V—was determined largely by power relations. To be loyal was to contribute actively to the power of the daimyo.

The existence of such power led the stronger fiefs to conquer, absorb, or subordinate the less powerful domains. In some cases the samurai of these conquered domains were tracked down and killed, some escaped to become *rōnin*—most of whom eventually entered the service of other lords. In other cases certain subgroups among the forces of a defeated lord were detached and made the retainers of the victorious daimyo. The hierarchic, pyramidal structure of loyalties provided a means by which formerly independent smaller fiefs could be assimilated into a larger unit. The emergence of a feudal bureaucracy provided the means by which increasingly larger units could be efficiently governed. The necessity to maximize power and the means by which it was achieved may be termed the centripetal forces in Warring States' society. It was these factors, abetted, perhaps, by the rise of a nation-wide net of trade, that led to the emergence of increasingly larger political aggregates, until Japan was finally centralized under Hideyoshi in the late sixteenth century.

But at the point at which it seemed that feudalism was about to be destroyed by the very success of these feudal centripetal forces, it was preserved and maintained for over 250 years in the form of the Tokugawa system. Why the first three shogun were willing to "freeze" the early seventeenth-century structure, rather than to push forward a radical program of centralization, is difficult to say at this stage in our knowledge. One reason, certainly, was that even after having conquered all of Japan the position of Tokugawa Ieyasu was far from perfect. His enemies subdued, Ieyasu had the appearance of absolute power; yet by this very token he was absolutely dependent on his vassals who were concerned for their own rights and semiprivate power. This is more than a verbal quibble.

One significant difference (among many) between England and France which led to limited monarchy in one and absolute monarchy in the other was that after the Conquest, English kings dealt with their vassals as a single group whereas the weaker French kings dealt with them individually as best they could. In the former case there existed little leverage, in the latter, much. Abstracting differences of culture and comparing for the moment the nature of power relations, the position of the Tokugawa vis-à-vis their *fudai* retainers was much closer to the English than to the French case. The *fudai* daimyo early emerged as a traditional collegial body, as a *de facto,* if not legal, check on the extension of central authority.

The strength of the *fudai* vis-à-vis the Tokugawa was further heightened by the existence of the *tozama* lords. The Tokugawa could not afford to alienate their *fudai* retainers since this would have left the Bakufu naked and vulnerable to *tozama* attack.

Real centralization (once the Tokugawa peace had been established) could only come about by two means: first, by the development of a nonfeudal Bakufu bureaucracy, by the destruction of feudal house government, and second, by Bakufu interference in the internal affairs of the han. Were such developments possible in Tokugawa Japan?

In 1600 Ieyasu confiscated the domains of some of his former enemies (the Chōsokabe), reduced those of others (the house of Mōri), and permitted a few to remain untouched (the Shimazu). Yet most of the land taken was merely given to others as fiefs: private power was upheld and real centralization thwarted. The bureaucratic principle was not sufficiently advanced for officials to be appointed to rule the conquered lands, as was done, for example, by the Ch'in state in Chou China.[5]

Merchants were politically insignificant in spite of the fact that commerce was more developed in the late sixteenth century than has usually been assumed. This may partially explain the inability of Hideyoshi or Ieyasu to create a bureaucracy of nonfeudal origin. Exceptions in their relations with merchants may be noted but these do not disprove the rule. Konishi Yukinaga, the second son of a merchant of Sakai, was made a daimyo and given domains of 240,000 *koku* by Hideyoshi. This, however, was a consequence of his ability

[5] The only exceptions to the rule were the Bakufu intendancies (*daikansho*) established in areas such as Nagasaki for special strategic or economic reasons.

and the esteem in which he was held: it is not subject to any simple economic interpretation. And more important, Konishi was a daimyo, not a bureaucrat. Other merchants and even some foreigners were appointed to fairly important economic positions by Ieyasu (who also numbered among his advisors Buddhist monks and Confucian scholars). None, however, could seriously compete with his house government of *fudai* lords.[6]

Moreover, not only were the nonfeudal classes weak, but the feudal ties, personal in nature, were strong and could not suddenly be replaced by a hired bureaucracy. Ieyasu spoke of this personal loyalty as follows:

I am loyal to Heaven by cherishing the loyal retainers of the Matsudaira house, and because of this I have received from Heaven the charge of the Empire of Japan. But if it is ruled improperly Heaven will take it away again. . . . One must consider the future and make no innovations or new families. And when the Empire is well ruled the military classes are obedient. When it is the reverse they are against the authorities. This has always been so and it is of Heaven.[7]

One is immediately struck by the congruence in Tokugawa Confucianism of Heaven and the Emperor from whom Ieyasu has

[6] Another tack by which the Tokugawa structure of government can be approached is to suggest that the *fudai* lords were themselves primarily bureaucratic in character, that they were structurally parallel to the higher ranks of samurai in a han. There are a number of arguments that can be adduced in support of this. The gradient between *gokenin, hatamoto,* and *fudai* daimyo was a slippery slope; it is difficult to say in terms of function where one leaves off and the other begins. All had been Tokugawa retainers before Sekigahara and lacked a tradition of autonomy; all could be employed at their respective levels within the Bakufu officialdom. The higher *hatamoto,* those with fiefs of over 3000 *koku,* had virtually as much control over their fiefs as did the smaller *fudai* daimyo. Unlike the *tozama* or *shimpan* lords, the *fudai* were, within limits, promoted or demoted according to their performance, and were more frequently shifted from one fief to another in order that status accord with function. All of this suggests that the Bakufu might be viewed as a giant han which contained within it (like *karō* fiefs in Chōshū) the *fudai* domains. I have not approached the question from this angle since I feel that to reduce them to samurai bureaucratic types plays down the autonomy which they did possess, and makes little of the fact that their bureaucratic position hinged on their status as lords of domains of over 10,000 *koku.*

[7] A. L. Sadler, *The Maker of Modern Japan, The Life of Tokugawa Ieyasu* (London, 1937), pp. 352–353.

received the mandate to rule. In good Neo-Confucian style both, while ultimately the source of all power, are to be controlled by the ruler (the shogun) through the maintenance of proper and virtuous relations. We note here that these are defined in terms of the shogun's reciprocal duty to the retainers of his house, the condition on which Heaven has seen fit to give him the mandate.

In the presence of such ethical notions and of *de facto* private power, a radical transformation of the principle of house government was impossible. The Bakufu bureaucracy continued, therefore, to be staffed by *fudai* lords and *hatamoto* whose bureaucratic advancement was accompanied by an increase in their domains, by a strengthening of their "feudal" private power. Men such as these were unlikely to support any general policy of interference in the affairs of the han—partly because any such policy once established could be extended to their own domains, but more important, because such interference could upset the equilibrium within which they held power.

In the early Tokugawa period the Bakufu had almost absolute power; the shogun or Tokugawa house did not. The Bakufu was a corporate affair in which power was widely diffused. Evidences of this are many. Crucial decisions were made largely by conferences of the top Bakufu officials, the *rōjū,* who were appointed from *fudai* daimyo (holders of castles, *jōshu*) with domains over 25,000 *koku.* Other of the more important officials were also *fudai* lords, while lesser officials were appointed from among the *hatamoto.* Throughout most of the Tokugawa period when their collective power was unchallenged, the *fudai* bureaucrats seem to have broken up into cliques. These cliques are still very much in evidence during the first few years after Perry; they provide the framework for the early period of Bakufu clique politics. Yet, with the entrance into Bakufu politics of *tozama* and *shimpan* daimyo, even during these early years, their underlying corporate consciousness emerged and they gradually coalesced into a single group. Most *fudai* daimyo supported Ii as *tairō,* they rejoiced collectively at the Satsuma-Aizu coup in 1863, they were the motive force behind the second Bakufu expedition against Chōshū, and they constituted the "conservative" power in Edo against which the *shimpan* leaders of the Bakufu, Tokugawa Keiki and Matsudaira Keiei, fought in vain.

Ieyasu, aware in 1600 that Bakufu power was diffuse, distributed most of the spoils of his victory at Sekigahara among his own family (founding the *shimpan* houses) and the pro-Tokugawa *tozama* daimyo (Yamanouchi or Kuroda, for example), in the hope of strengthening the Tokugawa against the *fudai* lords. Yet even this distribution of lands did not weaken the position of the *fudai* lords as the Bakufu officialdom; their small domains ensured their dependence on the Bakufu, but not their loyalty to the Tokugawa.

Earlier, speaking of Tokugawa loyalty, we found it to be unilateral and unconditional. This was true of the relation between samurai and their daimyo; it was not true of the relation between the *fudai* lords and the house of Tokugawa. Here "credit," support by the *fudai* daimyo, was still important; *fudai* loyalty was still, to a certain degree, reciprocal and conditional. "Uneasy lies the head that wears a crown" is particularly true in such a situation. With the examples of Oda Nobunaga and Hideyoshi before him, Ieyasu had good cause to worry concerning the fate of his progeny. That such unease had been warranted was strikingly confirmed by an event in 1680. At that time the *tairō* Sakai Tadakiyo, faced with the problem of choosing an heir for the failing shogun Ietsuna, proposed to the Council of *rōjū* (following the example of the Hōjō during the Kamakura period) that an Imperial prince, Arisugawa no Miya, be made the shogunal heir. Tadakiyo's scheme failed; his enemies within the Council of *rōjū* used this as a pretext to break his power within the Bakufu.

And yet, that he should have been judged primarily by other *fudai* daimyo (one *shimpan* lord seems to have played a role in this as well), by his peers, and not by the *shimpan* houses nor by a sick but absolute shogun, is significant. Power was institutional not familial. One is reminded of the feudal courts of medieval Europe—although the decision that foiled Tadakiyo's attempted usurpation of power was not strictly juridical. Had the situation been reversed, had a shogun attempted without a proper sanction to infringe upon the rights of a *fudai* daimyo, what would their judgment have been? That Tadakiyo could think of gaining for his own family the *de facto* hegemony over Japan only sixty-six years after the death of Ieyasu points up the Tokugawa weakness within the Bakufu structure.

How is this view to be reconciled with the usual view of the shogun as a figure possessing absolute power? In part this may be thought of as a two-phase process. When weak, sick, or in his minority, the shogun was impotent and all power was in the hands of his *fudai* bureaucracy. When strong and in his majority, then, as the lord to whom the *fudai* retainers owed allegiance, as the head of the Bakufu bureaucracy, and as overlord of all Japan, much of the power returned to his hands. The cyclic alternation of the two phases permitted, on the one hand, incidents such as that involving Sakai Tadakiyo, and, on the other, the direct rule of a strong figure such as Yoshimune. A further subpattern within such cyclic alternation, within the strong phase of a weak daimyo, was the rise of strong proxies such as Arai Hakuseki (the son of a *rōnin*) or Yanagizawa Yoshiyasu (the archetype of government by shogunal advisors, *soba yōnin seiji*), who with the backing of the daimyo could for a time challenge the hereditary *fudai* power. The advancement and use of a man such as Arai Hakuseki represents at the highest level of Bakufu government the furthest extent of bureaucratic rationalization of personnel. Yet even in the strong phase the shogun was absolute only within the system, he could effect changes only in certain boundary areas, and did not have, I feel, the power to change the constitution of the system itself.[8]

Another dimension of this constitution can be uncovered by asking whether the Tokugawa social structure or even the polity, its most stable aspect, was in fact "frozen" miraculously in the early seventeenth century. A more careful view of early Tokugawa history un-

[8] Admittedly, this is a distinction that blurs at times. Yet what would have happened if a shogun, even in his strong phase, had attempted to replace the *fudai* bureaucracy with one hired from other classes or groups? Was this within the absolute power of a strong shogun? Would not any advances in this direction have been repudiated by the Bakufu in the next weak shogunal phase, or might this not have had repercussions that would have endangered the Bakufu? In this sense I feel that the distinction is a valid one.

veils a motley array of succession disputes, power struggles among *rōjū,* pitched battles within certain han, *rōnin* uprisings, and so on. These disturbances appear to have been part of a settling down process, what Max Weber called the "routinization of charisma." Ieyasu had been a powerful and, in his own way, a principled ruler; he was responsible for the chain of victories that made the Tokugawa house the most powerful in Japan. This aura of success was further enhanced by the establishment of the Bakufu system and the Tokugawa peace; at the time of his demise Ieyasu was venerated in Japanese fashion as a god. For Ieyasu's immediate successors this charisma or right to power was still personal, but by the end of the seventeenth century it had been largely transferred to the Bakufu organization.

Viewed in this fashion, the turbulence characteristic of the early seventeenth century in Japan, the dispossession of the anti-Tokugawa *tozama* daimyo, the continuing warfare at Osaka and Shimabara, the *rōnin* uprisings of the mid-seventeenth century, and so on, may be regarded as belonging to the period before routinization took hold. Likewise, the attempted usurpation by Tadakiyo and other local uprisings may represent a conflict between the surviving personal charisma of han leaders and the traditional charisma embodied in the Bakufu organization. Only such an interpretation will explain why the period between 1700 and 1850, when the greatest erosion of the social foundation of Tokugawa society occurred, was the period of greatest political stability (except, perhaps, at the village level). By 1700 or so, even charisma within the han had been routinized.[9]

[9] There is no end of approaches from which this may be viewed. In a paper, "Skill and Intelligence in Tokugawa Society," given at the twelfth annual meeting of the Association for Asian Studies in New York in April, 1960, Ronald Dore spoke of an emphasis on "men of talent" in the 1615 *Buke shohatto* (Laws Governing the Military Households) that was omitted from later formulations. This mirrors the shift from ability to form implied by the concept of routinized authority. In the same paper Dore also spoke of the re-emergence of a quest for talent in Chōshū during the late Tokugawa period (seating in the han school by ability rather than rank, and so on). I feel that this later shift occurred within a situation in which authority was still firmly routinized and that it reflected 1) the tensions associated with the climate of reform, and 2) the particular adaptability of the Chōshū government.

This view of the settling-down process within Tokugawa society also tends to corroborate my views (see Chapter V) regarding the transition from Warring States to Tokugawa loyalty: the point at which the Bakufu-han structure was accepted as legitimate was also, inescapably, the point at which the loyalty of samurai to their daimyo became routinized, as well as the point at which loyalty became impersonal. A century is not too long a period for the change from the personal loyalty of the one period to the "han nationalism" of the other.

The system of Tokugawa rule is spoken of as centralized feudalism. This is proper; power was more centralized after 1600 than before. Yet the devices and institutions through which the Tokugawa ruled might, in the context in which they were formed, be more meaningfully viewed as having had a centrifugal force. The Tokugawa held back change. At a certain point they checked the process of centralization that had begun with Oda Nobunaga. As we have seen above, they purchased security at the price of administrative decentralization. Paradoxically, the Tokugawa preserved the earlier centripetal forces by preventing their fulfillment. The military power and the autonomy of the han under their own governments were preserved as a potential which could one day again be released. The hierarchic loyalty of the earlier period was not only preserved but its centripetal potential was heightened since, in its depersonalized and "free-floating" form, it could now be put, should the opportunity occur, to a nonfeudal use. Similarly, the advance of bureaucratic rationalization within the han (if not at the national level) had created greater concentrations of power at the "provincial" level. All of these opened the way in the nineteenth century for the establishment of the Meiji government with a type of bureaucracy which could not have been formed in 1600.

By 1700 a complex equilibrium had been established between the above latent centripetal forces and the centrifugal controls of the system, between the Tokugawa house and its *fudai* vassals, between the military power of the Bakufu heartland and the potential power of the *tozama* han. The same balance of elements can be seen in the stipulations of the Tokugawa code of loyalty. In all matters external to the han, samurai were obligated to conform to the rules set down by the Tokugawa hegemons, while internally they were expected to be frugal, to give service, to study, and to join in military exercises, all items designed to keep up the military strength of the han.

The arrival of Perry in 1853 destroyed the two bases on which the traditional charisma of the Bakufu had depended: putting an end to the security of the nation it vitiated the Tokugawa peace, and setting the Court against the Bakufu (after 1858) it subverted the shogun's authority as hegemon. Had the Bakufu resolutely decided that, after all, something was better than nothing, it could have compromised and the Tokugawa house could have dominated a new Imperial settlement. However, the insignificance of the *fudai* lords as territorial powers, one of the conditions which had enabled the Tokugawa to survive for 250 years, now proved their undoing. Having no other stake, the *fudai* lords felt compelled to maintain their power in the form of the traditional Bakufu autocracy. Their intransigence and unfounded arrogance alienated the powerful han and made conflict inevitable. In contrast to the *fudai,* the *shimpan* lords, equally realistic but better endowed, sought a compromise in which the power of their domains would be recognized. When rebuffed by the Bakufu they withdrew their support. In the end, its autocratic pretensions humiliated and without support, the Bakufu appointed a shogun who no longer really believed in the shogunate. The rapidity with which this house of cards collapsed astonished even its enemies.

From the viewpoint of the han the significant feature of foreign pressure was that it set in motion the forces that destroyed the centrifugal controls of the Bakufu system. This upset the Tokugawa equilibrium, releasing the centripetal potentialities of late Tokugawa loyalty. In this situation, for reasons given earlier, some han were unable to act. These continued to stress passive conformance to the ailing Tokugawa hegemony. Others began to act immediately, at first through Bakufu clique politics and subsequently on the national scene. The social reality behind the various attempts to "mediate" between Bakufu and Court was that of political action by autonomous han. The years of conflict between Chōshū and Satsuma, and even the character of their final *rapprochement,* bespeak the intensity of what I have, somewhat freely, labeled "han nationalism."

By 1863 political maneuvers had given way to military struggles for power. These Bakumatsu struggles were obviously not wars between unit-fiefs. To speak of the Restoration as the working out of the centripetal forces contained in the traditional society does not mean that they took the same form as those of the period of Warring States. Bakufu power continued to exist in some degree throughout the entire period; therefore the efforts by the han to increase their power unfolded on a highly structured political stage revolving around the Bakufu-Court axis. By 1868 a coalition of strong han had carried out the Restoration.

The second phase in this process of centralization, the Meiji revolution, took place after 1868 when the Restoration leaders, joining the infant Imperial government, transferred the power of the strong han to the new government, and then used this power to destroy the old han system. Edwin O. Reischauer has described the Tokugawa system as "basically reactionary even in the early seventeenth century."[10] That the evolving centripetal tendencies of Japanese feudalism should have produced a far greater degree of centralization two hundred years later is not surprising.

(2) A second crucial factor which added impetus to the process of centralization and prevented the rise of a new Bakufu was the adoption of Western military technology by the strong han. It was the use of this new type of power that enabled Chōshū to defeat the combined Bakufu army in 1866; it was this power that enabled Chōshū and Satsuma to triumph over numerically superior Bakufu forces at Fushimi and Toba in 1868; it was this power that maintained the Meiji government during its early years of trial. Without this strength the Bakufu could never have been overthrown by two han, whatever their traditional strength may have been. Without this factor of technology, the new government, whatever form it might have taken, would undoubtedly have been more of a compromise with the old Tokugawa institutions than was actually the case.

In Chōshū as in Japan the role of the new technology was im-

[10] Edwin O. Reischauer, *Japan Past and Present* (Cambridge, 1953), p. 108.

portant. The greatest concentration of this new power was in the *shotai,* which in 1865 furnished the model for a reform of the han military. The *shotai* were led by the loyalists, those who demanded the boldest program of political action. They functioned as schools in which samurai and non-samurai alike became imbued with the doctrines of the Restoration movement. Their in-group solidarity was primarily ideological, but was undoubtedly reinforced by their military effectiveness. The use of traditional forms within the *shotai,* the maintenance of class distinctions, cannot hide the fact that this was an institutional breakthrough of the first water, one that cut across and passed beyond the traditional, stratified structure of vassalage. That it should have occurred in the area where vested interests were strongest gives it a symbolic importance at least equal to that of the abolition by the Ch'ing of the civil service examination system in 1905. It was the existence of the *shotai* that enabled the loyalists to resist the Mukunashi government in 1864, and made possible the transfer of power to the "Court" in 1868.

Yet, the adoption of Western technology was possible only because of the strength of Chōshū's traditional society. The *shotai* were formed to increase the power of the han. The specific event leading to their formation was the 1863 Shimonoseki bombardment, a consequence of the han's attempts to contest the power of the West. Other han were as aware as Chōshū of the superiority of Western arms, but they were not as concerned with power. Just as Matsudaira Sadanobu had recognized the superiority of Western science and advocated its adoption, so now the Chōshū leaders grasped Western military technology as the "form and utensil" with which they could implement their goals. It was adopted not because it was Western, but in spite of being Western. The strength of political values in Chōshū not only made it financially able to buy the foreign weapons, but just as important, it gave to a majority in the han the will to use them.

(3) A third factor contributing to the formation of the centralized Meiji state, one more important than the immediate effects of the new technology on Japanese society, was the image of the West as a militarily superior power looming on the periphery of a weak and

backward Japan. This image was formed only gradually during the Bakumatsu period; it was virtually complete by 1868. It remained almost as a constant during the post-Restoration period.

Robert N. Bellah has suggested that the Tokugawa value system gave primacy to performance and the attainment of political goals.[11] The Meiji leaders, who had won out in the Bakumatsu struggle for power, represented in an extreme form this concern of the traditional society with power. In their eyes the existence on Japan's doorstep of the superior Western powers was intolerable. Thus, during the early Meiji period, the leaders of Japan were not concerned solely with consolidating their power against internal opposition—a process which would have demanded only a relatively small increase in political rationalization. On the contrary, their main task was to create a state sufficiently powerful to cope with the foreign powers of which they were so conscious. It was this latter challenge that led them to carry out the Meiji revolution; the centripetal forces of Tokugawa Japan (in the absence of external pressure) and even the adoption of Western arms during the Bakumatsu period cannot explain the post-Restoration destruction of the old society, the continuing demand for men of talent, or the borrowing of Western institutions.

In terms of the interplay between different types of values, these revolutionary reforms of the early Meiji period were in some sense the continuation of the changes instituted in Bakumatsu Chōshū. In both cases specific institutions were destroyed in order to achieve goals or values in terms of which the entire institutional order was justified. The formation of the *shotai* outside of the regular han military, the establishment in 1866 of a more rationalized system of rifle units including both *shotai* and vassal groups, the rise of men such as Takasugi, Ōmura, or Maebara, the reform of the han government were all designed to strengthen the han. Likewise, Chōshū's actions on the national scene, which at times involved relatively little risk (the 1861 mediation) but at other times endangered the very existence of the han (the 1864 counter-coup or the 1866 Chōshū-Bakufu war), were attempts to achieve greater han power. These can

[11] Robert N. Bellah, *Tokugawa Religion* (Glencoe, Illinois, 1957), pp. 14–15.

be thought of as the early beginnings—only partially supported by an Emperor-oriented concept of sovereignty—of the "revolution" which the Meiji oligarchs were to carry out for the same end of power.

That the values or goals of the earlier period continued to function with so little change explains the strength of Japan's obsession to equal the West. This was further bulwarked by the centripetal Tokugawa loyalty that was now extended from the daimyo who reigned but did not rule to an Emperor in much the same condition. These were central among the values that supported the nascent Meiji state. In many ways they resemble the nationalism which has appeared in other Asian countries only after decades of contact with the West. It was because Japan possessed such characteristics when first confronted by the West that it was able so early to achieve a part of the transformation which is the goal of other nations in Asia today. In Japan as in Chōshū it is in a large measure to the strength and not to the weaknesses of the traditional society that we must turn to comprehend its modern history.

BIBLIOGRAPHIC NOTE

The materials used in this study fall roughly into four categories: archival materials, general histories, biographies, and recent articles based on archival materials. The records and documents of Chōshū became, after the Restoration, the Chōshū Archives, which are now preserved in the Yamaguchi Prefectural Library. This certainly constitutes one of the most important remaining collections of materials relating to Tokugawa Japan. In addition to touching on all aspects of han organization and function, the Archives also contain shelf after shelf of narrative accounts of its late Tokugawa history and biographical materials regarding its leaders and the early lives of those who went on to become the Meiji oligarchs. During a short visit to Yamaguchi, however, I was able to peruse only several rosters of the ranking samurai (*shi*), to view a few manuscripts dealing with the Chōshū civil war, and to use certain books and transcriptions of interviews unavailable elsewhere.

In the second category, that of general histories, I should mention first the twelve-volume *Bōchō kaiten shi* (Restoration History of Chōshū). This constitutes the most important single source that I have drawn on in writing this book. It is an old-fashioned documentary history in which letters, records, memorials and the like are loosely strung together under an assortment of topics arranged more or less in chronological order. It is a history which, according to its introduction, "makes every effort to avoid criticisms, or conclusions, and attempts only to state the facts." Why this work the purpose of which is clearly to make known the Chōshū record should have taken this form is hard to say. In 1890 Itō Hirobumi, in reply to a proposal that a collection of materials relating to the history of the Restoration be made, responded as follows:

From the time of the battle at the palace gate [in 1864] various conflicts have arisen between Satsuma and Chōshū. The collection of materials regarding Restoration history is in one sense the collection of materials about Satsuma-Chōshū enmity. Thus, now, just when Satsuma and Chōshū, united, hope without incident to weather the first 1890 Imperial Diet, if unpleasant feelings were stirred up between the two it might have grave political consequences. I am very much in favor of the collection of materials relating to the Restoration, but it is not yet the time. [Tōyama Shigeki, *Meiji ishin* (Tokyo, 1951), p. 8]

Even twenty-one years later when the *Bōchō kaiten shi* was first published, such feelings may have influenced its mode of composition. Still another factor determining its form was that of traditional canons of historiography. Having spent so many hours with this work I find it difficult to avoid a feeling of ambivalence. Its shortcomings are obvious. Yet, all in all, its pastiche-like quality, its lack of index, and the fact that the same person is referred to (with complete accuracy) by different names at different stages of his career, are certainly outweighed by its rich array of valuable *sōrōbun* documents. After having read English and Japanese survey histories of the Restoration period in which the more enlightened statements of the future oligarchs are seized upon, magnified, and interpreted as the basic theme of the period, it is a sobering experience to delve in the *Bōchō kaiten shi* in which such statements are but rare flecks of froth in a torrent of Confucian rhetoric.

A second general history, an extremely useful one as an outline guide to the intricacies of pre-Restoration politics, is the six-volume *Ishin shi* (Restoration History) published by the Ministry of Education and the Office for Restoration Historiography in 1939–1941. In spite of the militarist era in which it appeared, this is a relatively impartial, well-organized, indexed, easily used, factual account of the most salient aspects of Restoration history. A third general work to which I am much indebted is the provocative *Meiji ishin* (Meiji Restoration) by Tōyama Shigeki. This is the best single volume history of the Restoration to have appeared since the end of the war;

I invariably learn more by disagreeing with Professor Tōyama than I do by agreeing with most other writers.

Biographies and the collected writings of individuals throw a great deal of light on those who died before the Restoration, but in general they have little of interest to say about the early lives of the Meiji oligarchs. In part this can be explained by the official character of the latter's biographies; in part this derives from the fact that most of them played only minor roles in the actual struggles of the Restoration period.

The fourth category, that of recent articles and monographs regarding Chōshū, is in its contribution to this book second only to the *Bōchō kaiten shi*. Much of what is new in this study, especially those sections concerning the class composition of government cliques, the social bases of military organization, and the character of the Chōshū Tempō Reform, is based on these works. Without them this book could not have taken the form that it now has. Most of this recent work is Marxist in its assumptions regarding history. All but one of the Japanese scholars who write on Chōshū are of this persuasion, and this distribution is not atypical for most areas of Restoration and Tokugawa history. When in Japan in 1956–1957 I was immersed in a sea of Marxist historiography. This is apparent in this book: even when not specifically attacking this interpretation I am often arguing against it. Yet, apart from its overly simple theoretical scheme, much of this work is extremely good. Young historians of this school have uncovered many new and valuable materials and meticulously documented many previously untouched areas of Japanese history. All who work in this field must use these materials if they are to participate in the discourse of Japanese history as it is presently carried on.

INDEX

INDEX

BIBLIOGRAPHY

GLOSSARY

BIBLIOGRAPHY

Western Sources

Abegglen, James C. The Japanese Factory, Aspects of its Social Organization. Glencoe, Illinois, 1958.

Allen, G. C. A Short Economic History of Modern Japan 1867-1937. London, 1946.

Beasley, W. G. Select Documents on Japanese Foreign Policy 1853-1868. Oxford, 1955.

Bellah, Robert N. Tokugawa Religion. Glencoe, Illinois, 1957.

Benedict, Ruth. The Chrysanthemum and the Sword. Boston, 1946.

Blacker, Carmen, tr. "Kyūhanjō by Fukuzawa Yukichi," Monumenta Nipponica, 9:304-329 (April 1953).

Brown, Devere Sidney. "Kido Takayoshi and the Meiji Restoration: A Political Biography, 1833-1877." Dissertation, University of Wisconsin, 1952.

Coulborn, Rushton. Feudalism in History. Princeton, 1956.

Craig, Albert M. "The Restoration Movement in Choshu," The Journal of Asian Studies, 18:187-198 (Feb. 1959).

Feuerwerker, Albert. China's Early Industrialization, Sheng Hsuan-huai (1844-1916) and Mandarin Enterprise. Cambridge, Mass., 1958.

Hackett, Roger. "Yamagata Aritomo 1838-1922: A Political Biography." Dissertation, Harvard University, 1955.

Jansen, Marius B. "Takechi Zuizan and the Tosa Loyalist Party," The Journal of Asian Studies, 18:199-212 (Feb. 1959).

Kautsky, Karl. Die Klassengegensätze im Zeitalter der Französischen Revolution. Stuttgart, 1919.

Keene, Donald. The Japanese Discovery of Europe: Honda Toshiaki and Other Discoverers 1720-1798. London, 1952.

Kyooka Eiichi, tr. The Autobiography of Fukuzawa Yukichi. Tokyo, 1934.

Nef, John U. War and Human Progress. Cambridge, Mass., 1952.

Norman, E. Herbert. "Andō Shōeki and the Anatomy of Japanese Feudalism," Transactions of the Asiatic Society of Japan, Third Series, 2:1-330 (Dec. 1949).

-------Japan's Emergence as a Modern State. New York, 1946.

-------Soldier and Peasant in Japan: The Origins of Conscription. New York, 1943.

Numata Jiro. "Acceptance and Rejection of Elements of European Culture in Japan," Journal of World History, 3.1:231-253 (1956).

Reischauer, Edwin O. Japan Past and Present. Rev. ed.; Cambridge, Mass., 1953.

-------The United States and Japan. Cambridge, Mass., 1950.

Sadler, A. L. The Maker of Modern Japan: The Life of Tokugawa Ieyasu. London, 1937.

Sansom, George B. The Western World and Japan. New York, 1951.

Satow, Ernest. A Diplomat in Japan. London, 1921.

Sayles, G. O. The Medieval Foundations of England. London, 1956.

Smith, Neil Skene. "Materials on Japanese Social and Economic History: Tokugawa Japan," Transactions of the Asiatic Society of Japan, Second Series, 14:1-176 (June 1937).

Smith, Thomas C. The Agrarian Origins of Modern Japan. Stanford, 1959.

-------"The Introduction of Western Industry to Japan during the Last Years of the Tokugawa Period," Harvard Journal of Asiatic Studies, 11:130-152 (June 1948).

-------Political Change and Industrial Development in Japan: Government Enterprise, 1868-1880. Stanford, 1955.

Takekoshi Yosaburo. The Economic Aspects of the History of the Civilization of Japan. 3 vols.; London, 1930.

Tsukahira, George Toshio. "The Sankin Kōtai System of Tokugawa Japan 1600-1868." Dissertation, Harvard University, 1951.

Tsunoda Ryusaku, William Theodore deBary and Donald Keene. Sources of the Japanese Tradition. Introduction to Oriental Civilizations Series, ed. William Theodore deBary. New York, 1958.

Veblen, Thorstein. Essays in Our Changing Order. New York, 1934.

Japanese Sources

Bōchō rekishi goyomi 防長歴史暦 [A historical calendar of Chōshū]. 3 vols.; Yamaguchi-ken, 1943.

Fujii Hokō 藤井葆光. Ishin shiryō: ōjōya Hayashi Yūzō 維新史料: 大庄屋林勇蔵 [Restoration materials: the life of Ōjōya Hayashi Yūzō]. Among papers titled Taishō ku sōisha ni kansuru torishirabe no ken kambō 大正九贈位者に関する取調べの件官房 in the Chōshū Archives (Mōri bunko 毛利文庫) of the Yamaguchi-ken Library.

Furushima Toshio 古島敏雄 "Kinsei ni okeru shōgyōteki nōgyō no tenkai" 近世に於ける商業的農業の展開 [The development of commercialized agriculture in the Tokugawa period]; Shakai kōsei shi taikei 社会構成史大系 [An outline history of social structure], ed.

Watanabe Yoshimichi 渡辺義通. Tokyo, 1950.

Furushima Toshio 古島敏雄 and Nagahara Keiji 永原慶二. Shōhin seisan to kisei jinushi sei 商品生産と寄生地主制 [Commercial production and the parasitic landlord system]. Tokyo, 1954.

Haraguchi Kiyoshi 原口清. "Chōshū han shotai no hanran" 長州藩諸隊の叛乱 [The rebellion of the Chōshū militia]; Meiji seiken no kakuritsu katei 明治政権の確立過程 [The establishment of Meiji political power]. Tokyo: Meiji shiryō kenkyū renraku kai 明治史料研究連絡会, 1957.

Hirano Shūrai 平野秋来. Chōshū no tenka 長州の天下 [The land of Chōshū]. 1912.

Honjō Eijirō 本庄栄治郎, ed. Kinsei Nihon no san dai kaikaku 近世日本の三大改革 [The three great reforms of modern Japan]. Tokyo, 1944.

-------Kinsei no keizai shisō 近世の経済思想 [Tokugawa economic thought]. 2 vols.; Tokyo, 1938.

-------Nihon keizai shisō shi gaisetsu 日本経済思想史概説 [An outline history of Japanese economic thought]. Tokyo, 1946.

-------Nihon keizai shisō shi kenkyū 日本経済思想史研究 [Studies in the history of Japanese economic thought]. 2 vols.; Tokyo, 1938.

Horie Hideichi 堀江英一 Hansei kaikaku no kenkyū 藩政改革の研究 [Studies of han reform]. Tokyo, 1955.

-------Meiji ishin no shakai kōzō 明治維新の社会構造 [Structure of the Meiji Restoration]. Tokyo, 1954.

Ienaga Saburō 家永三郎. Nihon dōtoku shisō shi 日本道徳思想史 [A history of Japanese moral thought]. Tokyo, 1954.

-------Nihon kindai shisō shi kenkyū 日本近代思想史研究 [Studies in the history of modern Japanese thought]. Tokyo, 1956.

Ikeda Toshihiko 池田俊彦. Shimazu Nariakira-kō den 島津斉彬公傳 [Life of Shimazu Nariakira]. Tokyo, 1954.

Inobe Shigeo 井野辺茂雄. Bakumatsu shi no kenkyū 幕末史の研究 [Studies in Bakumatsu history]. Tokyo, 1927.

Inoue Kiyoshi 井上清. "Bakumatsu ni okeru han shokuminchi-ka no kiki to tōsō" 幕末に於ける半植民地化の危機と斗争 [The crisis and struggle concerning semi-colonialization during the Bakumatsu period]; Rekishi hyōron 歴史評論, 33:1-16 (1951).

-------Nihon gendai shi I: Meiji ishin 日本現代史I:明治維新 [Modern Japanese history I: the Meiji Restoration]. Tokyo, 1951.

Inoue Tetsujirō 井上哲次郎. Shushigaku-ha no tetsugaku 朱子学派の哲学 [The philosophy of the Chu Hsi school]. Tokyo, 1911.

-------Yōmeigaku-ha no tetsugaku 陽明学派の哲学 [The philosophy of the Wang Yang-ming school]. Tokyo, 1903.

Ishii Ryōsuke 石井良助. Nihon hōsei shi gaisetsu 日本法制史概説 [An outline of Japanese legal history]. Tokyo, 1948.

Ishin shi 維新史 [Restoration history]. 6 vols.; Tokyo: Ishin shiryō hensan kai 維新史料編纂会, 1956.

Itō Hirobumi den 伊藤博文傳 [The life of Itō Hirobumi]. Tokyo: Shumpo-kō tsuishō kai 春畝公追頌会, 1943.

Itō Tasaburō 伊東多三郎. Bakuhan taisei 幕藩体制 [The Bakufu-han structure]. Tokyo, 1956.

-------"Sōmō no kokugaku" 草莽の国学 [Kokugaku in the commoner class]. MS.

Kada Tetsuji 加田哲二. Meiji shoki shakai keizai shisō shi 明治初期社会経済思想史 [A history of social and economic thought in the early Meiji period]. Tokyo, 1941.

Kagawa Seiichi 香川政一. Takasugi Shinsaku shōden 高杉晋作小傳 [A short biography of Takasugi Shinsaku]. Yamaguchi, 1936.

Kagoshima ken shi 鹿児島縣史 [A history of Kagoshima prefecture]. 5 vols.; Kagoshima: Kagoshima ken 鹿児島縣, 1940.

Kanjōtai shimatsu 干城隊始末 [An account of the Kanjōtai]. In the Chōshū Archives of the Yamaguchi-ken Library.

Kano Masanao 鹿野政直. Nihon kindai shisō no keisei 日本近代思想の形成 [The formation of modern Japanese thought]. Tokyo, 1956.

Kimura Motoi 木村礎. "Hagi han zaichi kashin dan ni tsuite" 萩藩在地家臣團について [Concerning the country samurai in Hagi han]; Shigaku zasshi 史学雑誌, 62.8:27-50 (1953).

Kobata Jun 小葉田淳, ed. Kinsei shakai 近世社会 [Tokugawa society] (Shin Nihon shi taikei 新日本史大系, vol. 4). Tokyo, 1952.

Konishi Shirō 小西四郎, ed. Meiji ishin 明治維新 [Meiji Restoration] (Shin Nihon shi taikei, vol. 5). Tokyo, 1952.

Koyama Yasuhiro 小山泰弘. "Iwakuni han no hansei kaikaku" 岩國藩の藩政改革 [The reform of the han government in Iwakuni]; Shigaku kenkyū 史学研究, 76:34-50 (May 1960).

Kumura Toshio 玖村敏雄. Yoshida Shōin 吉田松陰. Tokyo, 1944.

Maruyama Masao 丸山眞男. Nihon seiji shisō shi kenkyū 日本政治思想史研究 [Studies in the history of Japanese political thought]. Tokyo, 1952.

Meiji ishin shi kenkyū 明治維新史研究 [Studies in the Meiji Restoration]. Tokyo: Shigaku kai 史学会, 1933.

Meiji zen Nihon igaku shi 明治前日本医学史 [A history of Japanese medicine before the Meiji period]. Tokyo: Nihon gakushi in 日本学士院, 1955.

Misaka Keiji 三坂圭治. Hagi han no zaisei to buiku 萩藩の財政と撫育 [Finance and the buiku system in Hagi han]. Tokyo, 1944.

Murata Minejirō 村田峰次郎. Takasugi Shinsaku 高杉晋作. Tokyo, 1914.

Nakamura Kichiji 中村吉治. Nihon keizai shi gaisetsu 日本経済史概説 [An outline of Japanese economic history]. Tokyo, 1952.

Naramoto Tatsuya 奈良本辰也. "Bakumatsu ni okeru gōshi: chūnōsō no sekkyokuteki igi" 幕末に於ける郷士：中農層

の積極的意義 [The Bakumatsu gōshi: the positive significance of the middle stratum peasantry]; Rekishi hyōron, vol. 10 (1947).

-------"Kinsei hōken shakai ni okeru shōgyō shihon no mondai" 近世封建社会に於ける商業資本の問題 [The problem of commercial capital in Tokugawa feudal society]; Nihonshi kenkyū 日本史研究, vol. 5 (1947).

-------Kinsei hōken shakai shiron 近世封建社会史論 [A historical essay concerning Tokugawa feudal society]. Tokyo, 1952.

-------Yoshida Shōin 吉田松陰. Tokyo, 1955.

Nihon shi jiten 日本史辞典. Osaka: Kyōto daigaku bungakubu kokushi kenkyūshitsu 京都大学文学部國史研究室, 1954.

Niwa Kunio 丹羽邦男 and Unno Yoshio 海野由夫. "Bakumatsu yūhan no keizai kōzō 幕末雄藩の経済構造 [The economic structure of the active han in the Bakumatsu period]; Rekishi hyōron, 79:87-91 (Sept. 1956).

Nohara Yūsaburō 野原祐三郎. Bōchō ishin hiroku 防長維新秘録 [A secret history of restoration Chōshū]. Yamaguchi, 1937.

Nomura Kentarō 野村兼太郎. Gaikan Nihon keizai shisō shi 概観日本経済思想史 [An outline history of Japanese economic thought]. Tokyo, 1949.

Ogawa Gorō 小川五郎. "Chōshū han ni okeru shomin kinnō undō no tenkai to sono shisōteki haikei" 長州藩に於ける

庶民勤皇運動の展開とその思想的背景 [The development of the commoner pro-Emperor movement in Chōshū and its intellectual background]; Onoda kōtōgakkō kenkyū ronsō 小野田高等学校研究論叢, 8:3-19 (Dec. 1953).

Oka Mitsuo 岡光夫. "Chōshū han Setonai nōson ni okeru shōhin seisan no keitai" 長州藩瀬戸内農村に於ける商品生産の形態 [The form of merchant production in the inland sea villages of Chōshū]; Rekishi gaku kenkyū 歴史学研究, No. 159 (Sept. 1952).

Oka Yoshitake 岡義武. Kindai Nihon no keisei 近代日本の形成 [The formation of modern Japan]. Tokyo, 1947.

Osatake Takeshi 尾佐竹猛. Meiji ishin 明治維新 [The Meiji Restoration]. 4 vols.; Tokyo, 1942.

Oyama Hironari 小山博也. "Bakumatsu Satsuma han no ishin undō to sono haikei" 幕末薩摩藩の維新運動とその背景 [The Satsuma restoration movement and its background during the Bakumatsu period]; Saitama daigaku kiyō 埼玉大学紀要, 4.4:94-112 (1955).

Sakata Yoshio 坂田吉雄. "Meiji ishin to Tempō kaikaku" 明治維新と天保改革 [The Meiji Restoration and the Tempō reforms]; Jimbun gakuhō 人文学報, vol. 2 (1952).

-------"Nihon ni okeru kindai kanryō no hassei" 日本に於ける近代官僚の発生 [The emergence of a modern bureaucracy in Japan]; Jimbun gakuhō, 3:1-26 (Mar. 1953).

Seki Junya 関順也. "Bakumatsu ni okeru hansei kaikaku (Chōshū han): Meiji ishin seiritsuki no kiso kōzō" 幕末に於ける藩政改革(長州藩):明治維新成立期の基礎構造 [Han reform in Chōshū during the Bakumatsu period: the structural foundations of the Meiji Restoration]; Yamaguchi keizaigaku zasshi 山口経済学雑誌, 5:60-82, 6:58-80 (Mar., May, 1955).

-------"Bakumatsu ni okeru nōmin ikki: Chōshū han no baai" 幕末に於ける農民一揆:長州藩の場合 [Peasant rebellions during the Bakumatsu period: the case of Chōshū]; Shakai keizai shigaku 社会経済史学, 21.4:22-40 (1955).

-------Hansei kaikaku to Meiji ishin: han taisei no kiki to nōmin bunka 藩政改革と明治維新:藩体制の危機と農民分化 [Han reform and the Meiji Restoration: han crisis and peasant class differentiation]. Tokyo, 1956.

Shimazaki Tōson 島崎藤村. Yoakemae 夜明け前 [Before the dawn]. 2 vols.; Tokyo, 1936.

Shimmi Kichiji 新見吉治. Kakyū shizoku no kenkyū 下級士族の研究 [A study of lower samurai]. Tokyo, 1953.

Shōgiku Kido-kō den 松菊木戸公傳 [The life of Kido Kōin]. Tokyo: Kido-kō denki hensan kai 木戸公傳記編纂会, 1927.

Shunketsu Sakamoto Ryūma 雋傑坂本龍馬 [The life of Sakamoto Ryūma]. Tokyo: Sakamoto Nakaoka ryō sensei dōzō kensetsu kai 坂本中岡両先生銅像建設会, 1927.

Sōmi Kōu 相見香雨. "Watanabe Kazan" 渡辺華山; Dai jimmei jiten 大人名辞典, vol. 6. Tokyo, 1954.

Suematsu Kenchō 末松謙澄. Bōchō kaiten shi 防長回天史 [A history of Chōshū during the Restoration period]. 7 vols.; Tokyo, 1921.

Takashima Yanosuke 高島弥之助. Shimazu Hisamitsu-kō 島津久光公 [The life of Shimazu Hisamitsu]. Tokyo, 1937.

Takasu Yoshijirō 高須芳次郎. Kinsei Nihon jugaku shi 近世日本儒学史 [A history of Confucianism in Tokugawa Japan]. Tokyo, 1943.

Tanaka Akira 田中彰. "Bakumatsu-ki Chōshū han ni okeru hansei kaikaku to kaikakuha dōmei" 幕末期長州藩に於ける藩政改革と改革派同盟 [Han reform and the reformist clique in Chōshū during the Bakumatsu period]. Dissertation, Tokyo kyōiku daigaku bungakubu kenkyūka.

-------"Chōshū han kaikakuha no kiban: shotai no bunseki o tōshite mitaru" 長州藩改革派の基盤：諸隊の分析を通して見たる [The base of the Chōshū reformist clique: as seen through an analysis of the auxiliary militia]; Shichō 史潮, 51:9-23 (Mar. 1954).

-------"Chōshū han ni okeru hansei kaikaku to Meiji ishin" 長州藩に於ける藩政改革と明治維新 [Han reforms in Chōshū and the Meiji Restoration]; Shakai keizai shigaku, 22. 5:139-164 (1955).

-------"Chōshū han ni okeru Keiō gunsei kaikaku" 長州藩に於ける慶応軍政改革 [On the reformation of the military administration by Chōshū in the Keiō era]; Shirin 史林, 42.1:104-124 (1959).

-------"Chōshū han ni okeru Tempō ikki ni tsuite: Tempō kaikaku no zentei to shite" 長州藩に於ける天保一揆について：天保改革の前提として [The Tempō peasant rebellion in Chōshū: the background of the Tempō reform]; Shakai keizai shigaku, 21.4:89-131 (1955).

-------"Chōshū han no Tempō kaikaku" 長州藩の天保改革 [The Tempō reforms of Chōshū]; Historia, 8:15-39 (1957).

-------"Chōshū han o meguru Meiji ishin shi no sho mondai" 長州藩をめぐる明治維新史の諸問題 [The various problems of the Meiji Restoration history of Chōshū]; Yamaguchi-ken chihōshi kenkyū 山口縣地方史研究, 2:4-7 (1955).

-------"Meiji zettai shugi seiken seiritsu no ichi katei: dattai sōdō to nōmin ikki o megutte" 明治絶体主義政権成立の一過程：脱隊騒動と農民一揆をめぐって [One aspect of the establishment of the absolutist Meiji political power: concerning the discharged soldiers rebellion and peasant rebellion]; Rekishi hyōron, 75,4:1-19 (1956).

-------"Murata Seifū no kaibōron to yōgaku" 村田清風の海防論と洋学 [Murata Seifū's theory of maritime defense and Western studies]; Nihon rekishi 日本歴史, 93:22-27 (1956).

-------"Tōbakuha no keisei katei: Chōshū han bakumatsu hansei kaikaku to kaikakuha dōmei" 討幕派の形成過程：長州藩幕末藩政改革と改革派同盟 [The formation of the anti-Bakufu clique: han reforms and the reformist clique in Chōshū during the Bakumatsu period]; Rekishi gaku kenkyū, 205:1-17 (Mar. 1957).

Toba Masao 鳥羽正雄. "Edo jidai no rinsei" 江戸時代の林政 [Forest management during the Edo period]; Iwanami kōza Nihon rekishi 岩波講座日本歷史, vol. 9. Tokyo, 1934.

Tōkō sensei ibun 東行先生遺文 [The writings of Takasugi Shinsaku]. Tokyo: Tōkō sensei gojūnen sai kinen kai 東行先生五十年祭記念会, 1916.

Tokutomi Iichirō 德富猪一郎. Kinsei Nihon kokumin shi: Baku-Chō kōsen 近世日本國民史：幕長交戰 [A national history of Tokugawa Japan: the Bakufu-Chōshū war]. 2 vols.; Tokyo, 1937.

-------Kōshaku Yamagata Aritomo den 公爵山縣有朋傳 [The life of Prince Yamagata Aritomo]. 3 vols.; Tokyo, 1933.

Tōyama Shigeki 遠山茂樹. Meiji ishin 明治維新 [The Meiji Restoration]. Tokyo, 1951.

Toyoda Takeshi 豊田武. Nihon no hōken toshi 日本の封建都市 [The feudal city in Japan]. Tokyo, 1952.

Tsuchiya Takao 土屋喬雄. Hōken shakai hōkai katei no kenkyū 封建社会崩壞過程の研究 [Studies of the powers of decline of feudal society]. Tokyo, 1953.

Umaya Genjirō 馬屋原次郎. Bōchō jūgonen shi 防長十五年史 [Fifteen years of Chōshū history]. Tokyo, 1915.

Umetani Noboru 梅溪昇. "Meiji ishin shi ni okeru kiheitai no mondai" 明治維新史に於ける奇兵隊の問題 [The problem of the Kiheitai in Meiji Restoration history];

Jimbun gakuhō, 3:27-36 (Mar. 1953).

Wada Kenji 和田健爾 . Kusaka Genzui no seishin 久坂玄瑞の精神 [The spirit of Kusaka Genzui]. Tokyo, 1943.

Yoshida Shōin zenshū 吉田松陰全集 . 12 vols.; Tokyo: Yamaguchi-ken kyōiku kai 山口縣教育会 , 1940.

GLOSSARY

Personal Names

Abe Masahiro 阿部正弘
Abe Masatō 阿部正外
Aizawa Yasushi 会澤安
Akagawa Chūemon 赤川忠右衛門
Akagawa Kihei 赤川喜兵衛
Akagawa Naojirō 赤川直次郎
Akagawa Tarōemon 赤川太郎右衛門
Akane Taketo 赤根武人
Akimoto Shinzō 秋本新蔵
Andō Nobuyuki 安藤信行
Anenokōji Kintomo 姉小路公知
Aoyama Hanzō 青山半蔵
Arai Hakuseki 新井白石
Arima Shinshichi 有馬新七
Asano 浅野
Asano Shigenaga 浅野茂長
Ashikaga 足利
Ashikaga Takauji 足利尊氏
Ashikaga Yoshiakira 足利義詮
Ashikaga Yoshimitsu 足利義満
Awaya Hikotarō 粟屋彦太郎
Chōsokabe 長宗我部
Chu Hsi 朱熹

Date 伊達
Date Munenari 伊達宗城

Egawa Tarōzaemon 江川太郎左衛門
Etō Shimpei 江藤新平

Fujita 藤田
Fujita Tōko 藤田東湖
Fujita Yūkoku 藤田幽谷
Fukuhara Echigo 福原越後
Fukushima Masanori 福島正則
Fukuzawa Yukichi 福澤諭吉

Godaigo 後醍醐
Gotō Shōjirō 後藤象二郎

Hani Gorō 羽仁五郎
Hara Ichinoshin 原市之進
Hashimoto Sanai 橋本左内
Hayashi Suke 林助
Hayashi Yūzō 林勇蔵
Heizei 平城

Hideyoshi 秀吉
Hijikata Hisamoto 土方久元
Hirata Atsutane 平田篤胤
Hirosawa Saneomi 廣澤眞臣
Hitotsubashi 一橋
Hōjō 北條
Honda 本多
Honjō Munehide 本莊宗秀
Hori Jirō 堀次郎
Horie Hideichi 堀江英一
Horie Yasuzō 堀江保蔵
Hosokawa 細川
Hosokawa Yoshiyuki 細川慶順
Hotta Masayoshi 堀田正睦

Ibara Shukei 井原主計
Ii Naosuke 井伊直弼
Ikeda 池田
Ikeda Shigemasa 池田茂政
Ikeda Yoshinori 池田慶德
Ikeuchi Daigaku 池内大学
Inada 稲田
Inoue Kaoru 井上馨
Inoue Kiyoshi 井上清
Inoue Yoshirō 井上与四郎
Irie Sugizō 入江杉蔵
Isahaya 諫早
Ishikawa Kogorō 石川小五郎
Itakura Katsukiyo 板倉勝靜
Itō Hirobumi 伊藤博文
Itō Tasaburō 伊東多三郎
Iwakura Tomomi 岩倉具視

Jimbō Sōtoku 神保相德

Kamiyama Hōden 上山縫殿
Kaneshige Jōzō 兼重讓蔵
Katsu Awa 勝安芳
Kawakami Hanzō 河上範三
Kido Kōin 木戸孝允
Kihara Genemon 木原源右衛門
Kijima Matabei 來島又兵衛
Kikkawa Hiroie 吉川廣家
Kikkawa Tsunemoto 吉川経幹
Kodama Jochū 児玉如忠
Komai Chōon 駒井朝温
Komatsu Tatewaki 小松帯刀
Kōmei 孝明
Kondō Isamu 近藤勇
Konoe Tadahiro 近衛忠煕
Kujō Hisatada 久條尚忠
Kunishi Shinano 國司信濃
Kuroda 黒田

Kuroda Nagamasa 黒田長政
Kuruhara Ryōzō 來原良藏
Kusaka Genzui 久坂玄瑞
Kuze Hirochika 久世廣周

Maebara Issei 前原一誠
Maeda Magoemon 前田孫右衛門
Maki Yasuomi 眞木保臣
Manabe Akikatsu 間部詮勝
Masuda 益田
Masuda Danjō 益田彈正
Matsudaira 松平
Matsudaira Katamori 松平容保
Matsudaira Keiei 松平慶永
Matsudaira Mochiaki 松平茂昭
Matsudaira Sadanobu 松平定信
Matsushima Gōzō 松島剛藏
Minobe Matagorō 美濃部又五郎
Mitarai Kanichirō 御手洗幹一郎
Mitsukuri Gembo 箕作阮甫
Miyake Chūzō 三宅忠藏
Mizuno Tadakiyo 水野忠清
Mizuno Tadakuni 水野忠邦
Mōri 毛利
Mōri Chikuzen 毛利筑前
Mōri Dōshun 毛利洞春
Mōri Ise 毛利伊勢
Mōri Izumo 毛利出雲
Mōri Motonari 毛利元就
Mōri Motonori 毛利元德
Mōri Motosumi 毛利元純
Mōri Norito 毛利登人
Mōri Noto 毛利能登
Mōri Senjirō 毛利宣次郎
Mōri Takachika 毛利敬親
Mōri Terumoto 毛利輝元
Moriyama Tōen 森山棠園
Mukunashi Tōta 椋梨藤太
Murata Seifū 村田清風

Nagai Naomune 永井尚志
Nagai Uta 長井雅樂
Nagaya Tōbei 長屋藤兵衛
Nagoya Noboru 奈古屋登
Naitō Hyōe 内藤兵衛
Naitō Marisuke 内藤万里助
Nakamura Kyūrō 中村九郎
Nakamura Seiichi 中村誠一

Nakaoka Shintarō 中岡愼太郎
Nakatomi 中臣
Nakaya 中屋
Nakayama Tadayasu 中山忠能
Naramoto Tatsuya 奈良本辰也
Negoro Kazusa 根來上總
Niho Yaemon 仁保弥衛門
Nomura Wasaku 野村和作
Oda Nobunaga 織田信長
Ogasawara Nagamichi 小笠原長行
Ogawa Zenzaemon 小川善左衛門
Ōgimachi 正親町
Oguni Yūzō 小國融藏
Ogyū Sorai 荻生徂來
Ōhara Shigetomi 大原重德
Okamoto Sanuemon 岡本三右衛門
Okamoto Yoshinoshin 岡本吉之進
Ōkubo Toshimichi 大久保利通
Ōkura Genuemon 大倉源右衛門
Ōmura Masujirō 大村益次郎
Ōtani Bokusuke 大谷樸助
Ōuchi 大内
Rai Sanyō 賴山陽
Saigō Takamori 西鄉隆盛
Sakai 酒井
Sakai Tadakiyo 酒井忠清
Sakamoto Ryūma 坂本龍馬
Sakata Yoshio 坂田吉雄
Sakuma Shōzan 佐久間象山
Sakurai Shimpei 櫻井愼平
Sanjō Saneai 三條實愛
Sanjō Sanetomi 三條實美
Sasaki Otoya (Ikkan) 佐々木男也, 一貫
Satake 佐竹
Satō Shinuemon 佐藤新右衛門
Seki Junya 関順也
Shimazaki Tōson 島崎藤村
Shimazu 島津
Shimazu Hisamitsu 島津久光
Shimazu Nariakira 島津齊彬
Shimazu Narioki 島津齊興
Shimazu Shigehide 島津重豪
Shimazu Tadayoshi 島津忠義

Shimizu Seitarō 清水清太郎
Shinagawa Yajirō 品川弥二郎
Shiraishi Seiichirō 白石正一郎
Shishido Kurōbei 宍戸九郎兵衛
Shishido Tamaki 宍戸璣
Sufu Masanosuke 周布政之助
Sugawara no Michizane 菅原道真
Sugi Norisuke 杉德輔
Sugi Umetarō 杉梅太郎
Sugita Gempaku 杉田玄白
Suzuki Naomichi 鈴木直道
Suzuki Shigetane 鈴木重胤
Takashima Shūhan 高島秋帆
Takasugi Shinsaku 高杉晋作
Takechi Zuizan 武市瑞山
Tamaki Bunnoshin 玉木文之進
Tanuma Okitsugu 田沼意次
Tarusaki Yahachirō 樽崎弥八郎
Terashima Chūzaburō 寺島忠三郎
Tōdō 藤堂
Tōdō Takayuki 藤堂高猷
Tōjō Eian 東條英庵
Tokugawa 德川
Tokugawa Ienari 德川家斉
Tokugawa Iesada 德川家定
Tokugawa Ieyasu 德川家康
Tokugawa Keiki 德川慶喜
Tokugawa Keishō 德川慶勝
Tokugawa Mitsukuni 德川光圀
Tokugawa Mochitsugu 德川茂承
Tokugawa Nariaki 德川斉昭
Tokugawa Narinaga 德川斉修
Tokugawa Yoshitomi (Iemochi) 德川慶福, 家茂
Tokugawa Yoshiyori (Tayasu 田安 Yoshiyori) 德川慶頼
Tōyama Shigeki 遠山茂樹
Tsubaki 椿
Tsuboi Kuemon 坪井九右衛門
Tsukahara Masayoshi 塚原昌義
Umeda Umpin 梅田雲濱
Ura Yukie 浦靭負
Wang Yang-ming 王陽明
Watanabe Kazan (Noboru) 渡辺華山, 登
Watanabe Naizō 渡辺内蔵

Yamada Akiyoshi 山田顯義

Yamada Uemon 山田右衛門

Yamaga Sokō 山鹿素行

Yamagata Aritomo 山縣有朋

Yamagata Yoichihei 山縣与一兵

Yamanouchi 山内

Yamanouchi Yōdō 山内容堂

Yanagizawa Yoshiyasu 柳澤吉保

Yoshida Shōin 吉田松陰

Yoshida Tōyō 吉田東洋

Yoshimura Toratarō 吉村寅太郎

Yoshitomi Tōbei 吉富藤兵衛

Zusho Hirosato 調所廣鄉

Terms and Places

Aihonomura 秋穂村 — A village in Chōshū.

Aizu 會津 — A han in northern Honshū.

Akamanoseki 赤間関 — A village in Chōshū.

Aki 安藝 — A han in southwestern Honshū.

Akiragi 明木 — A village in Chōshū.

Amida-dera 阿彌陀寺 — A temple in Chōshū.

Ansei 安政 — Year period, 1845-1859.

Ansei no daigoku 安政の大獄 — The Purge of Ansei.

ashigaru 足輕 — A stratum of sotsu.

Awa 阿波 — A han in Shikoku.

baishin 陪臣 — Rear vassal.

Bakufu 幕府 — The government of the Tokugawa, the shogunate.

Bakumatsu 幕末 — The last period of Bakufu rule, usually 1853-1868.

Bansha 蠻社 — The Barbarian Society, a group including Watanabe Kazan which gathered to carry on Western studies.

Bitchū 備中 — A province in western Honshū.

Bizen 備前 — A province in western Honshū.

buikukyoku 撫育局 — A Chōshū government bureau for savings and investment.

Buke shohatto 武家諸法度 — Laws Governing the Military Houses, the Tokugawa code governing the daimyo.

bungen 分限	Duties of one's station.
Bunkyū 文久	Year period, 1861-1863.
bushi 武士	A common Japanese term for samurai.
bushidō 武士道	The code of ethics of the samurai.
chigyō 知行	Fief.
chika chisō 地下馳走	An extraordinary agricultural tax in Chōshū.
Chikuzen 筑前	A province in northern Kyūshū.
chinsei kaigi 鎮静会議	The Peace Assembly, a group of neutral samurai that formed in Chōshū at the time of the Chōshū civil war.
chō 町	A measure of land, 2.45 acres.
Chōfu 長府	One of the Chōshū branch han.
chōnin 町人	Townsmen.
Chōshū 長州	A han at the western tip of Honshū.
"chōtei e wa chūsetsu" 朝廷へは忠節	"Loyalty to the Court."
chūgen 中間	A stratum of <u>sotsu.</u>
chün tzu 君子	Gentleman.
daikan 代官	Either a Bakufu intendant or the samurai official in charge of a han district office.
daikansho 代官所	The office of the <u>daikan.</u>
daimyō 大名	Feudal lord of a han.
<u>Dai Nihon shi</u> 大日本史	The Mito history of Japan.

Dattai sōdō 脱隊騒動	The rebellion of discharged Chōshū militia.
Demmachō 傳馬町	A district in Edo.
Echizen 越前	A province in central Honshū on the Sea of Japan.
Edo 江戸	The Bakufu capital.
Edō 繪堂	A village in Chōshū.
Edo kachi 江戸徒士	A certain class of samurai who made up the cortege of the daimyo on his biennial trips to Edo.
Edo rusui karō 江戸留守居家老	Edo Resident Elder.
eidai karō 永代家老	A family with permanent Elder status.
enkintsuki 遠近附	A stratum of *shi* in Chōshū.
Enryōji 圓龍寺	A temple in Chōshū.
eta 穢多	See *semmin*.
fuchimai 扶持米	A form of samurai stipend.
fudai daimyō 譜代大名	See p. 17.
fudai hōkōnin 譜代奉公人	Hereditary servant, usually a hereditary cultivator in the service of a given family.
fudasashi 札差	Rice brokers who lent money to the *hatamoto*.
fukoku kyōhei 富國強兵	"Rich country, strong army," a slogan associated with the Restoration.
Fukui 福井	A han in central Honshū on the Sea of Japan.

Fukuoka 福岡	A han in northwestern Kyūshū.
funategumi 船手組	A stratum of *shi* in Chōshū.
Fushimi 伏見	A town near Kyoto.
gekokujō 下剋上	The overthrow of the high by the low, of lord by vassal.
Genroku 元禄	Year period, 1688-1704.
gijō 議定	Senior Councillor, one of the three offices in the earliest Meiji imperial government.
gisō 議奏	Court Councillor.
Giyūtai 義勇隊	One of the *shotai*.
gō 合	A measure of capacity, 0.318 pint.
gokenin 御家人	Direct vassals of the Tokugawa below the rank of *hatamoto*.
gōnō 豪農	Well-to-do peasant.
gontakure ごんたくれ	"Riff-raff," a pejorative localism used in southwestern Honshū.
gōshi 郷士	A samurai farmer.
gummugakari 軍務掛	Director of military affairs.
gun daikan 郡代官	District intendant.
gyōshō 行相	Accompanying Elder.
hachigumi 八組	A stratum of *shi* in Chōshū.
Hachimantai 八幡隊	One of the *shotai*.
Hagi 萩	The Chōshū castle town.
Hagi no ran 萩の乱	The Hagi Rebellion in 1876.
hakama 袴	A kind of Japanese kilt, an item of formal attire.

Hamada	濱田	A han in western Honshū on the Sea of Japan.
han	藩	A fief.
han nakama	藩仲間	A han-licensed guild.
hatamoto	旗本	Tokugawa vassals higher than *gokenin* but lower than *fudai* daimyo.
Heike monogatari	平家物語	"The Tale of the House of Taira," a military romance written during the Kamakura period.
Higo	肥後	A province.
Hikone	彦根	A han in Ōmi province.
Himeshima	姫島	An island off the coast of the Kyūshū province of Bungo.
Hiroshima	廣島	The castle town of Aki han.
Hokkyōdan	北強團	A militia unit formed by conservative retainers of the house of Masuda.
hōon	報恩	Repayment of an obligation.
Hōryaku	寶曆	Year period, 1751-1763.
hoshu kokusui	保守國粹	"Protect the national essence."
hōzōkin	寶藏金	See p. 45.
hsien	縣	The county office in Imperial China.
Hyōgo	兵庫	The Tokugawa name for what is today the port of Kobe.
ichidai karō	一代家老	Single Generation Elder.
ichimon	一門	The highest stratum of *shi* in Chōshū.

Ikedaya 池田屋	An inn in Kyoto at which loyalists from Chōshū and other han were attacked on 5/6/64. See p. 228.
ikki 一揆	A peasant rebellion.
Ikueikan 育英館	A school maintained by the house of Masuda (of Chōshū) for its retainers.
Ina 伊那	A valley on the Kiso Kaidō.
Inaba 稲葉	A han (Usuki) in northeastern Kyūshū.
Isa 伊佐	A village in Chōshū.
Ishindan 維新團	A militia unit formed by loyalists among the Masuda rear vassals.
Iwakuni 岩國	One of the Chōshū branch han.
jikachi 地徒士	Used in contrast to Edo *kachi*, troops on duty in the han.
jikimetsuke 直目附	Direct Inspector.
jikisanshi 直参士	Retainers with the right of an audience with their daimyo.
jinsō 陣僧	A stratum of *shi* in Chōshū.
jishagumi 寺社組	A stratum of *shi* in Chōshū.
jitsudaka 實高	See p. 10.
jiyū minken undō 自由民權運動	The Free People's Rights Movement of the early Meiji period.
jōi 攘夷	"Expel the barbarians."
junshi 准士	A stratum intermediate between *shi* and *sotsu* in Chōshū.
kabunakama 株仲間	A licensed guild.

Kaga 加賀	A han in central Honshū on the Sea of Japan.
Kagoshima 鹿児島	A han in southern Kyūshū, Satsuma.
kahanshū 加判衆	See karōshū.
kaibōron 海防論	"Argument for maritime defense."
kaikoku 開國	"Open the country."
Kaitengun 回天軍	One of the shotai.
Kaminoseki 上ノ関	A village in Chōshū.
kampaku 関白	The senior official at the Imperial Court.
kan 貫	A measure of weight, 8.27 lbs. Av. Used especially for silver.
Kanagawa Washin Jōyaku 神奈川和親條約	Kanagawa Treaty of Friendship.
Kanjōtai 干城隊	A military unit formed by the neutralist chinsei kaigi (Peace Assembly) at the time of the Chōshū civil war.
Kankō-maru 觀光丸	A Bakufu naval training boat.
Kansei 寛政	Year period, 1789-1800.
Kantō 関東	Japan east of the Hakone barrier, especially the plain on which Edo is situated.
Karatsu 唐津	A han in northwestern Kyūshū.
karō 家老	Elder, the senior minister of a daimyo.
karōshū 家老衆	Council of Elders.
Kawachi 河内	A village in Chōshū near Hagi.
keisotsu 軽卒	See sotsu.

kerai 家来	Vassals.
Kiheitai 奇兵隊	One of the Chōshū shotai.
Kii 紀伊	A province; also a han in central Honshū.
Kimmon no hen 禁門の變	The unsuccessful Chōshū counter-coup attempted in 1864.
kin 斤	A measure of weight, 1.32 lbs. Av.
Kinki 近畿	Osaka-Kyoto area.
kirimai 切米	A form of samurai stipend.
Kishū 紀州	See Kii.
kōbetsu 皇別	Descended from the Imperial line, one of the genealogical categories in traditional Japan.
Kōbōji 弘法寺	A temple near Hagi.
kōbugattai 公武合体	Union of Court and Camp, the moderate political movement headed by Satsuma.
kogakuha 古学派	See p. 156.
koihō 古医法	Ancient Medical Method.
Kōjōtai 鴻城隊	One of the shotai.
kōkoku 皇國	Imperial country.
koku 石	A measure of capacity, 4.96 bushels.
kokudaka 石高	See p. 9.
kokugaku 國学	National Studies.
kokuji goyōgakari 國事御用掛	Court Advisors on National Affairs, a special office created by extremists in 1863.

kokuji sansei 國事参政	Officials for Participation in National Affairs, a special office created by extremists in 1863.
kokuji yoriudo 國事寄人	Advisors on National Affairs, a special office created by extremists in 1863.
Kokura 小倉	A han in northwestern Kyūshū.
kokushō 國相	Han Administrative Elder.
kokutai 國体	The national polity.
Kōseikan 好生館	A school founded in Chōshū for the study of Western medicine.
koshinigata 越荷方	The Chōshū han *toiya*.
kōshōgakuha 考證学派	School of Textual Analysts.
Kumage 熊毛	A district in Chōshū.
Kumamoto 熊本	A han; also a city in west-central Kyūshū.
kunimoto rusui karō 国元留守居家老	Han Resident Elder.
kurairichi 藏入地	Land from which the revenues went to the han treasury.
kuramoto yaku 藏元役	Treasury official.
Kurashiki 倉敷	A Bakufu *daikansho.*
Kurume 久留米	A han in west-central Kyūshū.
Kurushima 久留島	Mori han in north-central Kyūshū.
Kuwana 桑名	A han in central Honshū.
Kyōhō 享保	Year period, 1716-1735.
Kyōhōji 教法寺	A temple in Chōshū.
Kyoto bugyō 京都奉行	A Bakufu official responsible for the administration of Kyoto.

Kyūhanjō 旧藩情	An essay by Fukuzawa Yukichi on conditions in the old han.
kyūri 究理	Natural philosophy.
kyūryō daikan 給領代官	Fief intendant.
Kyūshū 九州	One of the four main islands of Japan.
li 吏	Hereditary subofficials in the *hsien* office in Imperial China, particularly important during the Ming and Ch'ing periods.
Manji seihō 萬治制法	Statutes of the Manji Period.
Matsue 松江	A han on the Sea of Japan.
meibun 名分	Duties of one's station.
meigi 名義	Duty or honor.
Meiji 明治	Year period, 1868-1911.
Meirinkan 明倫館	The Chōshū college.
metsuke 目附	Censor, a Bakufu official.
Migita 右田	A village in Chōshū.
mikuni 御國	Honorable country.
Mino 美濃	A province in central Honshū.
Mitajiri 三田尻	A town in Chōshū on the Inland Se.
mitchoku 密勅	See p. 121.
Mito 水戸	A han immediately to the north-east of Edo.
mombatsu 門閥	"High-ranking," applied to samurai houses.
momme 匁	A measure of weight; 3.75 gramme.

Mori 森	A han in north-central Kyūshū.
muga ni 無我に	Selflessly.
mukyūdōri 無給通	A stratum of *shi* in Chōshū.
Myōenji 妙円寺	A temple near Kagoshima.
myōgakin 冥加金	Annual tax paid by a guild.
Nagasaki 長崎	A city under Bakufu control.
Nagasaki-maru 長崎丸	A Bakufu steamship sunk by Chōshū in 1863.
Nagato 長門	One of the two provinces in Chōshū.
naichoku 内勅	See p. 121.
naidaka 内高	See *jitsudaka*.
nakama 仲間	Guild.
Nihon gaishi 日本外史	"The Unofficial History of Japan," by Rai Sanyō.
Niigata 新潟	A port on the Sea of Japan.
nōheitai 農兵隊	Peasant militia.
ōbiromazume daimyō 大廣間詰大名	A group of daimyo who, when gathered in attendance on the shogun, occupied the same chamber at the palace of the shogun. It consisted of eleven *shimpan*, three of the more important *fudai*, and the twenty-four highest ranking *tozama* daimyo.
Ōda 大田	A village in Chōshū.
Ogōri 小郡	A town and a district in Chōshū.
ōgumi 大組	A stratum of *shi* in Chōshū.
ōjōya 大庄屋	The highest peasant official.

Okayama 岡山	A han in western Honshū on the Inland Sea.
ōmetsuke 大目附	Great Censor, a Bakufu official.
Ōmi 近江	A province in central Honshū.
Ōmori 大森	A Bakufu daikansho near Hamada han.
omotedaka 表高	The listed productive capacity of a han. See p. 10.
Ōno 大野	A village in Chōshū.
ōrōkanomazume kamon daimyō 大廊下間詰家門大名	The gosanke (Mito, Owari, and Kii) who occupied the Great Corridor attendance chamber in the shogun's palace.
Ōsaka 大阪	A city at the head of the Inland Sea.
Ōsaka sentō 大阪船頭	A hereditary rank in Chōshū.
Ōshima 大島	A district in Chōshū.
Ōsu 大洲	A han in Shikoku.
Ōtsu 大津	A district in Chōshū.
Owari 尾張	A han in central Honshū.
oyabun kobun 親分子分	The duty of the parent and of the child, used in speaking of the Japanese type of leader-follower relation.
rangaku 蘭学	Dutch studies.
rigaku 理学	Physics.
Rikishitai 力士隊	One of the Chōshū shotai.
rōjū 老中	The members of the highest Bakufu council.
rokudaka 禄高	The amount of a samurai's stipend or fief.

rōnin 浪人	Masterless samurai.
rōshin 老臣	High (literally, old) retainer.
ryō 兩	A unit of coinage, usually for gold.
ryūhei 流弊	Current evils.
Ryūkyū 琉球	Island chain to the south of Kyūshū.
sabaku 佐幕	"Support the Bakufu."
Saga 佐賀	A han in northwestern Kyūshū.
Sagami 相模	A province, a bay, in east-central Honshū.
saiban 宰判	The administrative districts within Chōshū.
Sakaimachi mon 堺町門	Sakaimachi Gate of the imperial palace in Kyoto.
samurai 侍	Feudal warrior.
Sanjō 三條	A street in Kyoto.
sanjūnindōri 三十人通	A stratum of <u>shi</u> in Chōshū.
sankin kōtai 参勤交代	The system of forced biennial residence of the daimyo in Edo.
sanyo 参與	Junior Councillor, one of three offices in the earliest Meiji imperial government.
sanyo daimyō 参豫大名	Participating daimyo.
sanzu no kawa 三途の河	An Oriental river Styx, a river to be crossed by the dead on their way to the Court of the King of Hell.
Sasanami 佐々並	A village in Chōshū.
satagaki 沙汰書	Official orders.
Satsuma 薩摩	A han in southern Kyūshū.

Seibutai 整武隊	One of the shotai.
seiji sōsai 政事總裁	An office created by the kōbugattai group in 1862 to strengthen their own power, roughly equivalent to that of tairō.
seinan sensō 西南戰爭	The war between the Meiji government and certain Satsuma troops in 1877.
Seiyō kibun 西洋紀聞	"Reports on the West," a work by Arai Hakuseki.
Sekigahara 関ヶ原	The battle in 1600 in which the Tokugawa gained their hegemony.
Sekigahara gunki 関ヶ原軍記	The Military Record of the Battle of Sekigahara.
semmin 賤民	The "unclean class" or outcastes of Tokugawa Japan.
Sempōtai 先鋒隊	The military of the conservative government in Hagi in 1864.
Sendai 仙台	A han in northeastern Honshū.
sensei 先生	Teacher.
sentō 船頭	A stratum of shi in Chōshū.
shi 士	Upper samurai as opposed to sotsu.
shihan 支藩	Branch han or collateral house.
shikan 士官	Officer.
shiko 士雇	See junshi.
shimbetsu 神別	A genealogical term referring to families descended from the various gods of the Shinto pantheon.
Shimonoseki 下ノ関	A city; also the straits between Honshū and Kyūshū.

Shimōsa 下總 — A province in east-central Honshū.

shimpan 親藩 — Tokugawa branch han.

shimpei 親兵 — Imperial troops.

Shimpūren 神風連 — The Divine Wind Company which rebelled in Kumamoto in 1876.

Shinano 信濃 — A province in central Honshū.

shingaku 心学 — A school of ethical teachings begun by Ishida Baigan, widespread among the Tokugawa merchant class.

shireikan 司令官 — Officer.

Shiyūtai 市勇隊 — One of the *shotai*.

shōgun 將軍 — The military ruler of feudal Japan.

Shōka sonjuku 松下村塾 — The school of Yoshida Shōin.

Shōkōji 清光寺 — A temple in Chōshū.

Shōkōkan 彰考館 — The Mito school.

Shōnai 庄内 — A han in northern Honshū.

shoshidai 所司代 — Kyoto *shoshidai*, the shogun's representative in Kyoto.

shotai 諸隊 — Auxiliary militia in Chōshū.

shōya 庄屋 — Village headman.

shōya dōmei 庄屋同盟 — Alliance of *shōya* formed in Chōshū to aid the loyalists in 1865.

shōyu 醤油 — Soy sauce.

Shūgitai 集義隊 — One of the *shotai*.

shuhan 主藩 — The main han as opposed to branch han.

soba yōnin seiji 側用人政治	Government by shogunal advisor; the soba yōnin was the Bakufu officer who, until the reforms of Yoshimune, acted as the intermediary between the shogun and the council of rōjū.
Sogekitai 狙撃隊	One of the shotai.
sonnō jōi 尊王攘夷	"Honor the Emperor; expel the barbarians."
sōsai 總裁	Supreme Executive Office, one of the three offices in the earliest Meiji imperial government.
sosen e kōdō 祖先へ孝道	"Filial duty to the ancestors (of the house of Mōri)."
sotsu 卒	Samurai below the class of shi.
Suō 周防	One of the two provinces in Chōshū.
Susamura 須佐村	A village in Chōshū.
Susa naikō 須佐内訌	An armed struggle between loyalist and conservative retainers of the house of Masuda.
tairō 大老	The highest office in the Bakufu, to which appointments were made only in time of crisis.
Tajima 但馬	A province in western Honshū on the Sea of Japan.
takajō 鷹匠	A stratum of samurai in Chōshū.
tama o ubau 玉を奪ふ	"Carry off the jewel (emperor)."
tamarinomazume daimyō 溜間詰大名	Five important shimpan daimyo and the six most important fudai daimyo.

Tamba 丹波	A province in western Honshū on the Sea of Japan.
tan 反	A measure of land, 0.245 acres.
Tanegashima 種ヶ島	An island off the southern coast of Kyūshū at which the Portuguese first landed in the middle of the sixteenth century.
Tatsuno 龍野	A han in western Honshū.
teikannoma daimyō 帝鑑間大名	A group of daimyo who, when gathered in attendance on the shogun, occupied the same chamber at the palace of the shogun. It consisted of the four least important *shimpan* and sixty middle grade *fudai* daimyo.
Tempō 天保	Year period, 1830-1844.
Teradaya 寺田屋	The Fushimi inn at which the Satsuma loyalists were attacked in 1862. See pp. 177-178.
to 斗	A measure of capacity, 3.97 gallons.
Toba 鳥羽	A district in Kyoto.
tōbaku 討(倒)幕	"Overthrow the Bakufu."
Tōhoku 東北	Northeastern Japan, the area around Sendai.
toimaru 問丸	The early form of the *toiya*.
toiya 問屋	Wholesale house.
Tokuji 徳地	A district in Chōshū.
Tokuyama 徳山	One of the Chōshū branch han.
Tosa 土佐	A han in Shikoku.
Toshima 当島	The Hagi district in Chōshū.
Tōshintai 東津隊	One of the *shotai*.

tōshoku 当職	Han Administrative Elder.
tōyaku 当役	See tōshoku.
tozama 外様	Outside lords.
Tsu 津	A han in west-central Honshū.
Tsūshō Jōyaku 通商條約	Commercial Treaty of 1858.
Tsuwano 津和野	A han in western Honshū.
Tsuyama 津山	A han in west-central Honshū.
ujō 鵜匠	A stratum of shi in Chōshū.
ukimai 浮米	A type of samurai stipend.
Uraga 浦賀	A village on Tokyo Bay.
Uwajima 宇和島	A han in Shikoku.
watari ni fune o esōrō kokochi 渡りに船を得候心地	"The feeling of having found a ferry [in time of need]."
yamabushi 山伏	Mountain monk, a devotee of a popular faith sect.
Yamaguchi 山口	City in Chōshū which became the castle town during the Bakumatsu period.
Yamashiro 山代	One of the districts in Chōshū.
Yamato 大和	A province in west-central Honshū.
Yanagawa 柳川	A han in northern Kyūshū.
yashiki 邸	A mansion or residence. During the Tokugawa period the mansions established by the daimyo in Edo were often quite extensive.
Yoakemae 夜明け前	"Before the Dawn," a novel by Shimazaki Tōson.

Yōchōtai	膺懲隊	One of the *shotai*.
Yonezawa	米澤	A han in northern Honshu.
yōnin	用人	The commercial agent of a han.
yorigumi	寄組	A high stratum of *shi* in Chōshū.
yōshi	養子	An adopted child, usually one adopted to carry on a family name.
Yoshida	吉田	A district in Chōshū.
Yūbikan	有備館	School established at the Chōshū residence in Edo.
Yuda	湯田	A village near Yamaguchi.
Yūgekitai	遊撃隊	One of the *shotai*.
za	座	Guild.
zempu	膳夫	A stratum of *shi* in Chōshū.
zokugan	俗眼	Pedestrian or vulgar eyes.
zokuron	俗論	Pedestrian view.
zokurontō	俗論黨	Party of the pedestrian view, the Tsuboi-Mukunashi clique.

HARVARD HISTORICAL MONOGRAPHS

** Out of Print*

1. Athenian Tribal Cycles in the Hellenistic Age. By W. S. Ferguson. 1932.
2. The Private Record of an Indian Governor-Generalship: The Correspondence of Sir John Shore, Governor-General, with Henry Dundas, President of the Board of Control, 1793–1798. Edited by Holden Furber. 1933.
3. The Federal Railway Land Subsidy Policy of Canada. By J. B. Hedges. 1934.
4. Russian Diplomacy and the Opening of the Eastern Question in 1838 and 1839. By P. E. Mosely. 1934.
5. The First Social Experiments in America: A Study in the Development of Spanish Indian Policy in the Sixteenth Century. By Lewis Hanke. 1935.*
6. British Propaganda at Home and in the United States from 1914 to 1917. By J. D. Squires. 1935.*
7. Bernadotte and the Fall of Napoleon. By F. D. Scott. 1935.*
8. The Incidence of the Terror during the French Revolution: A Statistical Interpretation. By Donald Greer. 1935.*
9. French Revolutionary Legislation on Illegitimacy, 1789–1804. By Crane Brinton. 1936.
10. An Ecclesiastical Barony of the Middle Ages: The Bishopric of Bayeux, 1066–1204. By S. E. Gleason. 1936.
11. Chinese Traditional Historiography. By C. S. Gardner. 1938. Rev. ed., 1961.
12. Studies in Early French Taxation. By J. R. Strayer and C. H. Taylor. 1939.
13. Muster and Review: A Problem of English Military Administration, 1420–1440. By R. A. Newhall. 1940.*
14. Portuguese Voyages to America in the Fifteenth Century. By S. E. Morison. 1940.*
15. Argument from Roman Law in Political Thought, 1200–1600. By M. P. Gilmore. 1941.*
16. The Huancavelica Mercury Mine: A Contribution to the History of the Bourbon Renaissance in the Spanish Empire. By A. P. Whitaker. 1941.*
17. The Palace School of Muhammad the Conqueror. By Barnette Miller. 1941.

18. A Cistercian Nunnery in Mediaeval Italy: The Story of Rifreddo in Saluzzo, 1220–1300. By Catherine E. Boyd. 1943.*
19. Vassi and Fideles in the Carolingian Empire. By C. E. Odegaard. 1945.*
20. Judgment by Peers. By Barnaby C. Keeney. 1949.
21. The Election to the Russian Constituent Assembly of 1917. By O. H. Radkey. 1950.
22. Conversion and the Poll Tax in Early Islam. By Daniel C. Dennett. 1950.*
23. Albert Gallatin and the Oregon Problem. By Frederick Merk. 1950.*
24. The Incidence of the Emigration during the French Revolution. By Donald Greer. 1951.*
25. Alterations of the Words of Jesus as Quoted in the Literature of the Second Century. By Leon E. Wright. 1952.*
26. Liang Ch'i Ch'ao and the Mind of Modern China. By Joseph R. Levenson. 1953.
27. The Japanese and Sun Yat-sen. By Marius B. Jansen. 1954.
28. English Politics in the Early Eighteenth Century. By Robert Walcott, Jr. 1956.
29. The Founding of the French Socialist Party (1893–1905). By Aaron Noland. 1956.
30. British Labour and the Russian Revolution, 1917–1924. By Stephen Richards Graubard. 1956.*
31. RKFDV: German Resettlement and Population Policy, 1939–1945. By Robert L. Koehl. 1957.
32. Disarmament and Peace in British Politics, 1914–1919. By Gerda Richards Crosby. 1957.
33. Concordia Mundi: The Career and Thought of Guillaume Postel (1510–1581). By W. J. Bouwsma. 1957.
34. Bureaucracy, Aristocracy, and Autocracy: The Prussian Experience, 1660–1815. By Hans Rosenberg. 1958.
35. Exeter, 1540–1640: The Growth of an English County Town. By Wallace T. MacCaffrey. 1958.
36. Historical Pessimism in the French Enlightenment. By Henry Vyverberg. 1958.
37. The Renaissance Idea of Wisdom. By Eugene F. Rice, Jr. 1958.
38. The First Professional Revolutionist: Filippo Michele Buonarroti (1761–1837). By Elizabeth L. Eisenstein. 1959.
39. The Formation of the Baltic States: A Study of the Effects of Great Power Politics upon the Emergence of Lithuania, Latvia, and Estonia. By Stanley W. Page. 1959.*
40. Conservation and the Gospel of Efficiency: The Progressive Conservation Movement, 1890–1920. By Samuel P. Hays. 1959.
41. The Urban Frontier: The Rise of Western Cities, 1790–1830. By Richard C. Wade. 1959.

42. New Zealand, 1769–1840: Early Years of Western Contact. By Harrison M. Wright. 1959.
43. Ottoman Imperialism and German Protestantism, 1521–1555. By Stephen A. Fischer-Galati. 1959.
44. Foch versus Clemenceau: France and German Dismemberment, 1918–1919. By Jere Clemens King. 1960.
45. Steelworkers in America: The Nonunion Era. By David Brody. 1960.
46. Carroll Wright and Labor Reform: The Origin of Labor Statistics. By James Leiby. 1960.
47. Chōshū in the Meiji Restoration. By Albert M. Craig. 1961.
48. John Fiske: The Evolution of a Popularizer. By Milton Berman. 1961.
49. John Jewel and the Problem of Doctrinal Authority. By W. M. Southgate. 1962.
50. Germany and the Diplomacy of the Financial Crisis, 1931. By Edward W. Bennett. 1962.
51. Public Opinion, Propaganda, and Politics in Eighteenth-Century England: A Study of the Jew Bill of 1753. By Thomas W. Perry. 1962.
52. Soldier and Civilian in the Later Roman Empire. By Ramsay MacMullen. 1963.
53. Copyhold, Equity, and the Common Law. By Charles Montgomery Gray. 1963.
54. The Association: British Extraparliamentary Political Association, 1769–1793. By Eugene Charlton Black. 1963.
55. Tocqueville and England. By Seymour Drescher. 1964.
56. Germany and the Emigration, 1816–1885. By Mack Walker. 1964.
57. Ivan Aksakov (1823–1886): A Study in Russian Thought and Politics. By Stephen Lukashevich. 1965.
58. The Fall of Stein. By R. C. Raack. 1965.
59. The French Apanages and the Capetian Monarchy, 1224–1328. By Charles T. Wood. 1966.
60. Congressional Insurgents and the Party System: 1909–1916. By James Holt. 1967.
61. The Rumanian National Movement in Transylvania, 1780–1849. By Keith Hitchins. 1969.
62. Sisters of Liberty: Marseille, Lyon, Paris and the Reaction to a Centralized State, 1868–1871. By Louis M. Greenberg. 1971.
63. Old Hatreds and Young Hopes: The French Carbonari against the Bourbon Restoration. By Alan B. Spitzer. 1971.
64. To the Maginot Line: The Politics of French Military Preparation in the 1920's. By Judith M. Hughes. 1971.
65. Florentine Public Finances in the Early Renaissance, 1400–1433. By Anthony Molho. 1971.
66. Provincial Magistrates and Revolutionary Politics in France, 1789–1795. By Philip Dawson. 1972.

67. The East India Company and Army Reform, 1783–1798. By Raymond Callahan. 1972.
68. Ireland in the Empire, 1688–1770: A History of Ireland from the Williamite Wars to the Eve of the American Revolution. By Francis Godwin James. 1973.
69. Industry and Economic Decline in Seventeenth-Century Venice. By Richard Tilden Rapp. 1976.
70. George Henry Lewes: A Victorian Mind. By Hock Guan Tjoa. 1977.